OXFORD IB DIPLOMA PROGRAMME

ENVIRONMENTAL SYSTEMS AND SOCIETIES

2015 EDITION

COURSE COMPANION

Jill Rutherford
Gillian Williams

OXFORD
UNIVERSITY PRESS

OXFORD
UNIVERSITY PRESS

Great Clarendon Street, Oxford, OX2 6DP, United Kingdom

Oxford University Press is a department of the University of Oxford. It furthers the University's objective of excellence in research, scholarship, and education by publishing worldwide. Oxford is a registered trade mark of Oxford University Press in the UK and in certain other countries

British Library Cataloguing in Publication Data

Data available

978-0-19-833256-5

15

Paper used in the production of this book is a natural, recyclable product made from wood grown in sustainable forests.
The manufacturing process conforms to the environmental regulations of the country of origin.

Printed and bound by CPI Group (UK) Ltd, Croydon, CR0 4YY

Acknowledgements

The publishers would like to thank the following for permissions to use their photographs:

Cover image: A MA/Shutterstock; p2: Public Domain; p6: SHONE/Gamma-Rapho/Getty Images; p16: C. M. Buddle; p18: Reto Stöckli, Nazmi El Saleous, and Marit Jentoft-Nilsen, NASA GSFC; p22: Roger Ressmeyer/CORBIS/ Image Library; p30: Shutterstock; p30: Shutterstock; p36: Tom Brakefield/Digital Vision/Getty Images; p51: Kim Kyung-Hoon/Reuters/Image Library; p53: iStock.com; p58: Tom Brakefield/CORBIS/Image Library; p60: Tim Graham/Corbis/Image Library; p61: William Ormerod/Visuals Unlimited, Inc.; p61: iStock.com; p62: Nigel Cattlin/Visuals Unlimited/Corbis/Image Library; p56 & 62: iStock.com; p67: Mark Medcalf/Shutterstock; p67: Shutterstock; p67: Paul Reeves Photography/Shutterstock; p67: iStock.com; p68: Mitrofanov Alexander/Shutterstock; p77: Sergey Uryadnikov/Shutterstock; p106: Shutterstock; p108: Leonardo Gonzalez/Shutterstock; p109: Shutterstock; p109: Florin Mihai/Shutterstock; p110: O. Alamany & E. Vicens/CORBIS/Image Library; p112: Norbert Wu/Minden Pictures/Getty Images; p112: Jurgen Freund/naturepl.com; p121: Adam Burton/Robert Harding World Imagery/Corbis/Image Libray; p127: NHBS, UK; p127: NHBS, UK; p130: Richard Hewitt Stewart/National Geographic Society/Corbis/Image Library; p132: Marlin E. Rice/Iowa State University Extension and Outreach; p137: TBD; p146: Menno Schaefer/Shutterstock; p146: Arthur van der Kooij/Shutterstock; p152: Shutterstock; p155: Public Domain; p158: Shutterstock; p159: Handout /Getty Images News/ Getty Images; p167: Christian Vinces/Shutterstock; p167: Peter J. Wilson/Shutterstock;p170: Stephane Bidouze/Shutterstock; p114&171: Shutterstock; p172: Shutterstock; p172: Public Domain; p172: James L. Stanfield/National Geographic/ Getty Images; p175: David Steele/Shutterstock; p175: Shutterstock; p176: Shutterstock; p177: Joyce Mar/Shutterstock; p177: Simon Bratt/Shutterstock; p178: John Carnemolla/Corbis/Image Library; p192: James king-holmes/Science photo library; p215: Papilio Photo Library/Photographersdirect.com; p231: Zhao jianpeng qd - Imaginechina/AP Images; p232: Shutterstock; p198 & 232: Pete Atkinson/Photographer's Choice/Getty Images p232: NASA; p246: Michal Madacky/Shutterstock; p247: Shutterstock; p252: Public Domain; p254: Shutterstock; p266: Nicolas Thibaut/Photononstop/Getty Images; p267: iStock.com; p268: Shutterstock; p268: Shutterstock; p268: Xinhua/Xinhua Press/Corbis/Image Library; p236 & 269: Jimmy Tran/Shutterstock; p269: Shutterstock; p271: iStock.com; p290: ChinaFotoPress /ChinaFotoPress/Getty Images; p291: iStock.com; p274 & 293: Lisa S./Shutterstock; p294: Shutterstock; p295: TopFoto; p302 & 306: Boarding1now/Dreamstime.com; p308: Bettmann/Corbis/Image Library; p316: Philip Lange/Shutterstock; p316: iStock.com; p326: Antonio V. Oquias/Shutterstock; p337: NASA; p344: Victor Habbick Visions/Science Photo Library; p372: iStock.com; p375: Lianem/Dreamstime.com; p378: Shutterstock; p378: James Mattil/Shutterstock; p378: Shutterstock; p381: Ulrich Mueller/Shutterstock; p350 & 386: Shutterstock; p389: Huguette Roe/Shutterstock; p390: Muellek Josef/Shutterstock; p390: iStock.com; p410: Shutterstock; p417: Shutterstock; p419: SRM.

Artwork by Six Red Marbles and OUP

The author and publisher are grateful for permission to reprint extracts from the following copyright material:

extracts from *A Sand Country Almanac* by Leopold (1968) from 'Foreword' and 'The Land Ethic', reprinted by permission of Oxford University Press, USA, (www.oup.com).

extracts from *Silent Spring* by Rachel Carson, copyright © 1962 by Rachel L. Carson, renewed 1990 by Roger Christie, reprinted by permission Houghton Mifflin Harcourt Publishing Company and Pollinger Limited (www.pollingerltd.com) on behalf of the Estate of Rachel Carson and Frances Collin, trustee, all rights reserved. All copying, including electronic, or re-distribution of this text, is expressly forbidden.

extract from article 'Here's the good news: 126 new species discovered in Greater Mekong' by AFP, 19 December 2012, copyright AFP, 2015, reprinted by permission.

extract from *A Short History of Nearly Everything* by Bill Bryson, copyright © 2003 by Bill Bryson, published by Doubleday, reprinted by permission of The Random House Group Limited and Broadway Books, an imprint of the Crown Publishing Group, a division of Random House LLC, all rights reserved.

from 'The Millennium Development Goals', from www.undp.org, reprinted by permission of UNDP Brazil.

from 'Neotropical plant diversity' by Andrew Henderson, Steven P. Churchill, James L. Luteyn, 2 May 1991, *Nature* Vol. 351, pp. 21-22, copyright 1991, reprinted by permission of Macmillan Publishers Ltd. via Copyright Clearance Center.

from *Biodiversity: an Introduction* by Kevin J. Gaston, John I. Spicer, reproduced with permission of Blackwell Publishing, Inc. in the format 'republish in a book' via Copyright Clearance Center.

extract from 'Foreword' from http://atlas.aaas.org, by Peter Raven, reprinted by permission of AAAS (American Association for the Advancement of Science).

table from *Environmental Science* by Kevin Byrne, (Nelson Thornes, 1997), reprinted by permission of the publishers, Oxford University Press.

from 'Water shortage looms in Israel after prolonged drought, but supply to Jordan continuing', Associated Press, 19 March 2008, reprinted by permission of The Associated Press Copyright © 2015, all rights reserved.

definition 'Aquaculture' from http://www.fao.org/aquaculture/en/, accessed 1 February 2015 © FAO 2015, Food and Agriculture Organisation of the United Nations, reprinted by permission.

from *Environmental Science* by Kevin Byrne, (Nelson Thornes, 1997), reprinted by permission of the publishers, Oxford University Press.

extract from IPCC, 2014: *Climate Change 2014: Impacts, Adaptation, and Vulnerability – Summary for Policymakers*. Contribution of Working Group II to the Fifth Assessment Report of the Intergovernmental Panel on Climate Change, Assessment, ...[Field, C.B, V.R. Barros, D.J. Dokken, K.J. Mach, M.D. Mastrandrea, T.E. Bilir, M. Chatterjee, K.L. Ebi, Y.O. Estrada, R.C. Genova, B. Girma, E.S. Kissel, A.N. Levy, S. MacCracken, P.R. Mastrandrea, and L.L. White (eds.)]. World Meteorological Organization, Geneva, Switzerland, reprinted by permission.

extract from IPCC, 2014: *Climate Change 2014: Mitigation of Climate Change – Summary for Policymakers*. Contribution of Working Group III to the Fifth Assessment Report of the Intergovernmental Panel on Climate Change, ...[Edenhofer, O., R. Pichs-Madruga, Y.Sokona, E. Farahani, S. Kadner, K. Seyboth, A. Adler, I. Baum, S. Brunner, P. Eickemeier, B. Kriemann, J. Savolainen, S. Schloemer, C. von Stechow, T. Zwickel and J.C. Minx (eds.)], Cambridge University Press, Cambridge, United Kingdom and New York, NY, USA, reprinted by permission of IPCC.

p. 399 extracts adapted from 'All About Recycling in Germany' from www.howtogermany.com, reprinted by permission.

'An exercise to teach bioscience students about plagiarism' by Chris J. R. Willmott and Tim M. Harrison, *Journal of Biological Education*, 1 June 2003, Taylor & Francis, reprinted by permission of the publisher (Taylor & Francis Ltd, www.tandfonline.com).

Course Companion definition

The IB Diploma Programme course books are resource materials designed to support students throughout their two-year Diploma Programme course of study in a particular subject. They will help students gain an understanding of what is expected from the study of an IB Diploma Programme subject while presenting content in a way that illustrates the purpose and aims of the IB. They reflect the philosophy and approach of the IB and encourage a deep understanding of each subject by making connections to wider issues and providing opportunities for critical thinking.

The books mirror the IB philosophy of viewing the curriculum in terms of a whole-course approach; the use of a wide range of resources, international mindedness, the IB learner profile and the IB Diploma Programme core requirements, theory of knowledge, the extended essay, and creativity, action, service (CAS).

Each book can be used in conjunction with other materials and indeed, students of the IB are required and encouraged to draw conclusions from a variety of resources. Suggestions for additional and further reading are given in each book and suggestions for how to extend research are provided.

In addition, the course books provide advice and guidance on the specific course assessment requirements and on academic honesty protocol. They are distinctive and authoritative without being prescriptive.

IB mission statement

The International Baccalaureate aims to develop inquiring, knowledgeable and caring young people who help to create a better and more peaceful world through intercultural understanding and respect.

To this end the IB works with schools, governments and international organizations to develop challenging programmes of international education and rigorous assessment.

These programmes encourage students across the world to become active, compassionate, and lifelong learners who understand that other people, with their differences, can also be right.

The IB Learner Profile

The aim of all IB programmes is to develop internationally minded people who, recognizing their common humanity and shared guardianship of the planet, help to create a better and more peaceful world. IB learners strive to be:

Inquirers   They develop their natural curiosity. They acquire the skills necessary to conduct inquiry and research and show independence in learning. They actively enjoy learning and this love of learning will be sustained throughout their lives.

Knowledgeable   They explore concepts, ideas, and issues that have local and global significance. In so doing, they acquire in-depth knowledge and develop understanding across a broad and balanced range of disciplines.

Thinkers   They exercise initiative in applying thinking skills critically and creatively to recognize and approach complex problems, and make reasoned, ethical decisions.

Communicators   They understand and express ideas and information confidently and creatively in more than one language and in a variety of modes of communication. They work effectively and willingly in collaboration with others.

Principled   They act with integrity and honesty, with a strong sense of fairness, justice, and respect for the dignity of the individual, groups, and communities. They take responsibility for their own actions and the consequences that accompany them.

Open-minded   They understand and appreciate their own cultures and personal histories, and are open to the perspectives, values, and traditions of other individuals and communities. They are accustomed to seeking and evaluating a range of points of view, and are willing to grow from the experience.

Caring   They show empathy, compassion, and respect towards the needs and feelings of others. They have a personal commitment to service, and act to make a positive difference to the lives of others and to the environment.

Risk-takers   They approach unfamiliar situations and uncertainty with courage and forethought, and have the independence of spirit to explore new roles, ideas, and strategies. They are brave and articulate in defending their beliefs.

Balanced   They understand the importance of intellectual, physical, and emotional balance to achieve personal well-being for themselves and others.

Reflective   They give thoughtful consideration to their own learning and experience. They are able to assess and understand their strengths and limitations in order to support their learning and personal development.

A note on academic honesty

It is of vital importance to acknowledge and appropriately credit the owners of information when that information is used in your work. After all, owners of ideas (intellectual property) have property rights. To have an authentic piece of work, it must be based on your individual and original ideas with the work of others fully acknowledged. Therefore, all assignments, written or oral, completed for assessment must use your own language and expression. Where sources are used or referred to, whether in the form of direct quotation or paraphrase, such sources must be appropriately acknowledged.

How do I acknowledge the work of others?

The way that you acknowledge that you have used the ideas of other people is through the use of footnotes and bibliographies.

Footnotes   (placed at the bottom of a page) or endnotes (placed at the end of a document) are to be provided when you quote or paraphrase from another document, or closely summarize the information provided in another document. You do not need to provide a footnote for information that is part of a 'body of knowledge'. That is, definitions do not need to be footnoted as they are part of the assumed knowledge.

Bibliographies   should include a formal list of the resources that you used in your work. 'Formal' means that you should use one of the several accepted forms of presentation. This usually involves separating the resources that you use into different categories (e.g. books, magazines, newspaper articles, Internet-based resources, CDs and works of art) and providing full information as to how a reader or viewer of your work can find the same information. A bibliography is compulsory in the extended essay.

What constitutes malpractice?

Malpractice is behaviour that results in, or may result in, you or any student gaining an unfair advantage in one or more assessment component. Malpractice includes plagiarism and collusion.

Plagiarism is defined as the representation of the ideas or work of another person as your own. The following are some of the ways to avoid plagiarism:

- Words and ideas of another person used to support one's arguments must be acknowledged.

- Passages that are quoted verbatim must be enclosed within quotation marks and acknowledged.

- CD-ROMs, email messages, web sites on the Internet, and any other electronic media must be treated in the same way as books and journals.

- The sources of all photographs, maps, illustrations, computer programs, data, graphs, audio-visual, and similar material must be acknowledged if they are not your own work.

- Works of art, whether music, film, dance, theatre arts, or visual arts, and where the creative use of a part of a work takes place, must be acknowledged.

Collusion is defined as supporting malpractice by another student. This includes:

- allowing your work to be copied or submitted for assessment by another student

- duplicating work for different assessment components and/or diploma requirements.

Other forms of malpractice include any action that gives you an unfair advantage or affects the results of another student. Examples include, taking unauthorized material into an examination room, misconduct during an examination, and falsifying a CAS record.

Contents

About the authors

Jill Rutherford has some 30 years of teaching, administrative and board experience within international and UK national schools. She is currently academic director of Ibicus International, offering workshops to IB teachers. She was the founding Director of the IB Diploma at Oakham School, England. She holds two degrees from the University of Oxford. Her passion lies in teaching and writing about the IB Environmental Systems and Societies course.

Gillian Williams graduated from Reading University and has taught Environmental Systems, Geography and TOK on the international circuit since 1993. In her international career Gillian has held various leadership position including Deputy Head, Head of Year and Head of Department. In 2011 she began advising on the IB Environmental Systems and Societies curriculum review. She is a workshop leader (online and face-to-face) and part of the IB Global Mentoring Team. Gillian is currently Experiential Education Director at Utahloy International School Zengcheng in China.

INTRODUCTION TO THE ESS COURSE

This book is a Course Companion to the IB Diploma Programme course – Environmental Systems and Societies. Although this course has had a history of some decades in several forms, the one that you are studying now is the culmination of the ideas and work of many teachers and their students. It introduces you to some big environmental issues facing humans and the world that we inhabit.

The IB mission statement, which is also expressed in the IB learner profile characteristics, is at the heart of this course. As you read and refer to this Course Companion, consider the examples, case studies and questions with reference to your characteristics as a learner and the characteristics of the learner profile. The earth faces many human-induced environmental issues. We must continue to enquire into and think about the environment and our actions within it so that we can build up knowledge across disciplines in order to solve problems. Governments, groups and individuals taking decisions on environmental issues must evaluate the different viewpoints with an open mind and balance the risks and benefits of their actions. We would not be adequate guardians of the planet unless we care about it, have principles by which we live, and accept accountability for our actions after due reflection. The maxim 'Think globally, act locally' is a driver of this course and of the IB Diploma Programme CAS requirement. If you are carrying out CAS activities, many of these could also involve protecting or repairing your environment, and we hope that you may gain some ideas for this from this book.

Writing this Course Companion would not have been possible without a team approach. Many IB Diploma Programme teachers have contributed in varying ways and the authors are most grateful to these busy people, living in different countries and biomes, ecosystems and environments. We believe that this team approach gives the book a truly international flavour, with case studies from many countries and viewpoints. Thank you to all these teachers. The book would not have been written at all were it not for past students who, with goodwill, enthusiasm and interest, were willing to get cold, wet and muddy for the purpose of collecting data and gaining understanding.

Any errors or omissions are entirely those of the authors and we welcome communication from you to point out where these are and to suggest improvements and updates for the next edition.

Aims of ESS

1. Acquire the knowledge and understanding of environmental systems at a variety of scales.

2. Apply the knowledge, methodologies and skills to analyse environmental systems and issues at a variety of scales.

3. Appreciate the dynamic interconnectedness between environmental systems and societies

4. Value the combination of personal, local and global perspectives in making informed decisions and taking responsible actions on environmental issues.

5. Be critically aware that resources are finite, and that these could be inequitably distributed and exploited, and that management of these inequities is the key to sustainability.

6. Develop awareness of the diversity of environmental value systems.

7. Develop critical awareness that environmental problems are caused and solved by decisions made by individuals and societies that are based on different areas of knowledge.

8. Engage with the controversies that surround a variety of environmental issues.

9. Create innovative solutions to environmental issues by engaging actively in local and global contexts.

Why the Big Questions?

The intention of the Big Questions (BQs) in the ESS course is to glue the topics of the course together by encouraging a holistic view of the subject. These questions are not examined but help link topics together and show the interconnectedness of ESS.

They may be weaved into the responses to the essay questions in Paper 2.

They stimulate an approach to ESS that should allow you to apply higher-order skills such as analysis, synthesis and evaluation in your (concept-based) learning and prevent the topics being addressed in isolation.

The Big Questions at the end of each topic allow you to reflect on your learning and to explore the connections between topics. Good teaching provides this anyway and these questions reinforce this for you.

The main themes in the Big Questions are:

1. Equilibrium: the systems approach

2. EVSs: environmental value systems

3. Sustainability

4. Strategy: management strategies – the effectiveness of human intervention in solving environmental issues

5. Biodiversity: a global viewpoint on the future of the Earth's environment.

The six Big Questions

A. Which strengths and weaknesses of the systems approach and of the use of models have been revealed through this topic?

B. To what extent have the solutions emerging from this topic been directed at **preventing** environmental impacts, **limiting** the extent of the environmental impacts, or **restoring** systems in which environmental impacts have already occurred?

C. What value systems are at play in the causes and approaches to resolving the issues addressed in this topic?

D. How does your personal value system compare with the others you have encountered in the context of issues raised in this topic?

E. How are the issues addressed in this topic relevant to sustainability or sustainable development?

F. In which ways might the solutions explored in this topic alter your predictions for the state of human societies and the biosphere decades from now?

LP links to ESS aims and Big Questions

The IB learner profile is a list of the characteristics that the IB expect you to develop in following an IB programme and it is linked to the aims of ESS.

The practical work and internal assessment assignment you undertake in this course should engage you in developing all the attributes of the learner profile; aim 4 should help you to develop your characteristics of being principled, open-minded, caring, risk-taking, balanced and reflective; aims 1 and 2 help you to become knowledgeable about ESS, apply and inquire into that knowledge and communicate it to others.

IB approaches to teaching and learning

There are five approaches to learning (p11 ESS guide). These are developing skills in:

- Thinking
- Social interactions
- Communication
- Self-management
- Research.

There are six approaches to teaching. These are:

- Inquiry-based
- Conceptually focused
- Contextualized
- Collaborative
- Differentiated
- Informed by assessment.

Teaching and learning in the IB should be approached in these ways and we should work together to try to make sense of the world. Knowing how the world works is essential to helping us face global challenges. The IB recognizes three phases of learning in the circle shown in figure I.1.

In the **inquiry phase**, we learn new things.

In the **reflection phase**, we make connections between the things we have learned and gain

▲ **Figure I.1**

a deeper understanding. Theory of knowledge (TOK) in the IB Diploma encourages reflection and you should have time in ESS to reflect on your knowledge, evaluate the evidence and recognize bias just as you do in TOK. The BQs should also help you reflect.

In the **action phase**, we learn by doing. This is principled action in which we make responsible choices with fairness, integrity and honesty.

How you might use the Big Questions

The ESS guide states that you might use the BQs:

- In learning each topic either at the start of a topic as an introduction or in group discussions and building on the last one or as revision themes.

- Or as student assignments – formative assessment.

How this book links the Big Questions, concepts, context and content

This book takes the BQs and applies them to each of the topics.

Each topic has BQs which should stimulate discussion, reflection or action. Sometimes you might use them as issues for debate in class, sometimes as ideas to develop in practical work or your IA. The questions in this book are not exhaustive but are indicative of many more that you will probably create as you go through the course.

The BQs are developed at three levels:

- Top level BQ in the ESS guide.

- Reflective BQ – knowledge questions based on content of the topic under study. Many of these reflective questions are TOK knowledge questions.

- Action BQ – practical work ideas are listed throughout the book and some of these could be developed for your individual investigation, practical scheme of work, or ESS extended essay topic.

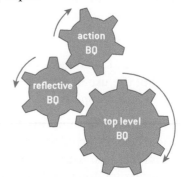

▲ **Figure I.2**

Topic	Big Questions					
	A	B	C	D	E	F
Foundations of environmental systems and societies	X		X	X	X	
Ecosystems and ecology	X				X	
Biodiversity and conservation		X	X	X	X	X
Water, food production systems and society	X	X			X	X
Soil systems, terrestrial food production systems and society	X	X	X		X	X
Atmospheric systems and society		X			X	X
Climate change and energy production	X	X	X	X	X	X
Human systems and resource use	X	X	X	X	X	X

▲ **Figure I.3** Big Questions relevant to ESS topics

Topic	Concepts				
	Systems approach	EVS	Sustainability	Management strategies	Global viewpoint
Foundations of environmental systems and societies	1.2, 1.3	1.1	1.4	1.5	1.1, 1.5
Ecosystems and ecology	2.1, 2.2, 2.3, 2.4			2.5	2.4
Biodiversity and conservation		3.1	3.4		3.2, 3.3
Water, food production systems and society	4.1, 4.3	4.2	4.3, 4.4	4.4	4.1, 4.2, 4.3, 4.4
Soil systems , terrestrial food production systems and society	5.1, 5.2	5.3	5.2	5.3	5.2, 5.3
Atmospheric systems and society	6.1	6.1		6.2, 6.3, 6.4	6.1, 6.2
Climate change and energy production	7.2	7.3	7.3	7.3	7.1, 7.2, 7.3
Human systems and resource use	8.1	8.3	8.4	8.3	8.1, 8.2, 8.3, 8.4

▲ **Figure I.4** ESS sub-topics particularly relevant to concepts

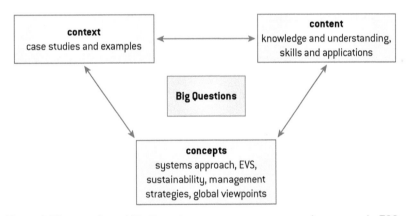

▲ **Figure I.5** Integration of Big Questions, content, context and concepts in ESS

The order in which the syllabus is arranged does **not** prescribe the order in which it is taught. It is up to individual teachers to decide on an arrangement that suits their circumstances. The diagram above shows how the inter-related strands of the subject can be integrated.

This ESS course is underpinned by a concept-based approach. The central concepts above should be revisited in each context through the 'Big Questions' that are given on page ii of this chapter.

You can only genuinely appreciate the overarching concepts and principles of ESS when the concepts are set in context. This course and the examination papers require you to explore the application of these concepts and principles in a wide range of situations.

The advantages of this concept-based approach are as follows:

- Facilitates disciplinary and interdisciplinary learning allowing for connections to be made with other subjects.

- Deepens your understanding of complex and dynamic ecosystems.

- Allows you to integrate new content into existing knowledge.

The key concepts are connected to the 'Big Questions'. These address the overarching issues that you will encounter throughout the duration of the ESS course.

1. Is it possible for individuals and societies to act in a sustainable manner?

2. To what extent are value systems held by individuals and societies dynamic?

3. To what extent are all ecosystems part of a global interconnected ecosystem?

4. Is it always possible to find appropriate solutions to environmental issues?

5. Do individuals and societies have effective strategies to address environmental issues?

6. To what extent does a systems approach enhance our understanding of environmental systems?

For each topic, examples of the reflective BQs are provided at the end of the topic.

For each topic, practical work ideas are included, which could for the basis of action BQs.

You will, no doubt, think of loads more BQs and make links that the authors have not yet thought about. Go for it and you should deepen your learning in ESS and do better in the assessment!

How to use this book

This book follows the sequence of topics in the ESS guide. This is not a teaching order though so you could read this book in any order that your teacher decides. There are many cross-references throughout the book because you will build up a bigger picture of the concepts, examples and themes as you follow the course.

At the end of each of the eight topics is a review section. For that topic, the review section contains:

- Big Questions
- Reflective, TOK questions
- A quick self-test of questions.

Answers to the quick review questions and many of the activities in the book can be found at www.oxfordsecondary.com/ib-ess.

1.1 Environmental value systems

Significant ideas:

→ Historical events, among other influences, affect the development of environmental values systems and environmental movements.

→ There is a wide spectrum of environmental value systems each with their own premises and implications.

Applications and skills:

→ **Discuss** the view that the environment can have its own intrinsic value.

→ **Evaluate** the implications of two contrasting environmental value systems in the context of given environmental issues.

→ **Justify** the implications using evidence and examples to make the justification clear.

Knowledge and understanding:

→ Significant historical influences on the development of the environmental movement have come from literature, the media, major environmental disasters, international agreements and technological developments.

→ **An environmental value system (EVS) is a worldview or paradigm that shapes the way an individual, or group of people, perceives and evaluates environmental issues**, influenced by cultural, religious, economic and socio-political contexts.

→ An EVS might be considered as a 'system' in the sense that it may be influenced by education, experience, culture and media (inputs) and involves a set of inter-related premises, values and arguments that can generate consistent decisions and evaluations (outputs).

→ There is a spectrum of EVSs from ecocentric through anthropocentric to technocentric value systems.

→ **An ecocentric viewpoint** integrates social, spiritual and environmental dimensions into a holistic ideal. It **puts ecology and nature as central to humanity** and emphasizes a less materialistic approach to life with greater self-sufficiency of societies. An ecocentric viewpoint prioritizes biorights, emphasizes the importance

of education and encourages self-restraint in human behaviour.

→ **An anthropocentric viewpoint argues that humans must sustainably manage the global system.** This might be through the use of taxes, environmental regulation and legislation. Debate would be encouraged to reach a consensual, pragmatic approach to solving environmental problems.

→ **A technocentric viewpoint argues that technological developments can provide solutions to environmental problems**. This is a consequence of a largely optimistic view of the role humans can play in improving the lot of humanity. Scientific research is encouraged in order to form policies and understand how systems can be controlled, manipulated or exchanged to solve resource depletion. A pro-growth agenda is deemed necessary for society's improvement.

→ There are extremes at either end of this spectrum (eg deep ecologists – ecocentric – to cornucopian – technocentric), but in practice EVSs vary greatly with culture and time and rarely fit simply or perfectly into any classification.

→ Different EVSs ascribe different intrinsic values to components of the biosphere.

Key term

An **environmental value system (EVS)** is a worldview or paradigm that shapes the way an individual or group of people perceive and evaluate environmental issues. This will be influenced by cultural, religious, economic and socio-political context.

To think about

Our environmental value systems will influence the way we see environmental issues.

1. List other value systems that influence how we view the world.

2. Outline one named global and one local environmental issue.

Describe your opinion on these issues and explain how your value systems influence it.

'Whatever befalls the Earth — befalls the sons of the Earth.

Humankind has not woven the web of life. We are but one thread within it. Whatever we do to the web, we do to ourselves. All things are bound together. All things connect.'

Attributed to Chief Seattle, 1855

▲ **Figure 1.1.1** The only known photo of Chief Seattle taken in the 1860s

Development of the environmental movement

The environmental movement as we know it originated in the 1960s BUT humans have been concerned about the effect we have on the environment for much longer.

- Romans reported on problems such as air and water pollution.

- Between the late 14th century and the mid 16th century, waste produced by humans was associated with the spread of epidemic disease in Europe.

- Soil conservation was practised in China, India and Peru as early as 2,000 years ago.

Such concerns did not really give rise to widespread public activism until recently. To understand modern environmentalism we must look back at the historical events which:

- caused concern over environmental impacts

- elicited the responses of individuals, groups of individuals, governments and the United Nations to these impacts.

Powerful individuals and independent pressure groups are now very influential though their use of media, and they have catalysed the movement to make it a people's or 'grass roots' movement. There has also been a continuing divide in philosophy between:

- those who see the reason for conserving nature as being to continue to supply goods and services to humankind in a sustainable way (environmental managers) and

- those who believe that we should conserve nature unconditionally, for its spiritual value (deep and self-reliance ecologists);

ie do we save it for **our** sake or for **its** sake?

Who is involved in the environmental movement?

It is probably fair to say that the majority of people in the world do not spend much time focusing on environmental issues unless they are brought to their attention or affect them directly. However, the activities of a number of groups have influenced

- norms of behaviour (eg purchasing choices such as dolphin-friendly tuna and recycling) and

- political choices (eg the successes of the 'Green Party').

Influential individuals often use media publications (eg Aldo Leopold's *A Sand County Almanac,* Rachel Carson's *Silent Spring,* Al Gore's *An Inconvenient Truth*) to raise issues and start the debate.

Independent pressure groups use awareness campaigns to effect a change (eg Greenpeace on Arctic exploration, World Wildlife Fund on saving tigers). They influence the public who then influence government and corporate business organizations. These groups are called non-governmental organizations (NGOs). 'Friends of the Earth' is another example.

Corporate businesses (especially multinational corporations – MNCs – and transnational corporations – TNCs) are involved since they are supplying consumer demand and in doing so using resources and creating environmental impact (eg mining for minerals or burning of fossil fuels).

Governments make policy decisions including environmental ones (eg planning permission for land use), and apply legislation (laws) to manage the country (eg emissions controls over factories). They also meet with other governments to consider international agreements (eg United Nations Environment Programme, UNEP). Different countries are at different stages of environmental awareness, as are different individuals. Legislating about emissions is important but so is making sure there is enough food for the population. While different countries may put environmental awareness at different levels of priority, all are aware of the issues facing the Earth and that all must be involved in finding solutions.

Intergovernmental bodies such as the **United Nations** have become highly influential in more recent times by holding Earth Summits to bring together governments, NGOs and corporations to consider global environmental and world development issues.

To research

Look up Chief Seattle on the web. His famous speech was in the Lushootseed language, translated into Chinook Indian trade language, and then into English. While he may not have said these exact words, does it matter?

'We abuse land because we regard it as a commodity belonging to us. When we see land as a community to which we belong, we may begin to use it with love and respect.'

Aldo Leopold,
A Sand County Almanac
(reprinted by permission of Oxford University Press, USA)

TOK

In 2013, 30 Greenpeace activists on board the Greenpeace ship *Arctic Sunrise* peacefully protested in Arctic international waters against the Russian Gazprom oil platform drilling for oil in the Arctic. They were arrested by armed Russian commandos and kept in prison for 100 days before being freed.

Read about this at www.greenpeace.org and news websites.

Do you agree with what the activists were doing or do you agree with the Russian authorities in stopping them?

Debate the issues in this with three teams: one represents Greenpeace views, one the Russian state and the other the Gazprom interests.

To what extent can we rely on reason to evaluate the Greenpeace approach to this issue?

The growth of the modern environmental movement in outline

Event	Impact
Neolithic Agricultural Revolution (10,000 years ago)	• Humans settled to become farmers instead of nomadic hunter-gatherers • Human population began to rise • Local resources (food, water, fuel) were managed sustainably from around the settlement
Industrial Revolution (early 1800s)	• Population growth and resource usage escalated • Large scale production of goods and services for all • Burning of large amounts of fuel in the form of trees and coal • Mining of minerals from the earth to produce metals to make machines • Limestone quarried for cement production • Land was cleared, natural waterways polluted, cities became crowded and smoky • Our urban consumer society arose
Green Revolution of the 1940s to 1960s	• Mechanized agriculture and boosted food production massively • Required the building of machinery and burning of enormous amounts of fossil fuels such as oil • Technology was applied to agriculture • New crop varieties were developed and fertilizer and pesticide use rose sharply • The world population grew to about 3 billion • Our resource use and waste production rocketed
Modern environmental movement (1960s onwards)	• The impacts became more global: collapsing fish stocks, endangered species, pesticide poisoning, deforestation, nuclear waste, ozone layer depletion, global warming, acid precipitation, etc. • A new breed of **environmentalists** surfaced who had scientific backgrounds and spearheaded the modern environmental movement • Greenpeace founded 1971 • Influential individuals wrote books (eg Rachel Carson's *Silent Spring*) • NGOs campaigned and the media reported • Governments formed nature reserves and put environmental issues on their agenda • Some businesses marketed themselves as environmentally friendly • UNEP organized Earth Summits on the environment • The movement became public and gained momentum
Environmentalism today	• More research on loss of biodiversity and climate change leading to more action to protect the environment and encourage sustainability from governments, corporations and individuals • Small number of climate sceptics voice doubts over climate change • Discovery of fracking process to release shale gas and oil shale reserves increases tensions between technocentrists and ecocentrists

'And this is why I sojourn here,
Alone and palely loitering,
Though the sedge is wither'd from the lake,
And no birds sing.'

From *La Belle Dame Sans Merci* by John Keats

Case studies – historical influences on the environmental movement

There is general agreement that the modern environmental movement was catalysed by Rachel Carson's book, *Silent Spring*, published in 1962. The title comes from the John Keats poem (right). Carson warned of the effects of pesticides on insects, both pests and others, and how this was being passed along the food chain to kill others, including birds (hence a silent spring). What really gained people's attention was her belief that pesticides such as DDT (**d**ichloro**d**iphenyl**t**richloroethane, a persistent, synthetic insecticide) were finding their way into people and accumulating in fatty tissues, causing higher risks of cancer. Chemical industries tried to ban the book but many scientists shared her concerns and when an investigation, ordered by US president John F. Kennedy, confirmed her fears, DDT was banned.

In the decades since the publication of *Silent Spring*, it has been criticized as scaremongering without enough scientific evidence. The banning of DDT may have caused more harm than good (see 2.2) by allowing the mosquitoes that carry malaria to survive and so spread the disease causing millions of deaths.

Al Gore, former US vice-president, was heavily influenced by the book to become involved in environmental issues, particularly with his documentary on climate change *An Inconvenient Truth*, 2006. This raised awareness of climate change – then called global warming – and clearly stated that global climate change was a result of greenhouse gases released by human activities and that we had to act as this is a moral issue. George Bush's response to the documentary when he was president of the USA was 'Doubt it' and he later said that we should focus on technologies that enable us to live better lives and protect the environment.

Mercury is a heavy metal and is poisonous to animals. It affects the nervous system causing loss of vision, hearing and speech and lack of coordination in arms and legs. Severe poisoning causes insanity or death. Mercury was used in the hat-making industry into the 20th century. Hat makers were known to often suffer mental illnesses although the source of such illnesses was unknown. This is the basis of the name of the 'Mad Hatter' character in Lewis Carroll's *Alice in Wonderland* and the phrase 'mad as a hatter'.

The Chisso Corporation built a chemicals factory in Minamata, Japan and was very successful for years. But a by-product was methylmercury which bioaccumulated in the bodies of humans, causing mercury poisioning (see 2.2).

In the early hours of the morning of 3 December 1984, in the centre of the city of Bhopal, India, in the state of Madhya Pradesh, a Union Carbide pesticide plant released 40 tonnes of methyl isocyanate (MIC) gas, immediately killing nearly 3,000 people and ultimately causing at least 15,000–22,000 total deaths. This has been called the **Bhopal Disaster** and is considered to be the world's worst industrial disaster. The world was in shock.

In 1986, at **Chernobyl**, the worst nuclear disaster ever occurred. This was a few miles north of Kiev, the capital of Ukraine (then part of the USSR) where an explosion and then fire resulted in a level 7 event (the highest) in reactor number 4. The reactor vessel containing the uranium radioactive material split so exposing the graphite moderator to air which caused it to catch fire. The reactor went into uncontrollable meltdown and a cloud of highly radioactive material from this drifted over much of Russia and Europe as far west as Wales and Scotland. Fission products from the radioactive cloud, eg isotopes

'For the first time in the history of the world, every human being is now subjected to contact with dangerous chemicals, from the moment of conception until death.'

Rachel Carson, *Silent Spring*, 1962

'Now I truly believe that we in this generation must come to terms with nature, and I think we're challenged, as mankind has never been challenged before, to prove our maturity and our mastery, not of nature but of ourselves.'

Rachel Carson, *Silent Spring*, 1962

of caesium, strontium and iodine, have a long half-life and were accumulated in food chains. In 2009, there were still restrictions on selling sheep from some Welsh farms due to their levels of radiation. There is much debate about how many people have been affected by the radiation as long-term effects, such as cancers and deformities at birth, are difficult to link to one event. 31 workers died of radiation sickness as they were exposed to high levels in trying to shut down the reactor and some had a lethal dose of radiation within one minute of exposure. Estimates of later deaths vary but some state about 1,000 extra cases of thyroid cancer and 4,000 other cancers caused by the fallout cloud. Other estimates state that 1 million people will have died as a result of the disaster.

The authorities of the day did not announce the disaster but it was picked up in Sweden when fallout was found on the clothing of workers at one of their nuclear plants.

Even today, the reactor is still dangerous. It was encased in a concrete shell but the other reactors continued to run until 2000. Now, a metal arch is being built as the concrete shell only has a lifetime of 30 years but estimates of the date of completion have been put back to 2016.

In 2011, there was another nuclear accident at the **Fukushima Daiichi** nuclear plant in Japan. An earthquake set off a tsunami which caused damage resulting in meltdown of 3 reactors in the plant. The water flooding these became radioactive and will take many years to remove. Although the radiation leak was only about 30% that of Chernobyl and radiation levels in the air low, one third of a million people were evacuated as the plant was sited in a densely populated area. Later reports showed the accident was caused by human error – it was not built to withstand a tsunami even though it was close to the sea in an earthquake zone. The plant is still not secured.

After the disaster, there were anti-nuclear demonstrations in other countries and Germany announced it was closing older reactors and phasing out nuclear power generation. France, Belgium, Switzerland all had public votes to reduce or stop nuclear power plants. In other countries, plans for nuclear plants were abandoned or reduced.

▲ **Figure 1.1.2** The Chernobyl nuclear reactor plant after the explosion in 1986

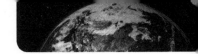

A review of major landmarks in environmentalism

Years	Events	Significance
10,000 yrs BP	**Neolithic agricultural revolution**	Settlements, population increase, local resource management began.
Early 1800s	**Industrial revolution in Europe**	Increased urbanization, resource usage and pollution.
Late 1800s	**Influential individuals such as Thoreau and Muir write books on conservation**	First conservation groups form and nature reserves established. NGOs form (RSPB, NT).
1914	**Once the most prolific bird, the passenger pigeon becomes extinct**	Conservation movement grows. Concern for tigers, rhinoceros, etc.
1930s and 1940s	**Dustbowl in North America**	Recognition that agricultural practices may affect soils and climate.
1940s	**Green Revolution – intensive technological agriculture**	Resource use (especially fossil fuel use) and pollution increased. Human population rises sharply.
1949	**Leopold writes *A Sand County Almanac***	Concept of 'stewardship' is applied to nature.
1951	**UK's ten National Parks are established**	Recognition of need to conserve natural areas.
1956 to 1968	**Minamata Bay Disaster**	Emphasizes the ability of food chains to accumulate toxins into higher trophic levels, including into humans.
1962	**Rachel Carson publishes *'Silent Spring'***	General acceptance of dangers of chemical toxins affecting humans. The pesticide DDT is banned.
1960s and early 1970s	**NGOs gain greater following**	Public awareness grows. WWFN, Greenpeace, Friends of the Earth all formed.
1972	**First Earth Summit – UN Conference on the Human Environment**	Declaration of UN conference. Action Plan for the Human Environment. Environment Fund established. Formation of UN Environment Programme (UNEP). Earth Summits planned at ten-year intervals.
1975	**C.I.T.E.S. formed by IUCN**	Endangered species protected from international trade.

Mid 1970s	Environmental philosophy established	Recognition that nature has its own intrinsic value. Stewardship ethic grows.
1979	James Lovelock publishes 'Gaia – A new look at life on Earth' and presents the 'Gaia hypothesis'	Systems approach to studying the environment begins. Nature seen as self regulating.
1982	Nairobi Earth Summit	Ineffective.
1983	UN World Commission on Environment and Development publishes the Brundtland Report	Sustainability established as the way forward.
1984	Bhopal Disaster	World's worst industrial disaster.
1986	Chernobyl Disaster	Nuclear fallout affects millions.
Mid 1980s	British Antarctic Survey Team detects ice sheets thinning and ozone hole	Public awareness of ozone depletion and risks of skin cancer.
1987	Montreal Protocol	Nations agree to reduce CFC use.
1980s	Green political parties form around the world	Political pressure placed on governments.
1988	IPCC formed by UNEP	Advises governments on the risks of climate change.
1992	Rio Earth Summit and Kyoto Protocol	Agreement to reduce carbon (CO_2) emissions to counter enhanced greenhouse effect and global warming. Agenda 21.
1990s	Green awareness strengthens	Environmentally friendly products, recycling and ecotourism become popular.
2002	Johannesburg Earth Summit	Plans to globally improve: Water and sanitation Energy supply issues Health Agricultural abuse Biodiversity reduction.
2005	Kyoto protocol becomes a legal requirement	174 countries signed and are expected to reduce carbon emissions to some 15% below expected emissions in 2008. It expires in 2012.
2006	Film 'An Inconvenient Truth' released	Documentary by Al Gore, former US vice-president, describing global warming.
2007	Nobel Peace Prize	Awarded in 2007 jointly to Al Gore and the IPCC for their work on climate change.
	IPCC release 4th assessment report in Nov 07	Report states that 'Warming of the climate system is unequivocal' and 'Most of the observed increase in globally averaged temperatures since the mid-20th century is *very likely* due to the observed increase in anthropogenic greenhouse gas concentrations.'
	UN Bali meeting Dec 07	187 countries meet and agree to open negotiations on an international climate change deal.
2012	Rio +20	Paper 'The Future We Want' published.

To do

Environmental headlines

9 March 2014 Lucy Hornby, Beijing in *Financial Times*

'Only three Chinese cities meet air quality standards'

11 March 2014 Suzanne Goldenberg, *The Guardian*

'California drought: authorities struggle to impose water conservation measures'

A. Look at newspaper headlines for one week.

Copy out the headlines that refer to environmental issues.

Put these in a table or on a notice board.

Good news	Bad news

B. Discuss with your fellow students what the environmental headlines may be in 2020 and 2050.

To research

Look up these people who were involved in environmentalism and write three sentences on each.

| Mahatma Mohandas Gandhi |
| Henry David Thoreau |
| Aldo Leopold |
| John Muir |
| E O Wilson |

To do

Find a local environmental issue where a pressure group is fighting for a cause.

Describe the issue and state the argument of the pressure group. What are the opposing arguments to their case? These may be economic, aesthetic, socio-political or cultural. State your own position on this issue and defend your argument.

'Nothing in this world is so powerful as an idea whose time has come.'

Victor Hugo 1802–1885

To do

Earth Days

This is just one of many images promoting Earth Day – what else can you find?

In the late 1960s, after *Silent Spring*, environmentalism turned into action, particularly in North America. Founded then were 'Earth Days' to encourage us all to be aware of the wonder of life and the need to protect it. There are two different ones. The UN Earth Day each year is on the Spring equinox (so in March in the Northern hemisphere and September in the Southern when the Sun is directly above the equator). John McConnell, an activist for peace, drove this concept.

The other Earth Day is always on 22 April each year, and was founded by US politician Gaylord Nelson as an educational tool on the environment. Up to 500 million people now take part in its activities worldwide each year.

Some are critical of Earth Days as marginalized activities that do not change the actions of politicians. Do you think they have an effect?

The spectrum of environmental value systems

- Different societies hold different environmental philosophies and comparing these helps explain why societies make different choices.

- The EVS we each hold will be influenced by cultural, religious, economic and socio-political contexts.

- The environment or any organism can have its own intrinsic value regardless of its value to humans. How we measure this value is a key to understanding the value we place on our environment.

For much of history, our viewpoint has been that the Earth's resources are unlimited and that we can exploit them with no fear of them running out. And for much of history that has been true. A much smaller human population in the past has been just one species among many. The words and phrases we use describe how we have seen the environment: 'fighting for survival', 'battle against nature', 'man or beast', 'conquering Everest', 'beating the elements'. It has only been in very recent times that humans have been able to control our environment and even think about terraforming (altering conditions to make it habitable for humans) on Mars. The Industrial Revolution heralded the arrival of the 'unbound Prometheus' of technological development when we were driven to explore, conquer and subdue the planet to the will of industrial growth. This ideology has reigned in the industrial world with the worldview that economic growth improves the lot of us all. But now it is clearer that the Earth's resources are not limitless as the Earth is not limitless. Humans may be the first species to change the conditions on Earth and so make it unfit for human life.

What is your environmental worldview?

You have a view of the world that is formed through your experiences of life – your background, culture, education and the society in which you live. This is your paradigm or worldview. You may be optimistic or pessimistic in outlook – see the glass as half full or half empty.

▲ **Figure 1.1.3** Is this half full or half empty?

To do

Environmental attitudes questionnaire

Consider these statements and decide if you agree strongly, agree, don't know, disagree or disagree strongly with each.

1. Humans are part of nature.
2. Humans are to blame for all the world's environmental problems.
3. We depend on the environment for our resources (food, water, fuel).
4. Nomadic and indigenous peoples live in balance with their environment.
5. Traditional farming methods do not damage the environment.
6. Nature will make good any damage that humans do to the Earth.
7. Humans have every right to use all resources on the planet Earth.
8. Technology will solve our energy crisis.
9. We have passed the tipping point on climate change and the Earth is warming up and we cannot stop it.
10. Animals and plants have as much right to live on Earth as humans.
11. Looking at a beautiful view is not as important as economic progress.
12. Species have always become extinct on Earth and so it does not matter that humans are causing extinctions.

Discuss your responses with your colleagues. Do they have different ones? Why do you think this is?

Consider these words:

- Environment
- Natural
- Nature.

Think about what they mean to you. Write down your responses.

Now discuss what you wrote with two of your classmates. Do you agree?

What have you written that is similar or different?

Why do you think your responses may be different?

How different do you think the responses of someone from a different century or culture may be? Discuss some examples.

Our relationship with the Earth

Can you think of other phrases that describe our relationship with nature and the Earth?

The words we use are often evaluative and not purely descriptive.

How do you think our language has influenced human perspectives on the environment?

How does the language we use influence your viewpoint?

A classification of different environmental philosophies

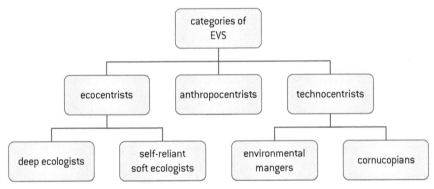

▲ **Figure 1.1.4** EVS categories

Humans like to classify and categorize, and environmental philosophies are no exception to this. The major categories of EVSs are:

- The **ecocentric** worldview – **puts ecology and nature as central to humanity** and emphasizes a less materialistic approach to life with greater self-sufficiency of societies. Is life-centred – which respects the rights of nature and the dependence of humans on nature so has a holistic view of life which is earth-centred. Extreme ecocentrists are **deep ecologists**.

- The **anthropocentric** worldview **– believes humans must sustainably manage the global system**. This might be through the use of taxes, environmental regulation and legislation. Is human-centred – in which humans are not dependent on nature but nature is there to benefit humankind.

- The **technocentric** worldview **– believes that technological developments can provide solutions to environmental problems**. **Environmental managers** are technocentrists. Extreme technocentrists are **cornucopians**.

Many in the industrial world have an **anthropocentric** (human-centred) or **technocentric** (planetary management) worldview. In this humans are seen as the dominant species on Earth and we can manage the environment to suit our needs. Other species only then have value if they are useful to us. This can be summarized as:

- We are the Earth's most important species, we are in charge.

- There will always be more resources to exploit.

- We will control and manage these resources and be successful.

- We can solve any pollution problem that we cause.

- Economic growth is a good thing and we can always keep the economy growing.

- In summary – whatever we do, we can solve it.

Cornucopians include those people who see the world as having infinite resources to benefit humanity. Cornucopians think that through technology and our inventiveness, we can solve any environmental problem and continually improve our living standards. For them, it is growth that will provide the answers and wealth to improve the lot of all and nothing should stand in the way of this. This paradigm sees a free-market economy – capitalism with minimal government control or interference – as the best way to manage markets and the planet. Some see the Earth as a spaceship and we are its captain and crew. If we understand the machine, we can steer it.

Environmental managers see the Earth as a garden that needs tending – the **stewardship** worldview. We have an ethical duty to protect and nurture the Earth. Environmental managers hold the view that there are problems and we need governments to legislate to protect the environment and resources from overexploitation and to make sustainable economies. We may need to compensate those who suffer from environmental degradation and the state has a duty to intervene. Environmental managers believe that if we look after the planet, it looks after us.

The **ecocentric** worldview believes that the views above are too simplistic. We do not even know what species are alive on Earth at the moment and certainly do not know how they interact so it is arrogant of us to think that we can manage it all. To think that we can continue economic growth until every person alive has as high a standard of living as the most affluent is just not possible and so we shall either fall off the treadmill of growth or find it stops beneath us.

Biocentric (life-centred) thinkers see all life as having an inherent value – a value for its own sake, not just for humans. So animals are not just for hunting and eating, trees for logging, lakes for fishing. We should not cause the premature extinction of any other species, whether it does us harm or good or neither. An extreme view of this is that we should not cause the harm of any individual of a species, which is what animal rights activists believe. Others who also call themselves ecocentric (earth-centred) broaden this out to the protection of ecosystems and habitats in which the species live. If we can preserve the ecological integrity and complexity of systems, then life will thrive. To broaden this further, some emphasize the holistic nature of our ethical obligation to the Earth. We are just one species, no more important than the others. Because we are sentient beings and can alter our environment, it is our duty to

restore degraded ecosystems, remove pollution and deal with global environmental problems.

To summarize the ecocentric view:

- The Earth is here for all species.

- Resources are limited.

- We should manage growth so that only beneficial forms occur.

- We must work with the Earth, not against it.

- We need the Earth more than it needs us.

Ecocentrists believe in the importance of small-scale, local community action and the actions of individuals making a difference. They view materialism and our need for more as wrong and do not like centralized decision-making.

At the end of the continuum are the **deep ecologists** who put more value on nature than humanity. They believe in biorights – universal rights where all species and ecosystems have an inherent value and humans have no right to interfere with this. Deep ecologists would like policies to be altered to reduce our impact on the environment, which includes a decrease in the human population and consuming less. Deep ecology is not an ecoreligion but a set of guidelines and values to help us think about our relationship with the Earth and our obligations towards it.

Another way of looking at these environmental value systems is to consider them as nurturing (ecocentric) and intervening or manipulative (technocentric/anthropocentric). These are two extremes of the spectrum on environmental values but most of us also think in both ecocentric and technocentric ways about issues and we may change our minds depending on various factors and as we get older. It is too simplistic to say that we fit into one or the other group all the time.

As we can only experience the world through our human perceptions, our views of the environment are biased by this viewpoint. We talk of animal rights but can only discuss these using our anthropocentric viewpoint. Most of us will take an accommodating view of the environment ('light-green') – faith in the ability of our institutions to adapt to environmental demands and changes and in communities to work together to reduce resource use (eg bottle banks, recycling aluminium cans) – and so be classified as environmental managers in figure 1.1.6. Some of us are cornucopians ('bright-green') with faith in the appliance of science to solve environmental problems and very few are deep ecologists ('deep-green' or 'dark-green') who believe in green rights and the survival of the Earth above the survival of the human species.

'A thing is right when it tends to preserve the integrity, stability and beauty of the biotic community. It is wrong when it tends otherwise.'

Aldo Leopold
(reprinted by permission of Oxford University Press, USA)

To think about

Cost-benefit analysis and the environment

Environmental economists working in industry may be asked how much pollution should be removed from a smokestack of a chimney before the waste is released to the atmosphere. All the pollutant could be removed but at a high cost financially and, in doing so, the company may not be able to afford cleaning up the outflow of heavy metals into a nearby ditch. The opportunity cost of the action is high. There are limited funds and unlimited demands on those funds. Usually costs are passed on to the consumer. So decisions may have to be made that mean some pollution escapes but both demands are met to some extent. Often a cost-benefit analysis is carried out to trade off the costs and benefits.

But valuing the environmental cost is very difficult and it can be argued that cost-benefit analysis cannot apply to these nonmarket effects. How do you value an undisturbed ecosystem or a wild animal or human health? Cost-benefit analysis is still used in decision-making for industry as it is transparent but it may not be the best way. Later in this book, we talk more about how to value the environment, but do be aware that an environmentalist may not always promote the total clean-up or eliminate solution if the opportunity cost is too high. When you add in questions of ethical practice and what is fair to do, you can see how complex this can become.

Practical Work

* For a named local environmental issue, investigate the relationship between position in society and EVS.

* Investigate the relationship between age and environmental attitudes. Investigate the relationship between gender and environmental attitudes.

* Ecosystems may often cross national boundaries and this may lead to conflict arising from the clash of different value systems about exploitation of resources. For one named example (eg ocean fishing, whaling, tropical rainforest exploitation, Antarctica), research the issue and consider the actions taken by different countries in the exploitation of the resources.

To do

1. Draw a table with two columns labelled 'Ecocentric' and 'Anthropocentric/Technocentric'.

2. Put each of the words or phrases below in one of these columns. Don't think for too long about each one. Go with your instinct now you have read about environmental value systems.

Aesthetic	Earth-centred	Managerial
Animal rights	Ecology	Manipulative
Authoritarian	Economy	Nurturing
Belief in technology	Feminist	Participatory
Capitalism	Global co-existence	Preservation
Centralist	Holistic	Reductionist
Competitive	Human-centred	Seeking progress
Consumerism	Individual	Seeking stability
Cooperative	Intervening	Utilitarian

Then put a tick next to the words that best describe your environmental viewpoint.

Draw a line with ecocentric on the left hand side and technocentric/anthropocentric on the right.

Put a cross which you think gives your position and get all your classmates to mark their own as well.

Review this at the end of the course and see if you have moved along the line – to left or right – or moved relative to your classmates.

To do

Copy and complete this table to show the main points of the different environmental philosophies.

Environmental value system	Ecocentric	Anthropocentric	Technocentric
Environmental management strategies			
Environmental philosophies			
Labels and characteristics			
Social movements			
Politics			

Various environmental worldviews

Communism and capitalism in Germany

After the Iron Curtain and Berlin Wall fell in Germany in 1989, western journalists rushed to see East Germany and report upon it. Communism was seen as the antidote to capitalist greed and communists claimed that their system could produce more wealth than capitalism and distribute it more evenly, in the process curing social ills including environmental degradation. But journalists reported on a polluted country in East Germany with the Buna chemical works dumping ten times more mercury into its neighbouring river in a day than a comparable West German plant did in a year. And the smoky two-stroke Trabant cars emitting one hundred times as much carbon monoxide as a western car with a catalytic converter. The message was that capitalism would clean up the industry – but it was not such a non-polluter itself. In some ways the paternalistic communist state had protected the interests of primary producers like farmers and fishermen and so the environment. There was a law that made smelters shut down and so not pollute in spring when crops were growing.

Native American environmental worldview

While there are many native American views, a broad generalization of their views is that they tend to hold property in common (communal), have a subsistence economy, barter for goods rather than use money, and use low-impact technologies. Politically, they come to consensus agreements by participation in a democratic process. The laws are handed down by oral tradition. Most communities have a matrilineal line (descent follows the female side) as opposed to patriarchal, with extended families and low population density. In terms of religion, they are polytheistic (worshipping many gods) and hold that animals and plants as well as natural objects have a spirituality.

Worldviews of Christianity and Islam

The two religions on Earth with the most adherents are Christianity and Islam, together numbering some 3.6 billion. They share the belief in a separation of spirit and matter or body and soul and a notion of 'dominion' or mastery over the Earth. But the ancient Greek view of citizenship and democracy, the Judaic notion of the covenant and the Christian view of unconditional love are examples which have perhaps been distorted in the anthropocentric views of the West. In the biblical book of Genesis, God commands humans to 'replenish the earth, and subdue it; and have dominion over it' (Genesis 1:28). But what does this mean? Are humans to be masters or stewards of the Earth? Do stewards own something or just look after it?

The Quran states that the Earth (and its bounty) has been given to humans for their sustenance. The Quran does differentiate from the Judeo-Christian model in a number of areas however.

'The earth shall rise on new foundations: We have been nought, we shall be all!'

Taken from the Internationale, the anthem of international socialists and communists

15

Shades of green: Where are we now?

In any political movement, there will be changes and developments. It is now difficult to avoid marketing that is based on environmental well-being, often related to human well-being. Organic, biotic, low emissions, energy-saving, sustainable, free-range, green credentials are all terms used in green marketing of products although exactly what they mean and how we perceive them is questionable. 'Greenwash' and 'Green sheen' are terms that describe activities that are not as good for the environment as the producer would like us to believe.

A way of classifying environmentalists today is as dark greens, light greens and bright greens.

Dark greens are dissenters seeking political change in a radical way as they believe that economic development and industrial growth are not the answer. They see a change in the status quo and a reduction in the size of the human population as the way to go. Light greens are individuals who do not want to work politically for change but change their own lifestyles to use fewer resources. For them, it is an individual choice. Bright greens want to use technological developments and social manipulation to make us live sustainably and believe that this can be done by innovation. For bright greens, economic growth may be beneficial if it means more of us live in efficient cities, use more renewable energy and reduce the size of our ecological footprints while increasing our standard of living. For them, we can have it all. The viridian design movement is a spin-off from the bright greens and is about global citizenship and improved design of green products.

What shade of green are you?

- Humans are not given mastery or dominion over the Earth but rather have been granted it as a gift or inheritance. This is a significant difference as it implies caretaker status of God's work not rulers over it.

- The Quran also recognises that the animal world is a community equal to the human one.

- There is more emphasis on the trustee status of human beings and thus the imperative towards charity (the 3rd pillar of Islam).[1]

Another layer comes from ecofeminism as an environmental movement in which ecofeminists argue that it is the rise of male-dominated societies since the advent of agriculture that has led to our view of nature as a foe to be conquered rather than a nurturing Earth mother.

Buddhism's environmental worldview – a religious ecology

Buddhism has evolved over 2,500 years to see the world as conjoined in four ways – morally, existentially, cosmologically and ontologically. Buddhists believe that all sentient beings share the conditions of birth, old age, suffering and death and that every living thing in the world is co-dependent. Buddhist belief teaches that as we are all dependent on each other, whether plant or animal, we are not autonomous and humans cannot be more important than other living things and must extend loving-kindness and compassion not just to life but to the Earth itself.

---

[1] Personal communication from Kosta Lekanides to the authors.

1.2 Systems and models

Significant ideas:

→ A systems approach can help in the study of complex environmental issues.

→ The use of models of systems simplifies interactions but may provide a more holistic view than reducing issues to single processes.

Applications and skills:

→ **Construct** a system diagram or a model from a given set of information.

→ **Evaluate** the use of models as a tool in a given situation, eg for climate change predictions.

Knowledge and understanding:

→ A **systems approach** is a way of visualizing a complex set of interactions which may be ecological or societal.

→ These interactions produce the emergent properties of the system.

→ The concept of a system can be applied to a range of scales.

→ A system is comprised of **storages and flows**.

→ The flows provide inputs and outputs of energy and matter.

→ The flows are processes and may be either **transfers** (a change in location) or **transformations** (a change in the chemical nature, a change in state or a change in energy).

→ In **system diagrams**, storages are usually represented as rectangular boxes, and flows as arrows with the arrow indicating the direction of the flow. The size of the box and the arrow may represent the size/magnitude of the storage or flow.

→ An **open system** exchanges both energy and matter across its boundary while a **closed system** only exchanges energy across its boundary.

→ An **isolated system** is a hypothetical concept in which neither energy nor matter is exchanged across the boundary.

→ **Ecosystems are open systems**. Closed systems only exist experimentally although the global geochemical cycles approximate to closed systems.

→ A **model** is a simplified version of reality and can be used to understand how a system works and predict how it will respond to change.

→ A model inevitably involves some approximation and loss of accuracy.

Why systems?

A system can be living or non-living.

- Systems can be on any scale – small or large. A cell is a system as are you, a bicycle, a car, your home, a pond, an ocean, a smart phone and a farm.

- Open, closed and isolated systems exist in theory though most living systems are open systems.

'Nature does nothing uselessly.'

Aristotle (384–322 BC)

- Material and energy undergo transfers and transformations in flowing from one storage to the next.

- Models have their limitations but can be useful in helping us to understand systems.

This course is called Environmental Systems and Societies and not Environmental Science or Studies. Have you considered why this is? There is a difference in emphasis. In the systems approach, the environment is seen as a set of complex systems: sets of components that function together and form integrated units. You study plants, animals, soils, rocks or the atmosphere not separately (as is sometimes the case in other sciences such as biology, geology or geography), but together as the component parts of complex environments. You also study them in relation to other elements of the system of which they are a part. The course takes an integrated view, and this emphasis on relationships and linkages distinguishes the systems approach. We consider ecosystems in this book and they can be on many scales from a drop of pond water to an ocean, a tree to a forest, a coral reef to an island continent. A biome can be seen as an ecosystem, though it helps if an ecosystem has clear boundaries. The whole biosphere is an ecosystem as well.

We also consider other systems such as the social and economic systems that make our human world work. Decisions about the environment are rarely simply decisions based on ecology or science. We may want to save the tigers but will be constrained by economics, society and political systems which all influence decisions we make.

A system may be an abstract concept as well as something tangible. It is a way of looking at the world. Usually, we can draw a system as a diagram. The environmental value system that you hold consists of your opinions on the environment and how you evaluate it.

A system may remain stable for a long time or may change quickly. Systems occur within their own environment which may be made up of other systems or ecosystems, and they usually exchange inputs and outputs – energy and matter in living systems, information in non-living ones – with their environment. Systems are all more than the sum of their parts, for example a computer is more than the materials used to make it.

Key term

A **system** is a set of inter-related parts working together to make a complex whole.

▲ **Figure 1.2.1** The Earth and its Moon viewed from space

The human place in the biosphere

The biosphere is a fragile skin on the planet Earth. It includes the air (atmosphere), rocks (lithosphere) and water (hydrosphere) within which life occurs. Humans and all other organisms live within this thin layer yet we know little about how it is regulated or self-regulates, or about the effects the human species is having upon it.

Types of system

Systems can be thought of as one of three types: open, closed and isolated.

An **open system** exchanges matter and energy with its surroundings (see figure 1.2.3).

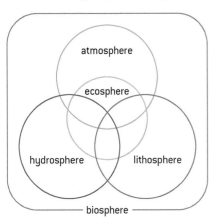

biosphere = atmosphere + lithosphere + hydrosphere + ecosphere

Figure 1.2.2 Relationships within the biosphere

All systems have:	Represented by:
STORAGES or stores of matter or energy	a box
FLOWS into, through and out of the system	arrows
INPUTS	arrows in
OUTPUTS	arrows out
BOUNDARIES	lines
PROCESSES which transfer or transform energy or matter from storage to storage	Eg respiration, precipitation, diffusion

Figure 1.2.3 Systems terminology

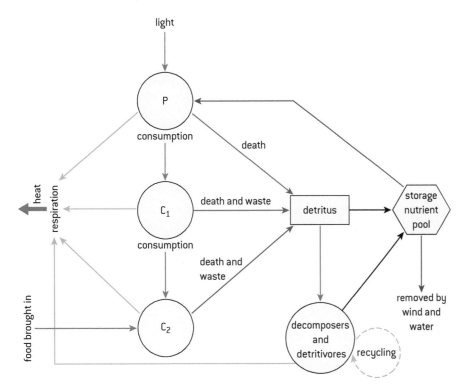

Figure 1.2.4 Energy and matter exchange in an immature forest ecosystem

Transfers and transformations

Both matter (or material) and energy move or flow through ecosystems as:

- transfers: water moving from a river to the sea, chemical energy in the form of sugars moving from a herbivore to a carnivore or:
  - the movement of material through living organisms (carnivores eating other animals)
  - the movement of material in a non-living process (water being carried by a stream)
  - the movement of energy (ocean currents transferring heat).

Practical Work

* Create a model ecosystem in a plastic soda bottle (sub-topic 5.4).

* Construct a model of your home, with storages and flows.

* Evaluate climate change models (7.2).

▶ **Figure 1.2.5** The Biogeochemical Cycle illustrating the general flows in an ecosystem. Energy flows from one compartment to another, eg in a food chain. But when one organism eats another organism, the energy that moves between them is in the form of stored chemical energy: the body of the prey organism

- transformations: liquid to gas, light to chemical energy:
  - matter to matter (soluble glucose converted to insoluble starch in plants)
  - energy to energy (light converted to heat by radiating surfaces)
  - matter to energy (burning fossil fuels)
  - energy to matter (photosynthesis).

Both types of flow require energy; transfers, being simpler, require less energy and are therefore more efficient than transformations.

Flows and storages

Both energy and matter flow (as inputs and outputs) through ecosystems but, at times, they are also stored (as storages or stock) within the ecosystem.

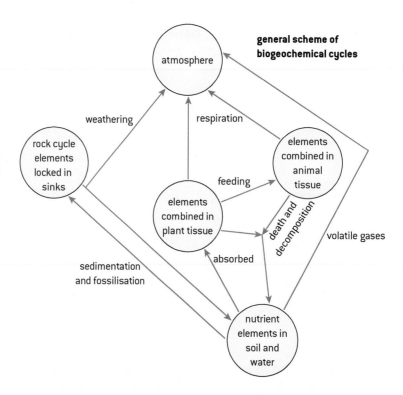

More on systems

Examples of systems

An ecosystem is a good example of a 'system'.

Using the model below, draw your own systems diagram for:

a) A candle

b) A mobile phone

c) A green plant

d) You

e) Your school

f) A lake

Label the inputs, outputs, storages and flows.

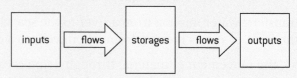

▲ **Figure 1.2.6**

Most systems are **open systems**. All ecosystems are open systems exchanging matter and energy with their environment.

In forest ecosystems:

- Plants fix energy from light entering the system during photosynthesis.
- Nitrogen from the air is fixed by soil bacteria.
- Herbivores that live within the forest may graze in adjacent ecosystems such as a grassland, but when they return they enrich the soil with feces.
- Forest fires expose the topsoil which may be removed by wind and rain.
- Mineral nutrients are leached out of the soil and transported in groundwater to streams and rivers.
- Water is lost through evaporation and transpiration from plants.
- Heat is exchanged with the surrounding environment across the boundaries of the forest.

Open system models can even be applied to the remotest oceanic island – energy and matter are exchanged with the atmosphere, surrounding oceans and even migratory birds.

A **closed system** exchanges energy but not matter with its environment.

Closed systems are extremely rare in nature. However, on a global scale, the hydrological, carbon, and nitrogen cycles are closed – they exchange only energy and no matter. The planet itself can be thought of as an 'almost' closed system.

Light energy in large amounts enters the Earth's system and some is eventually returned to space as long-wave radiation (heat). (Because a small amount of matter is exchanged between the Earth and space, it is not truly a closed system. What types of matter can you think of that enter the Earth's atmosphere and what types that leave it?)

Most examples of closed systems are artificial, and are constructed for experimental purposes. An aquarium or terrarium may be sealed so that only energy in the form of light and heat but not matter can be exchanged. Examples include bottle gardens or sealed terraria but they usually do not survive for long as the system becomes unbalanced, for example not enough food for the animals, or not enough oxygen or carbon dioxide, and organisms die. An example of a closed system that went wrong is Biosphere 2 (see p 22). An example of a closed system that is in equilibrium is at http://www.dailymail.co.uk/sciencetech/article-2267504/The-sealed-bottle-garden-thriving-40-years-fresh-air-water.html

An **isolated system** exchanges neither matter nor energy with its environment. Isolated systems do not exist naturally though it is possible to think of the entire universe as an isolated system.

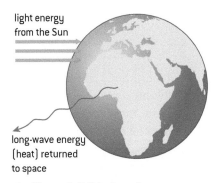

▲ **Figure 1.2.7** A closed system – the Earth

System	Energy exchanged	Matter exchanged
Open	Yes	Yes
Closed	Yes	No
Isolated	No	No

Biosphere 2

Biosphere 2, a prototype space city, was a human attempt to create a habitable closed system on Earth. Built in Arizona at the end of the 1980s, Biosphere 2 was a three-hectare greenhouse intended to explore the use of closed biospheres in space colonization. Two major 'missions' were conducted but both ran into problems. The Biosphere never managed to produce enough food to adequately sustain the participants and at times oxygen levels became dangerously low and needed augmenting – they opened the windows so making it an open system.

▲ **Figure 1.2.8** Biosphere 2

Inside were various ecosystems: a rainforest, coral reef, mangroves, savanna, desert, an agricultural area and living quarters. Electricity was generated from natural gas and the whole building was sealed off from the outside world.

For two years, eight people lived in Biosphere 2 in a first trial. But oxygen levels dropped from 21% to 14% and of the 25 small animal species put in, 19 became extinct, while ants, cockroaches and katydids thrived. Bananas grew well but there was not enough food to keep the eight people from being hungry. Oxygen levels gradually fell and it is thought that soil microbes respired much of this. Carbon dioxide levels fluctuated widely. A second trial started in 1994 but closed after a month when two of the team vandalized the project, opening up doors to the outside. Cooling the massive greenhouses was an issue, using three units of energy from air conditioners to cool the air for the input of every one unit of solar energy. So there were social, biological and technological problems with the project as the team split into factions and questions were asked as to whether this was a scientific, business or artistic venture.

The result was to show how difficult it is to make a sustainable closed system when the complexities of the component ecosystems are not fully understood.

To do

There is a TED talk about this http://www.ted.com/talks/jane_poynter_life_in_biosphere_2. Watch it.

Questions

1. Why do you think this was called Biosphere 2?

2. Biosphere 2 has been described as a 'closed system'. What does this mean?

3. Biosphere 2 was designed to include some of the major ecosystems of the Earth.

4. List the ecosystems and divide them into terrestrial (land based) and marine (sea-water based).

To think about

Atoms

All matter is made up of atoms. You are taught this in some of your first science lessons. Living things are made up of atoms, grouped into molecules and macromolecules, organelles, cells, tissues, organs and systems.

Read these two excerpts and think about what makes you you.

From 'Quantum theory and relativity explained' (Daily Telegraph 20th November 2007)

'Quantum theory has made the modern world possible, giving us lasers and computers and iPod nanos, not to mention explaining how the sun shines and why the ground is solid.

Take the fact that you are constantly inhaling fragments of Marilyn Monroe. It is stretching it a bit to say that this is a direct consequence of quantum theory.

Nevertheless, it is connected to the properties of atoms, the Lego bricks from which we are all assembled, and quantum theory is essentially a description of this microscopic world.

The important thing to realize is that atoms are small. It would take about 10 million of them laid end to end to span the full stop at the end of this sentence. It means that every time you breathe out, uncountable trillions of the little blighters spread out into the air.

Eventually the wind will spread them evenly throughout the Earth's atmosphere. When this happens, every lungful of the atmosphere will contain one or two atoms you breathed out.

So, each time someone inhales, they will breathe in an atom breathed out by you — or Marilyn Monroe, or Alexander the Great, or the last *Tyrannosaurus rex* that stalked the Earth.'

From Bill Bryson's 'A Short History of Nearly Everything'

'Why atoms take this trouble is a bit of a puzzle. Being you is not a gratifying experience at the atomic level. For all their devoted attention, your atoms do not actually care about you—indeed, they do not even know that you are there. They don't even know that they are there. They are mindless particles, after all, and not even themselves alive. (It is a slightly arresting notion that if you were to pick yourself apart with tweezers, one atom at a time, you would produce a mound of fine atomic dust, none of which had ever been alive but all of which had once been you.) Yet somehow for the period of your existence they will answer to a single overarching impulse: to keep you you.

The bad news is that atoms are fickle, and their time of devotion is fleeting indeed. Even a long human life adds up to only about 650,000 hours, and when that modest milestone flashes past, for reasons unknown, your atoms will shut you down, silently disassemble, and go off to be other things. And that's it for you. Still, you may rejoice that it happens at all. Generally speaking in the universe, it doesn't...so far as we can tell.'

Bill Bryson continues to say that 'life is simple in terms of chemicals – oxygen, hydrogen, carbon, nitrogen make up most of all living things and a few other elements too – sulfur, calcium and some others. But in combination and for a short time, they can make you and that is the miracle of life.'

Models of systems

A model is a simplified version of the real thing. We use models to help us understand how a system works and to predict what happens if something changes. Systems work in predictable ways, following rules, we just do not always know what these rules are. A model can take many forms. It could be:

- a physical model, for example a wind tunnel or river, a globe or model of the solar system, an aquarium or terrarium

- a software model, for example of climate change or evolution (Lovelock's Daisyworld)

- mathematical equations
- data flow diagrams.

Models have their limitations as well as strengths. While they may omit some of the complexities of the real system (through lack of knowledge or for simplicity), they allow us to look ahead and predict the effects of a change to an input to the system.

To do

Compare these two models

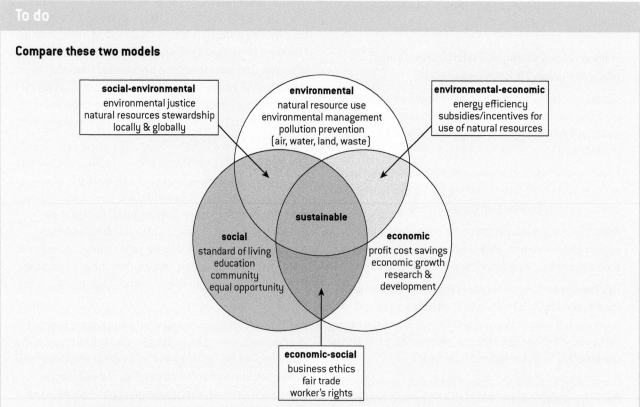

▲ **Figure 1.2.9** The spheres of a sustainable model. Only when all three overlap is there sustainability

Why are any of these circles in the Venn diagram outside the environment?

Is culture relevant to these models of sustainability? Where would you draw it in?

Does the model change how we treat our environment?

Evaluate these models. (Consider their strengths and weaknesses.)

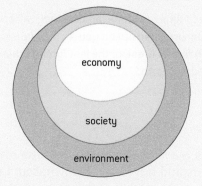

▲ **Figure 1.2.10** An alternative model of sustainability representing all other systems within the environmental system

The strengths of models are:

- Easier to work with than complex reality.
- Can be used to predict the effect of a change of input.
- Can be applied to other similar situations.
- Help us see patterns.
- Can be used to visualize really small things (atoms) and really large things (solar system).

The weaknesses of models are:

- Accuracy is lost because the model is simplified.
- If our assumptions are wrong, the model will be wrong.
- Predictions may be inaccurate.

Sustainable development modelling

See 1.4 for more on sustainable development and sustainability.

> **To do**
>
> http://gingerbooth.com/ flash/daisyball/ links to the Daisyworld game. Have a go.

To think about

Gaia – a model of the Earth

The 'Great Aerial Ocean' was Alfred Russel Wallace's description of the atmosphere.

'You can cut the atmosphere with a knife' is a common saying. If we could see the atmosphere, perhaps we would consider it and look after it more. As we cannot, perhaps we take it for granted.

In 1979, James Lovelock published *The Gaia Hypothesis*. In it he argued that the Earth is a planet-sized organism and the atmosphere is its organ that regulates it and connects all its parts. (Gaia is an Ancient Greek Earth goddess.) Lovelock argued that the biosphere keeps the composition of the atmosphere within certain boundaries by negative feedback mechanisms.

He based his argument on these facts:

1. The temperature at the Earth's surface is constant even though the Sun is giving out 30% more energy than when the Earth was formed.

2. The composition of the atmosphere is constant with 79% nitrogen, 21% oxygen and 0.03% carbon dioxide. Oxygen is a reactive gas and should be reacting but it does not.

3. The oceans' salinity is constant at about 3.4% but rivers washing salts into the seas should increase this.

He was much criticized over this hypothesis but Lynn Margulis who worked with him also supported his views though uses less emotive language about the Earth as an organism. Lovelock has defended his hypothesis for 30 years and many people are now accepting some of his views. He developed a Daisyworld as a mathematical simulation to show that feedback mechanisms can evolve from the activities of self-interested organisms – black and white daisies in this case.

In Lovelock's 2007 book, *The Revenge of Gaia*, he makes a strong case for the Earth being an 'older lady' now, more than half way through her existence as a planet and so not being able to bounce back from changes as well as she used to. He suggests that we may be entering a phase of positive feedback when the previously stable equilibrium will become unstable and we will shift to a new, hotter equilibrium state. Controversially, he suggests that the human population will survive but with a 90% reduction in numbers.

To do

1. Define a system

2. Fill in the gaps

The terms 'open', 'closed' and 'isolated' are used to describe particular kinds of systems. Match the above names to the following definitions:

- A _____ system exchanges matter and energy with its surroundings (eg an ecosystem).

- A _____ system exchanges energy but not matter (The 'Biosphere 2' experiment was an attempt to model this. These systems do not occur naturally on Earth, although the biosphere (or Gaia) itself can be considered a _____ system.)

- A _____ system exchanges neither matter nor energy. (No such systems exist, with the possible exception of the entire cosmos.)

All ecosystems are _____ systems, because of the input of _____ energy and the exchange of _____ with other ecosystems.

3. Systems circus

Look at the following simple systems and complete the table:

	Burning candle	Boiling kettle	A plant	Animal population
Inputs				
Outputs				
Energy and material transfers				
Energy and material transformations				

1.3 Energy and equilibria

Significant ideas:

→ The laws of thermodynamics govern the flow of energy in a system and the ability to do work.

→ Systems can exist in alternative stable states or as equilibria between which there are tipping points.

→ Destabilizing positive feedback mechanisms will drive systems toward these tipping points, whereas stabilizing negative feedback mechanisms will resist such changes.

Applications and skills:

→ **Explain** the implications of the laws of thermodynamics to ecological systems.

→ **Discuss** resilience in a variety of systems.

→ **Evaluate** the possible consequences of tipping points.

Knowledge and understanding:

→ The **first law of thermodynamics** is the **principle of conservation of energy**, which states that energy in an isolated system can be transformed but cannot be created or destroyed.

→ The principle of conservation of energy can be modelled by the energy transformations along food chains and energy production systems.

→ The **second law of thermodynamics** states that the entropy of a system increases over time. **Entropy** is a measure of the amount of disorder in a system. An increase in entropy arising from energy **transformations** reduces the energy available to do work.

→ The second law of thermodynamics explains the inefficiency and decrease in available energy along a food chain and energy generation systems.

→ As an open system, an ecosystem, will normally exist in a **stable equilibrium**, either a **steady-state** or one developing over time (eg succession), and maintained by stabilizing negative feedback loops.

→ **Negative feedback loops** (stabilizing) occur when the output of a process inhibits or reverses the operation of the same process in such a way to reduce change – it counteracts deviation.

→ **Positive feedback loops** (destabilizing) will tend to amplify changes and drive the system toward a tipping point where a new equilibrium is adopted.

→ The **resilience** of a system, ecological or social, refers to its tendency to avoid such tipping points and maintain stability.

→ Diversity and the size of storages within systems can contribute to their resilience and affect the speed of response to change (time lags).

→ Humans can affect the resilience of systems through reducing these storages and diversity.

→ The delays involved in feedback loops make it difficult to predict **tipping points** and add to the complexity of modelling systems.

Energy in all systems is subject to the laws of thermodynamics.

According to the **first law of thermodynamics**, energy is neither created nor destroyed. What this really means is that the total energy in any isolated system, such as the entire universe, is constant. All that can happen is that the form the energy takes changes. This first law is often called the **principle of conservation of energy.**

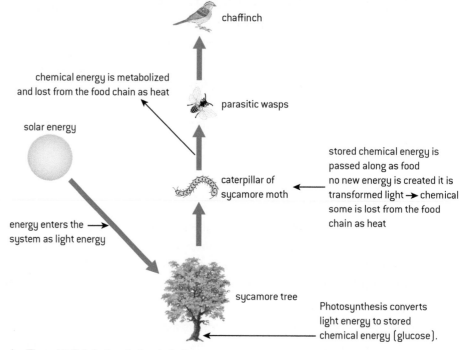

▲ **Figure 1.3.1** A simple food chain

In a power station, one form of energy (from eg coal, oil, nuclear power, moving water) is converted or transformed into electricity.

In your body, food provides chemical energy which you convert into heat or kinetic energy.

If we look at the sunlight falling on Earth, not all of it is used for photosynthesis.

The **second law of thermodynamics** states that the **entropy** of an isolated system not in equilibrium will tend to increase over time.

- Entropy is a measure of disorder of a system and it refers to the spreading out or dispersal of energy.

- More entropy = less order.

- Over time, all differences in energy in the universe will be evened out until nothing can change.

- Energy conversions are never 100% efficient.

- When energy is used to do work, some energy is always dissipated (lost to the environment) as waste heat.

This process can be summarized by a simple diagram showing the energy input and outputs.

Key terms

The **first law of thermodynamics** is the **principle of conservation of energy**, which states that energy in an isolated system can be transformed but cannot be created or destroyed.

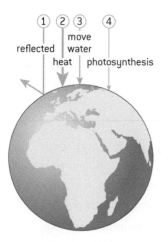

▲ **Figure 1.3.2** The fate of the Sun's energy hitting the Earth. About 30% is reflected back into space (1), around 50% is converted to heat (2), and most of the rest powers the hydrological cycle: rain, evaporation, wind, etc (3). Less than 1% of incoming light is used for photosynthesis (4).

Key terms

The **second law of thermodynamics** refers to the fact that energy is transformed through energy transfers. **Entropy** is a measure of the amount of disorder in a system. An increase in entropy arising from energy **transformations** reduces the energy available to do work.

energy = work + heat (and other wasted energy)

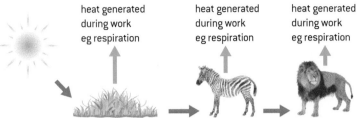

heat generated during work eg respiration

heat generated during work eg respiration

heat generated during work eg respiration

▲ **Figure 1.3.4** Loss of energy to the environment in a food chain

In the example in figure 1.3.4, the energy spreads out so the useful energy consumed by one trophic level is less than the total energy at the level below.

- Depending on the type of plant, the efficiency at converting solar energy to stored sugars is around 1–2%.

- Herbivores on average only assimilate (turn into animal matter) about 10% of the total plant energy they consume. The rest is lost in metabolic processes and escaping from the carnivore. This changes the stored chemical energy in its cells into useful work (running). But during its attempted escape some of the stored energy is converted to heat and lost from the food chain.

- A carnivore's efficiency is also only around 10% (see 2.3). As with the herbivore they metabolize stored chemical energy, in this case trying to catch the herbivore.

- So as energy is dispersed to the environment, there will always be a reduction in the amount of energy passed on to the next trophic level.

- That means the carnivore's total efficiency in the chain is 0.02 × 0.1 × 0.1 = 0.0002%.

- This means the carnivore loses most of its energy as heat into the surrounding environment.

Life is a battle against entropy and, without the constant replenishment of energy, life cannot exist. Consider this pictorial view (figure 1.3.5) of paddling upstream. Stop for a moment and you are swept back downstream by the current of entropy.

Simple example of entropy:

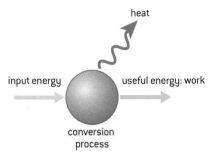

heat

input energy

useful energy: work

conversion process

▲ **Figure 1.3.3** The second law of thermodynamics

Key term

Entropy is a measure of the amount of disorder in a system.

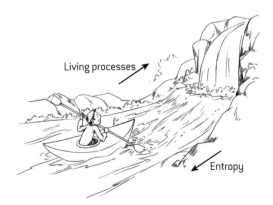

Living processes

Entropy

▲ **Figure 1.3.5** A representation of life against entropy

Tidy room has order: low entropy
Does this happen naturally without the input of energy?

Untidy room has disorder: high entropy

▲ **Figure 1.3.6** Which is your room?

The situation depicted in figure 1.3.6 obeys the second law of thermodynamics, since the tidy room of low entropy becomes untidy, a situation of high entropy. In the process, entropy increases spontaneously.

Solar energy powers photosynthesis. Chemical energy, through respiration, powers all activities of life. Electrical energy runs all home appliances. The potential energy of a waterfall turns a turbine to produce electricity. These are all high-quality forms of energy, because they power useful processes. They are all ordered forms of energy. Solar energy reaches us via photons in solar rays; chemical energy is stored in the bonds of macromolecules like sugars; the potential energy of falling water is due to the specific position of water, namely that it is high and falls. These ordered forms have low disorder, so low entropy.

On the contrary, heat may not power any process; it is a low-quality form of energy. Heat is simply dispersed in space, being capable only of warming it up. Heat dissipates to the environment without any order; it is disordered. In other words, heat is a form of energy characterized by high entropy.

To think about

Implications of the second law for environmental systems

We experience the second law in our everyday lives. All living creatures die and in doing so:

- entropy or disorder tends to increase

- the creatures move from order to disorder

- but organisms manage to 'survive' against the odds, that is against the second law of thermodynamics

- living creatures manage to maintain their order and defy entropy to stay alive by continuous input of energy by continuously getting chemical energy from organic compounds via respiration

- energy is even required at rest – if they do not respire they die.

This is the same as the example of the room; the only way to keep the room tidy is to continuously clean it, that is to expend energy.

In any process, some of the useful energy turns into heat:

- Low-entropy (high-quality) energy degrades into high-entropy (low-quality) heat.

- So the entropy of the living system stays low, whilst the entropy of the environment is increasing.

- Photosynthesis and respiration are good examples.

  - Low-entropy solar energy turns into higher-entropy chemical energy.

  - Chemical energy turns into even higher-entropy mechanical energy and is 'lost' as heat (low-quality, high-entropy).

- This increases the entropy of the environment, in which heat dissipates.

- As a consequence, no process can be 100% efficient.

Key term

Efficiency is defined as the useful energy, the work or output produced by a process divided by the amount of energy consumed being the input to the process:

efficiency = work or energy produced / energy consumed

efficiency = useful output/ input

Multiply by 100%, if you want to express efficiency as a percentage.

Complexity and stability

Most ecosystems are very complex. There are many feedback links, flows and storages. It is likely that a high level of complexity makes for a more stable system which can withstand stress and change better than a simple one can, as another pathway can take over if one is removed. Imagine a road system where one road is blocked by a broken-down truck; vehicles can find an alternative route on other roads. If a community has a number of predators and one is wiped out by disease, the others will increase as there is more prey for them to eat and prey numbers will not increase. If on the other hand systems are simple they may lack stability.

- Tundra ecosystems are fairly simple and thus populations in them may fluctuate widely, eg lemming population numbers.

- Monocultures (farming systems in which there is only one major crop) are also simple and thus vulnerable to the sudden spread of a pest or disease through a large area with devastating effect. The spread of potato blight through Ireland in 1845–8 provides an example; potato was the major crop grown over large areas of the island, and the biological, economic and political consequences were severe.

Key term

Negative feedback loops are stabilizing and occur when the output of a process inhibits or reverses the operation of the same process in such a way to reduce change – it counteracts deviation.

Equilibrium

Equilibrium is the tendency of the system to return to an original state following disturbance; at equilibrium, a state of balance exists among the components of that system.

We can think of systems as being in dynamic (steady-state) or static equilibria as well as in stable or unstable equilibria. We discuss each of these here. Note that the term steady-state equilibrium is used instead of dynamic equilibrium in this book.

Open systems tend to exist in a state of balance or stable equilibrium. Equilibrium avoids sudden changes in a system, though this does not mean that all systems are non-changing. If change exists it tends to exist between limits.

A **steady-state equilibrium** is a characteristic of open systems where there are continuous inputs and outputs of energy and matter, but the system as a whole remains in a more-or-less constant state (eg a climax ecosystem).

Key term

A **steady-state equilibrium** is a characteristic of open systems where there are continuous inputs and outputs of energy and matter, but the system as a whole remains in a more-or-less constant state (eg a climax ecosystem).

Negative feedback stabilizes steady-state equilibria. It tends to damp down, neutralize or counteract any deviation from an equilibrium, and it stabilizes systems or results in steady-state (dynamic) equilibrium. It results in self-regulation of a system.

In a steady-state equilibrium there are no long-term changes but there may be small fluctuations in the short term, eg in response to weather changes, and the system will return to its previous equilibrium condition following the removal of the disturbance.

Some systems may undergo long-term changes to their equilibrium as they develop over time while retaining integrity to the system. Successions (see 2.4) are good examples of this.

Examples of a steady-state equilibrium

1. A water tank. If it fills at the same rate that it empties, there is no net change but the water flows in and out. It is in a steady state.

2. In economics, a market may be stable but there are flows of capital in and out of the market.

3. In ecology, a population of ants or any organism may stay the same size but individual organisms are born and die. If these birth and death rates are equal, there is no net change in population size.

4. A mature, climax ecosystem, like a forest, is in steady-state equilibrium as there are no long-term changes. It usually looks much the same for long periods of time, although all the trees and other organisms are growing, dying and being replaced by younger ones. However, there are flows in and out of the system – light inputs from the sun, energy outputs as heat lost through respiration; matter inputs in rainwater and gases, outputs in salts lost in leaching and rain washing away the soil. However, over years, the inputs and outputs balance.

5. Another example of a steady-state equilibrium is people maintaining a constant body weight, thus 'burning' all the calories (energy) we get from our food. In cases of increasing or decreasing body weight there is no steady state.

6. The maintenance of a constant body temperature is another example. We sweat to cool ourselves and shiver to warm up but our body core temperature is about 37 °C.

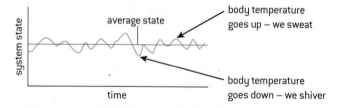

▲ **Figure 1.3.7** Steady-state equilibrium

Maintenance of a steady-state equilibrium is achieved through negative feedback mechanisms, as we shall see later.

Static equilibrium

Another kind of equilibrium is called a **static equilibrium**, in which there is no change over time, eg a pile of books which does not move unless toppled over. When a static equilibrium is disturbed it will adopt a new equilibrium as a result of the disturbance. A pile of scree material (a mass of weathered rock fragments) piled up against a cliff could be said to exist in static equilibrium. The forces within the system are in balance, and the components (the rock fragments, the cliff and the valley floor) remain unchanged in their relationship to one another for long periods of time.

Most non-living systems like a pile of rocks or a building are in a state of static equilibrium. This means that they do not change their position or state, ie they look the same for long periods of time and the rocks or bricks stay in the same place.

This cannot occur in living systems as life involves exchange of energy and matter with the environment.

Unstable and stable equilibria

Systems can also be **stable** or **unstable**.

In a stable equilibrium the system tends to return to the same equilibrium after a disturbance.

In an unstable equilibrium the system returns to a new equilibrium after disturbance.

Possibly this is happening to our climate and the new state will be hotter.

Feedback loops

Systems are continually affected by information from outside and inside the system. Simple examples of this are:

1. If you start to feel cold you can either put on more clothes or turn the heating up. The sense of cold is the information, putting on clothes is the reaction.

2. If you feel hungry, you have a choice of reactions as a result of processing this 'information': eat food, or do not eat and feel more hungry.

Natural systems act in exactly the same way.

Feedback loop mechanisms can either be:

- Positive:
  - Change a system to a new state.
  - Destabilizing as they increase change.
- Negative
  - Return it to its original state.
  - Stabilizing as they reduce change.

▲ **Figure 1.3.8** Static equilibrium

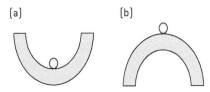

▲ **Figure 1.3.9** Diagrams of (a) stable and (b) unstable equilibrium

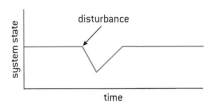

▲ **Figure 1.3.10** Stable equilibrium

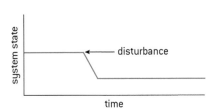

▲ **Figure 1.3.11** Unstable equilibrium

Key term

A **feedback loop** is when information that starts a reaction in turn may input more information which may start another reaction.

Examples of negative feedback

1. Your body temperature starts to rise above 37 °C because you are walking in the tropical sun and the air temperature is 45 °C. The sensors in your skin detect that your surface temperature is rising so you start to sweat and go red as blood flow in the capillaries under your skin increases. Your body attempts to lose heat.

2. A thermostat in a central heating system is a device that can sense the temperature. It switches a heating system on when the temperature decreases to a predetermined level, and off when it rises to another warmer temperature. So a room, a building, or a piece of industrial plant can be maintained within narrow limits of temperature.

3. Global temperature rises causing ice caps to melt. More water in the atmosphere means more clouds, more solar radiation is reflected by the clouds so global temperatures fall. But compare this with figure 1.3.14 which interprets it differently.

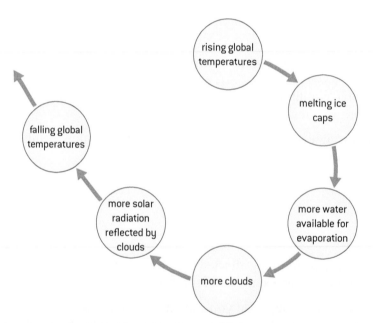

▲ **Figure 1.3.12** Negative feedback dampening change

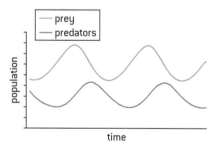

▲ **Figure 1.3.13** Cycles of predator and prey in the Lotka–Volterra model

4. Predator–prey interactions. The Lotka–Volterra model (proposed in 1925 and 1926) is also known as the predator–prey model and shows the effect of changing numbers of prey on predator numbers. When prey populations (eg mice) increase, there is more food for the predator (eg owl) so they eat more and breed more, resulting in more predators which eat more prey so the prey numbers decrease. If there are fewer prey, there is less food and the predator numbers decrease. The change in predator numbers lags behind the change in prey numbers. The snowshoe hare and Canadian lynx is a well-documented example of this (see box, p36).

5. Some organisms have internal feedback systems, physiological changes occurring that prevent breeding when population densities are high, promoting breeding when they are low. It is negative feedback loops such as these that maintain 'the balance of nature'.

Positive feedback results in a further increase or decrease in the output that enhances the change in the system. It is destabilized and pushed to a new state of equilibrium. The process may speed up, taking ever-increasing amounts of input until the system collapses. Alternatively, the process may be stopped abruptly by an external force or factor. Positive feedback results in a 'vicious circle'.

Examples of positive feedback

1. You are lost on a high snowy mountain. When your body senses that it is cooling below 37 °C, various mechanisms such as shivering help to raise your body core temperature again. But if these are insufficient to restore normal body temperature, your metabolic processes start to slow down, because the enzymes that control them do not work so well at lower temperatures. As a result you become lethargic and sleepy and move around less and less, allowing your body to cool even further. Unless you are rescued at this point, your body will reach a new equilibrium: you will die of hypothermia.

2. In some developing countries poverty causes illness and contributes to poor standards of education. In the absence of knowledge of family planning methods and hygiene, this contributes to population growth and illness, adding further to the causes of poverty: 'a vicious circle of poverty'.

3. Global temperature rises causing ice caps to melt. Dark soil is exposed so more solar radiation is absorbed. This reduces the **albedo** (reflecting ability of a surface) of Earth so global temperature rises. Compare this with figure 1.3.12 and you can see that the same change can result in positive or negative feedback. This is one reason that predicting climate change is so difficult.

> ### Key term
>
> **Positive feedback loops** (destabilizing) will tend to amplify changes and drive the system toward a tipping point where a new equilibrium is adopted.

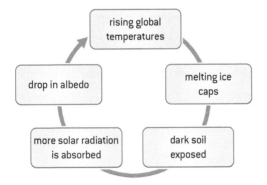

Figure 1.3.14 Positive feedback in global warming

Whether a system is viewed as being in static or steady-state equilibrium may be a matter of the timescale. An ecosystem undergoing succession (see 2.4) is in a state of flux – it changes constantly. In succession, the system undergoes long-term changes. However, the system retains its long-term integrity, since it is functioning properly, in a balanced, natural way. A better way to describe this situation is that the system shows stability and all systems in nature show **stability** by default.

Both natural and human systems are regulated by feedback mechanisms. Generally, we wish to preserve the environment in its present state, so negative feedback is usually helpful and positive feedback is usually undesirable. However there are situations where change is needed and positive feedback is advantageous, eg if students enjoy their Environmental Systems and Societies lessons, they want to learn more, so attend classes regularly and complete assignments. Consequently they move to a new equilibrium of being better educated about the environment.

We shall come back to feedback loops in various sections of this book, particularly in climate change and sustainable development.

To do

Predator–prey interactions and negative feedback

▲ **Figure 1.3.15** Canadian lynx chasing snowshoe hare

The Hudson Bay Trading Company in Northern Canada kept very careful records of pelts (skins) brought in and sold by hunters over almost a century. This is a classic set of data and shows this relationship because the hare is the only prey of the lynx and the lynx its only predator. Usually things are more complicated.

Figure 1.3.16 (adapted from Odum, *Fundamentals of Ecology*, Saunders, 1953) shows a plot of that data.

We have to assume that the numbers of animals trapped were small compared to the total populations and that the numbers trapped were roughly proportional to total population numbers. Also assumed is the prey always has enough food so does not starve. Given that, the cycles are remarkably constant with the lynx populations always smaller than and lagging behind the hare ones.

1. On average, what was the cycle length of the lynx population?

2. On average, what was the cycle length of the hare population?

3. Why do lynx numbers lag behind hare numbers?

4. Why are lynx numbers smaller than hare numbers?

Things are never as straightforward in ecology as we expect though. In regions where lynx died out, hare populations still continued to fluctuate. Why do you think this was?

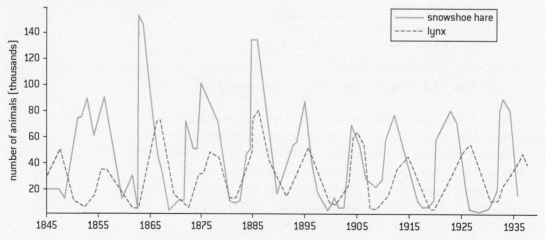

▲ **Figure 1.3.16** Snowshoe hare and Canadian lynx population numbers from 1845 to 1940

To do

Here are a number of examples of how both positive and negative feedback mechanisms might operate in the physical environment. No one can be sure which of these effects is likely to be most influential, and consequently we cannot know whether or not the Earth will manage to regulate its temperature, despite human interference with many natural processes.

Label each example as either positive or negative feedback.

Draw diagrams of one example of positive feedback and one example of negative feedback using the examples given, to show how feedback affects a system. Include feedback loops on your diagrams.

1. As carbon dioxide levels in the atmosphere rise the temperature of the Earth rises.

   As the Earth warms the rate of photosynthesis in plants increases, more carbon dioxide is therefore removed from the atmosphere by plants, reducing the greenhouse effect and reducing global temperatures.

2. As the Earth warms:

   Ice cover melts, exposing soil or water.

   Albedo decreases (albedo is the fraction of light that is reflected by a body or surface).

   More energy is absorbed by Earth's surface.

   Global temperature rises.

   More ice melts.

3. As Earth warms, upper layers of permafrost melt, producing waterlogged soil above frozen ground.

   Methane gas is released in an anoxic environment.

   The greenhouse effect is enhanced.

   Earth warms, melting more permafrost.

4. As Earth warms, increased evaporation produces more clouds.

   Clouds increase albedo, reflecting more light away from Earth.

   Temperature falls.

   Rates of evaporation fall.

5. As Earth warms, organic matter in soil is decomposed faster:

   More carbon dioxide is released.

   Enhanced greenhouse effect occurs.

   Earth warms further.

   Rates of decomposition increase.

6. As Earth warms, evaporation increases:

   Snowfall at high latitudes increases.

   Icecaps enlarge.

   More energy is reflected by increased albedo of ice cover.

   Earth cools.

   Rates of evaporation fall.

7. As Earth warms, polar icecaps melt releasing large numbers of icebergs into oceans.

   Warm ocean currents such as Gulf Stream are disrupted by additional freshwater input into ocean.

   Reduced transfer of energy to poles reduces temperature at high latitudes.
   Ice sheets reform and icebergs retreat.

   Warm currents are re-established.

Resilience of systems

The resilience of a system measures how it responds to a disturbance. The more resilient a system, the more disturbance it can deal with. Resilience is the ability of a system to return to its initial state after a disturbance. If it has low resilience, it will enter a new state – see figure 1.3.17.

Resilience is generally considered a good thing, whether in a society, individual or ecosystem as it maintains stability of the system.

In eucalypt forests of Australia, fire is seen as a major hazard. But eucalypts have evolved to survive forest fires. Their oil is highly flammable and the trees produce a lot of litter which also burns easily. But the trees regenerate quickly after a fire because they have buds within their trunks

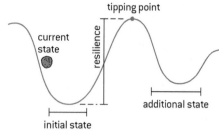

Figure 1.3.17 Resilience can be modelled as a ball in a bowl. If the ball is pushed upwards, it returns to the bottom of the bowl – its initial state. But if it is pushed enough, it will leave the bowl and settle elsewhere – in an additional state. The higher the walls of the bowl, the more resilience the system has as the more energy you need to push it out of the bowl

and plants that would have competed with them are destroyed. The eucalypts are resilient. But when the indigenous eucalypts are replaced by tree species that cannot withstand fire, it can be devastating.

In managed systems, such as agriculture, we want stability so we can predict that the amount of food we grow is about the same each year. If this does not happen, there can be disastrous consequences, for example the Irish potato famine or the Sahel drought and famine.

But resilience is not always good, eg a pathogenic bacterium causing a fatal disease could be very resilient to antibiotics which means it will kill many people so, in this case, its resilience is not so good for us.

Factors affecting ecosystem resilience

- The more diverse and complex an ecosystem, the more resilient it tends to be as there are more interactions between different species.

- The greater the species biodiversity of the ecosystem, the greater the likelihood that there is a species that can replace another if it dies out and so maintain the equilibrium.

- The greater the genetic diversity within a species, the greater the resilience. A monoculture of wheat or rice can be wiped out by a disease if none of the plants have resistance which is more likely in a diverse gene pool.

- Species that can shift their geographical ranges are more resilient.

- The larger the ecosystem, the more resilience as animals can find each other more easily and there is less edge-effect.

- The climate affects resilience – in the Arctic, regeneration of plants is very slow as the low temperatures slow down photosynthesis and so growth. In the tropical rain forests, growth rates are fast as light, temperature and water are not limiting.

- The faster the rate at which a species can reproduce means recovery is faster. So r-strategists (2.4) with a fast reproductive rate can recolonize the system better than slowly reproducing K-strategists.

- Humans can remove or mitigate the threat to the system (eg remove a pollutant, reduce an invasive species) and this will result in faster recovery.

Tipping points

Small changes occur in systems and may not make a huge difference. But when these changes tip the equilibrium over a threshold, known as a tipping point, the system may transform into a very different one. Then positive feedback loops drive the system to a new steady state.

An ecological **tipping point** is reached when an ecosystem experiences a shift to a new state in which there are significant changes to its biodiversity and the services it provides.

Characteristics of tipping points:

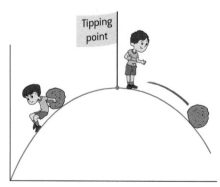

Figure 1.3.18 Illustrating a tipping point

- They involve positive feedback which makes the change self-perpetuating; eg deforestation reduces regional rainfall, which increases fire risk, which causes forest dieback.

- There is a threshold beyond which a fast shift of ecological states occurs.

- The threshold point cannot be precisely predicted.

- The changes are long-lasting.

- The changes are hard to reverse.

- There is a significant time lag between the pressures driving the change and the appearance of impacts, creating great difficulties in ecological management.

Examples of tipping points

1. **Lake eutrophication –** if nutrients are added to a lake ecosystem, it may not change much until enough nutrients are added to shift the lake to a new state – then plants grow excessively, light is blocked by decomposing plant material, oxygen levels fall and animals die. The lake becomes eutrophic and it takes a great effort to restore it to the previous state (4.4).

2. **Extinction of a keystone species** (eg elephants) from a savanna ecosystem can transform it to a new state which cannot be reversed.

3. **Coral reef death** – if ocean acidity levels rise enough, the reef coral dies and cannot regenerate.

Tipping points are well-known in local or regional ecosystems but there is debate about whether we are reaching a global tipping point. Some people say that climate change caused by human activities will force the Earth to a new, much warmer state – as much as 8 °C warmer than today. But evidence is that we see warming in one region and cooling in others, wetter in some and drier in others. The global system is so complex and ecosystems respond differently, often independently of other ecosystems.

If there were to be global tipping points, there are major implications for decision-makers. Some may think that below this point, not much would change while, once it is reached, all is lost as society could not respond fast enough. That could lead to inaction or despair – the 'what's the point, there is nothing we can do now' point of view.

The best approach we can have may be the precautionary one where we don't know what will happen exactly but can take steps to modify what we do in case. Such risk management is the responsible route to take.

Practical Work

* Create a model of a feedback loop.

* Create a model of a food web.

To do

1. Negative and positive feedback control. Look at this example of feedback control.

   a) How is the growth of the animal population regulated in the diagram?

   b) Explain why it is an example of negative feedback control.

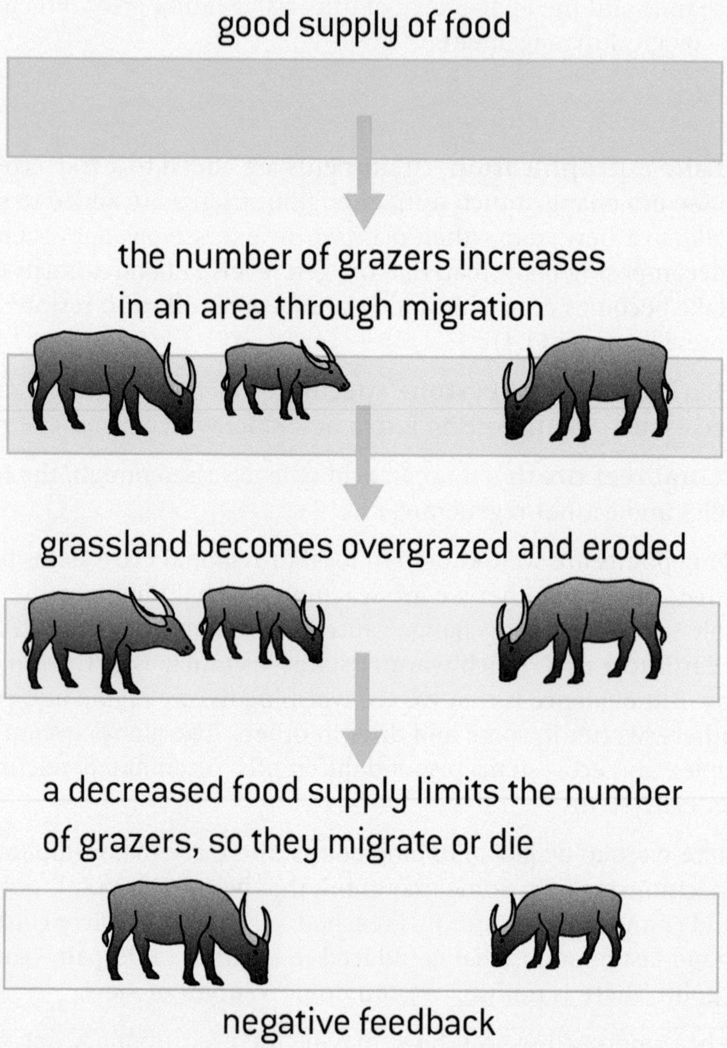

good supply of food

the number of grazers increases
in an area through migration

grassland becomes overgrazed and eroded

a decreased food supply limits the number
of grazers, so they migrate or die

negative feedback

▲ **Figure 1.3.19** Negative feedback amongst grazing animals

2. Explain with a named example how positive feedback may contribute to global warming.

3. Complete the diagram of a generalized ecosystem showing inputs, outputs and stores. Remember to add in human activities.

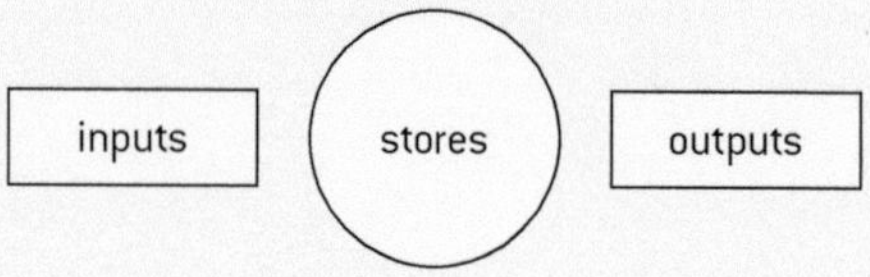

| inputs | (stores) | outputs |

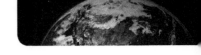

1.4 Sustainability

Significant ideas:

→ All systems can be viewed through the lens of sustainability.

→ Sustainable development meets the needs of the present without compromising the ability of future generations to meet their own needs.

→ Environmental indicators and ecological footprints can be used to assess sustainability.

→ Environmental Impact Assessments (EIAs) play an important role in sustainable development.

Applications and skills:

→ **Explain** the relationship between natural capital, natural income and sustainability.

→ **Discuss** the value of ecosystem services to a society.

→ **Discuss** how environmental indicators can be used to evaluate the progress of a project to increase sustainability, eg Millennium Ecosystem Assessment.

→ **Evaluate** the use of EIAs.

→ **Explain** the relationship between ecological footprint (EF) and sustainability.

Knowledge and understanding:

→ **Sustainability** is the use and management of resources that allows full natural replacement of the resources exploited and full recovery of the ecosystems affected by their extraction and use.

→ **Natural capital** is a term used for natural resources that can produce a sustainable natural income of goods or services.

→ **Natural income** is the yield obtained from natural resources

→ Ecosystems may provide life-supporting services such as water replenishment, flood and erosion protection, and goods such as timber, fisheries and agricultural crops.

→ Factors such as biodiversity, pollution, population or climate may be used quantitatively as environmental indicators of sustainability. These factors can be applied on a range of scales from local to global. The **Millennium Ecosystem Assessment** gave a scientific appraisal of the condition and trends in the world's ecosystems and the services they provide using environmental indicators, as well as the scientific basis for action to conserve and use them sustainably.

→ **Environmental Impact Assessments** (EIAs) incorporate baseline studies before a development project is undertaken. They assess the environmental, social and economic impacts of the project, predicting and evaluating possible impacts and suggesting mitigation strategies for the project. They are usually followed by an audit and continued monitoring. Each country or region has different guidance on the use of EIAs.

→ EIAs provide decision makers with information in order to consider the environmental impact of a project. There is not necessarily a requirement to implement an EIA's proposals and many socio-economic factors may influence the decisions made.

→ Criticisms of EIAs include the lack of a standard practice or training for practitioners, the lack of a clear definition of system boundaries and the lack of inclusion of indirect impacts.

→ An **ecological footprint** (EF) is the area of land and water required to sustainably provide all resources at the rate at which they are being consumed by a given population. Where the EF is greater than the area available to the population, this is an indication of unsustainability.

'The ultimate test of a moral society is the kind of world that it leaves to its children.'

Dietrich Bonhoeffer,
German theologian

'He who slaughters his cows today shall thirst for milk tomorrow.'

Muslim proverb

Key terms

Sustainability is the use and management of resources that allows full natural replacement of the resources exploited and full recovery of the ecosystems affected by their extraction and use.

The term **'sustainable development'** has been defined as 'development that meets the needs of the present without compromising the ability of future generations to meet their own needs.' (From *Our Common Future*, the report of the World Commission on Environment and Development, 1987).

Sustainability

Sustainability means living within the means of nature, on the 'interest' or sustainable natural income generated by natural capital. But sustainability is a word that means different things to different people. Economists have a different view from environmentalists about what sustainable means. The word sustainable is often used as an adjective in front of words such as resource, development and population.

Any society that supports itself in part by depleting essential forms of natural capital is unsustainable. There is a finite amount of materials on Earth and we are using much of it unsustainably – living on the capital as well as the interest. Our societies and economies cannot grow or make progress outside of environmental limits (figure 1.4.1).

Ecological overshoot

According to UN data (figure 1.4.1) humanity has overshot its sustainable level of resource exploitation.

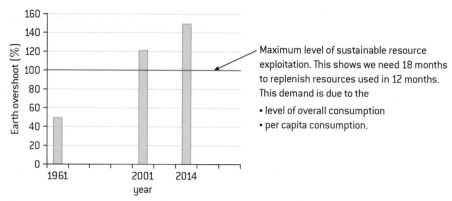

Maximum level of sustainable resource exploitation. This shows we need 18 months to replenish resources used in 12 months. This demand is due to the

• level of overall consumption
• per capita consumption.

▲ **Figure 1.4.1** Ecological overshoot

It is more in some parts of the world and cannot continue indefinitely.

Sustainability indicators

How we measure sustainability is crucial and there are many indices we can use together, both ecological and socio-economic. These could be anything from air quality, environmental vulnerability and water poverty to US$ GDP (Gross Domestic Product) per capita, life expectancy or gender parity. We can also measure sustainability on scales from local to global. The smaller the scale, the more accurate it can be but we also need a global measurement to get the whole picture.

To think about

The **Millennium Ecosystem Assessment (MEA)**, funded by the UN and started in 2001, is a research programme that focuses on how ecosystems have changed over the last decades and predicts changes that will happen. In 2005, it released the results of its first four-year study of the Earth's natural resources. It was not happy reading.

The report said that natural resources (food, freshwater, fisheries, timber, air) are being used in ways that degrade them so make them unsustainable in the longer term.

Key facts reported are:

- 60% of world ecosystems have been degraded.
- About 25% of the Earth's land surface is now cultivated.
- We use 40–50% of all available surface freshwater and water withdrawals from underground sources have doubled over the past 40 years
- Over 25% of all fish stocks are overharvested.
- Since 1980, about 35 % of mangroves have been destroyed.
- About 20% of corals have been lost in 20 years and another 20% degraded.

- Nutrient pollution has led to eutrophication of waters and dead coastal zones.
- Species extinction rates are now 100–1,000 times above the background rate.
- We have had more effect on the ecosystems of Earth in the last 50 years than ever before.

Some recommendations were to:

- Remove subsidies to agriculture, fisheries and energy sources that harm the environment.
- Encourage landowners to manage property in ways that enhance the supply of ecosystem services, such as carbon storage and the generation of fresh water.
- Protect more areas from development, especially in the oceans.

To think about

Using the figure below think about the following questions. Are you optimistic or pessimistic about the results of the impact of humans on the Earth?

What evidence are you using for your decision?

If you had the power, what actions would you force governments to take now to safeguard the environment but also protect humans from suffering? Give your reasons.

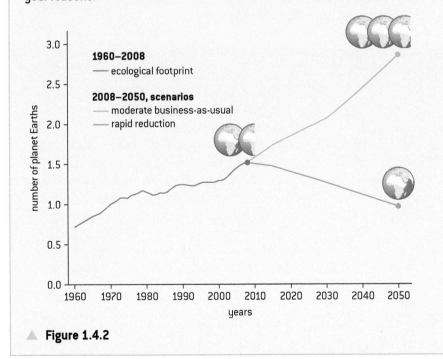

▲ Figure 1.4.2

But how can we change?

You may wonder why this continues if we all know it to be so. It is perhaps due to many factors including:

- Inertia: when changing what we do seems too difficult.

- The result of the '**tragedy of the commons**' (4.3), when many individuals act in their own self-interest to harvest a resource but destroy the long-term future of that resource so there is none for anyone. It may be obvious that this will happen, but each individual benefits from taking the resource in the short term so they continue to do so. For example, hunting an endangered species may result in its extinction but if your family are starving and it is the only source of food, you will probably hunt it to eat it.

Some people think that the real worth of natural capital is about the same as the value of the gross world product (total global output) – about US$65 trillion per year, yet we are only just beginning to give economic value to soil, water and clean air and to measure the cost of loss of biodiversity (see Topic 8).

Natural capital and natural income

Capital is what economists term the means of production – factories, tools, machines – and is used to create goods which provide income.

Natural capital is the goods and services that the environment provides humans with in order to provide natural income. For example, a forest (natural capital) provides timber (natural income); a shoal of fish or an agricultural crop provides food for us. Natural capital also provides services, for example erosion control, water management, recycling waste. See in more detail in Topic 8.

Key terms

Natural capital is a term used for natural resources that can produce a sustainable **natural income** of goods or services.

To think about

The Millennium Development Goals
http://www.undp.org/mdg/
The Millennium Development Goals (MDGs) are eight goals to be achieved by 2015 that respond to the world's main development challenges. The MDGs are drawn from the actions and targets contained in the **Millennium Declaration** that was adopted by 189 nations and signed by 147 heads of state and governments during the **UN Millennium Summit** in September 2000.
Goal 1: Eradicate extreme poverty and hunger
Goal 2: Achieve universal primary education

Goal 3: Promote gender equality and empower women
Goal 4: Reduce child mortality
Goal 5: Improve maternal health
Goal 6: Combat HIV/AIDS, malaria and other diseases
Goal 7: Ensure environmental sustainability
Goal 8: Develop a Global Partnership for Development
Are we on target to reach these goals? Research what actions have been taken since 2000. (Try searching the web for Millenium Development Goals BBC and you should find some BBC webpages with an update.)
Do you think these were attainable goals or too ambitious?

Environmental impact assessments

An environmental impact assessment or EIA is a report prepared **before** a development project to change the use of land, for example to plant a forest or convert fields to a golf course. An EIA weighs up the relative advantages or disadvantages of the development. It is therefore necessary to establish how the abiotic environment and biotic community would change if a development scheme went ahead. An EIA will try to quantify changes to microclimate, biodiversity, scenic and amenity value resulting from the proposed development. These measurements represent the production of a **baseline** study.

EIAs look at what the environment is like now and forecast what may happen if the development occurs. Both negative and positive impacts are considered as well as other options to the proposed development. While often EIAs have to deal with questions about the effect on the natural environment they can also consider the likely effects on human populations. This is especially true where a development might have an effect on human health or have an economic effect for a community.

What are EIAs used for?

EIAs are often, though not always, part of the planning process that governments set out in law when large developments are considered. They provide a documented way of examining environmental impacts that can be used as evidence in the decision-making process of any new development. The developments that need EIAs differ from country to country, but certain types of developments tend to be included in the EIA process in most parts of the world. These include:

- major new road networks
- airport and port developments
- building power stations
- building dams and reservoirs
- quarrying
- large-scale housing projects.

Where did EIAs come from?

In 1969, the US Government passed the National Environmental Policy Act (NEPA). NEPA made it a priority for federal agencies to consider the natural environment in any land use planning. This gave the natural environment the same status as economic priorities. Within 20 years of NEPA becoming law in the US, many other countries also included EIAs as part of their planning policy. In the US, environmental assessments (EA) are carried out to determine if an EIA (called EIS – environmental impact statement) needs to be undertaken and filed with the federal agencies.

What does an EIA need in it?

There is no set way of conducting an EIA, but various countries have minimum expectations of what should be included in an EIA. It is possible to break an assessment down into three main tasks:

- Identifying impacts (scoping).
- Predicting the scale of potential impacts.
- Limiting the effect of impacts to acceptable limits (mitigation).

There is always a non-technical summary so that the general public can understand the issues.

Weaknesses of EIAs

Different countries have different standards for EIAs which makes it hard to compare them. Also, it is hard to determine where the boundary of the investigation should be. How large an area, how many variables, how much does the EIA cost? It is also very difficult to consider all indirect impacts of a development so some may be missed.

To think about

EIAs are models of the system under study and allow us to predict the effects of the proposed change. A model is only as good as its parameters and asking the right questions is crucial. A change of land use will always have an effect but whether this is a net positive or negative one depends on the criteria used to measure it. Simplistically, if a factory blocks your view of the mountains that may be a loss to you but it may bring employment to the area, produce goods that would otherwise be imported and reduce the country's ecological footprint.

Cost-benefit analysis measures impacts of a development or change of land use translated into monetary values. In theory, this puts all costs into the same units of measure – money – so they can be assessed. Of course, how the assessment is made is critical to the values assigned and there are several ways to do this. For example, it may be based on the cost of restoring the environment to its previous state (eg after an open cast mine operation) or ask people which of several options they would select or be prepared to pay for.

Strategic environmental assessment tries to measure the social and environmental costs of a development but this can be subjective or a not very accurate prediction. Does it also depend on the environmental worldview of those planning the assessment?

Imagine a development or change of land use in an area near to your school or home. Decide amongst your class what this will be (it may be an actual one that is about to happen or has happened) and discuss:

- What criteria you would use to select the factors you think will change (eg number of jobs provided, net profit, land degradation, habitat loss, pollution).

- How you value these (is there another way of measuring them apart from financial?)

- How you weigh up the evidence to make a decision on whether the project should proceed or proceed in a modified state.

Practical Work

* Investigate what Ecological Footprint modelling can tell us about resource use.

* Consider whether sustainable development is a term that contradicts itself.

* Can sustainability agreements only be international? What is the point of a nation being sustainable if the rest of the world is not?

Key term

An **ecological footprint (EF)** is the area of land and water required to sustainably provide all resources at the rate at which they are being consumed by a given population.

Ecological footprints

EF is a **model** used to estimate the demands that human populations place on the environment. The measure takes into account the area required to provide all the resources needed by the population, and the assimilation of all wastes. Where the EF is greater than the area available to the population, this is an indication of unsustainability as the population exceeds the carrying capacity (8.4) of the area.

EFs may vary significantly from country to country and person to person and include aspects such as lifestyle choices (EVS), productivity of food production systems, land use and industry.

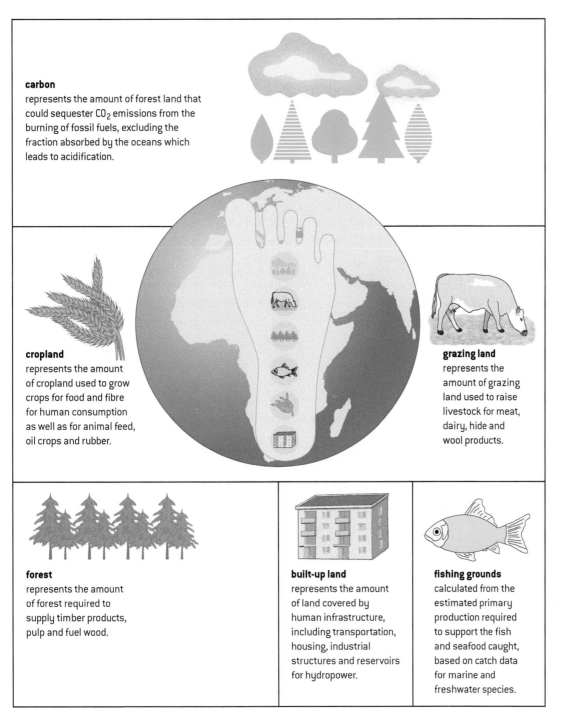

carbon
represents the amount of forest land that could sequester CO_2 emissions from the burning of fossil fuels, excluding the fraction absorbed by the oceans which leads to acidification.

cropland
represents the amount of cropland used to grow crops for food and fibre for human consumption as well as for animal feed, oil crops and rubber.

grazing land
represents the amount of grazing land used to raise livestock for meat, dairy, hide and wool products.

forest
represents the amount of forest required to supply timber products, pulp and fuel wood.

built-up land
represents the amount of land covered by human infrastructure, including transportation, housing, industrial structures and reservoirs for hydropower.

fishing grounds
calculated from the estimated primary production required to support the fish and seafood caught, based on catch data for marine and freshwater species.

Figure 1.4.3 Ecological footprint (EF) (© WWF)

1.5 Humans and pollution

Significant ideas:

→ Pollution is a highly diverse phenomenon of human disturbance in ecosystems.

→ Management strategies can be applied at different levels.

Applications and skills:

→ **Construct** systems diagrams to show the impact of pollutants.

→ **Evaluate** the effectiveness of each of the three different levels of intervention, with reference to figure 1.5.6.

→ **Evaluate** the use of DDT.

Knowledge and understanding:

→ Pollution is the addition of a substance or an agent to an environment by human activity, at a rate greater than that at which it can be rendered harmless by the environment, and which has an appreciable effect on the organisms in the environment.

→ Pollutants may be in the form of organic/inorganic substances, light, sound or heat energy, or biological agents/invasive species, and derive from a wide range of human activities including the combustion of fossil fuels.

→ Pollution may be non-point or point source, persistent or biodegradable, acute or chronic.

→ Pollutants may be primary (active on emission) or secondary (arising from primary pollutants undergoing physical or chemical change).

→ Dichlorodiphenyltrichloroethane (DDT) exemplifies a conflict between the utility of a 'pollutant' and its effect on the environment.

Key term

Pollution is the addition of a substance or an agent to an environment by human activity, at a rate greater than that at which it can be rendered harmless by the environment, and which has an appreciable effect on the organisms within it.

Figure 1.5.1 Does the earth need a gas mask?

Pollutants and pollution

Pollutants are released by human activities and may be:

- matter (gases, liquids or solids) which is organic (contains carbon atoms) or inorganic

- energy (sound, light, heat)

- living organisms (invasive species or biological agents).

There are:

- **primary pollutants** which are active on emission eg carbon monoxide from the incomplete combustion of fossil fuels, which causes headaches and fatigue and can kill

- **secondary pollutants** which are formed by primary pollutants undergoing physical or chemical changes eg sulphuric acid forms when sulphur trioxide reacts with water.

Photochemical smog is a mixture of primary and secondary pollutants (see sub-topic 6.3).

Since humans have been on Earth, we have polluted to a greater or lesser extent. Sewage and waste are products of human communities and burning wood and then coal has caused air pollution for 1,000 years. When population levels were lower, the environment could cope with these pollutants. However, pollution may be an inevitable side-effect of the economic development that has given most humans a far higher standard of living than we would otherwise have had. Since the Industrial Revolution pollution has increased but how we deal with it has also improved as we monitor industries and legislate against excessive pollution.

Major sources of pollutants

Figure 1.5.3 lists some major sources of pollutants. We shall be considering some of these later on.

To think about

It is sometimes said that a pollutant is a substance in the wrong place, in the wrong amount or at the wrong time. Could this be true of carbon dioxide, ozone or nitrate?

АЫХАНЬЕ СОВЕТСКОИ РОССИИ

▲ **Figure 1.5.2** Poster from the USSR before 1950 encouraging production by saying that the smoke from chimneys is the breath of Soviet Russia

Major source	Pollutant	Effect
Combustion of fossil fuels	Carbon dioxide	Greenhouse gas – climate change
	Sulphur dioxide	Acid deposition – tree and fish death, respiratory disease in humans
	Nitrogen oxides	Respiratory infections, eye irritation, smog
	Photochemical smog including tropospheric ozone, PANs, VOCs (volatile organic compounds)	Secondary pollutants (formed from others in the atmosphere) – damage to plants, eye irritation, respiratory problems in humans
	Carbon monoxide	Binds with haemoglobin in red blood cells instead of oxygen – can lead to death by suffocation
Domestic waste	Organic waste (food and sewage)	Eutrophication, waterborne diseases
	Waste paper	Volume fills up landfill sites, forests cut to produce it
	Plastics – containers, packaging	Volume fills up landfill sites, derived from oil
	Glass	Energy required to manufacture it (as with all products), can be recycled but most goes into landfill sites
	Tins/cans	Can be recycled but also goes into landfill
Industrial waste	Heavy metals	Poisoning, eg mercury, lead, cadmium
	Fluorides	Poisoning
	Heat	Reduces solubility of gases in water, so less oxygen so organisms may die
	Lead	Disabilities in children
	Acids	Corrosive
Agricultural waste	Nitrates	Eutrophication
	Organic waste	Eutrophication, disease spread
	Pesticides	Accumulate up food chains

▲ **Figure 1.5.3** Major sources of pollutants and their effects

Point source and non-point source pollutants

Non-point source (NPS) pollution:

- Is the release of pollutants from numerous, widely dispersed origins, for example gases from the exhaust systems of vehicles, chemicals spread on fields.

- May have many sources and it may be virtually impossible to detect exactly where it is coming from.

- Rainwater can collect nitrates and phosphates which are spread as fertilizer as it infiltrates the ground or as runoff on the surface. It may travel many kilometres before draining into a lake or river and increasing the concentration of nitrates and phosphates so much that eutrophication occurs. It would not be possible to say which farmer spread the excess fertilizer.

- Air pollution can be blown hundreds of kilometres and chemicals released from open chimneys mix with those from others.

So one solution is to set limits for all farmers and all industries to reduce emissions and then monitor what they actually do.

Point source (PS) pollution:

- Is the release of pollutants from a single, clearly identifiable site, for example a factory chimney or the waste disposal pipe of a sewage works into a river.

- Is easier to see who is polluting – a factory or house.

- Is usually easier to manage as it can be found more easily.

Persistent organic pollutants (POPs) and biodegradable pollutants

POPs were often manufactured as pesticides in the past. They are resistant to breaking down and remain active in the environment for a long time. Because of this, they bioaccumulate in animal and human tissues and biomagnify in food chains (see 2.2) and can cause significant harm.

Examples of these are DDT (see 2.2), dieldrin, chlordane and aldrin. Other POPs are polyvinyl chloride (PVC), polychlorinated biphenyls (PCBs) and some solvents. They have similar properties:

- high molecular weight

- not very soluble in water

- highly soluble in fats and lipids – which means they can pass through cell membranes

- halogenated molecules, often with chlorine.

PCBs were widely used in electrical apparatus and as coolants since the 1930s but banned by 2001. They cause cancers and disrupt hormone functions and have a similar structure and action in animals to dioxin which is one of the most deadly chemicals that humans have made. Because they are so persistent, PCBs are found everywhere in water as well as in animal tissues, even in the Arctic Circle.

Figure 1.5.4

Practical Work

* Create a poster/website/ wiki on the benefits and disadvantages of using DDT.

* Construct a systems diagram to show pollution of a local ecosystem.

Biodegradable pollutants do not persist in the environment and break down quickly. They may be broken down by decomposer organisms or physical processes, eg light or heat. Examples are soap, domestic sewage, degradable plastic bags made of starch. One common herbicide is glyphosate which farmers use to kill weeds. It is degraded and broken down by soil organisms.

Acute and chronic pollution

Acute pollution is when large amounts of a pollutant are released, causing a lot of harm. An example of this was when the chemical aluminium sulphate was accidently tipped into the wrong place in a water treatment works in Cornwall in the UK in 1988 and many people drank water which poisoned them. Another example was in the Bhopal Disaster of 1984 in India (1.1).

Chronic pollution results from the long-term release of a pollutant but in small amounts. It is serious because:

* often it goes undetected for a long time

* it is usually more difficult to clean it up

* it often spreads widely.

Air pollution is often chronic causing non-specific respiratory diseases, for example asthma, bronchitis, emphysema. Beijing's poor air quality is an example of chronic air pollution.

▲ **Figure 1.5.5** Chronic air pollution in Beijing 2014

To think about

The Prisoner's Dilemma

A big question about us is whether we are, by nature, loving or aggressive, noble or selfish, nice or nasty. Do we not steal or cheat because we may be found out or because we know it is wrong. Is it our default position to be kind and helpful to each other or to be top even if, or particularly if, it hurts someone else? Scientists, sociologists, philosophers, politicians and all thinking people want to know about our innate nature and why we react as we do.

There is a type of game that you can play as an example of Game Theory and it is called the Prisoner's Dilemma. Here is a version of it.

Two people A and B are suspected of a crime and arrested. There is not enough evidence to convict them unless they confess. The police separate them and offer each one the same deal. If one admits that they both did the crime and betrays the other, that one goes free and the other goes to prison for 10 years. If both stay silent, they both go to prison for a year. If both confess, they both go to prison for 5 years. What should they do? The best scenario for one is to confess and the other stays silent. But they don't know what the other will do. What has this to do with pollution? Quite a lot.

The best economic scenario for a polluter is to keep polluting as long as he/she is not found out. Not to confess. The cost of the pollution is then shared between everyone and the polluter does not have to spend money reducing their own personal or business pollution. If the polluter confesses, they may be punished by a fine, imprisonment or having to spend money in reducing the pollution.

But, just as in the Prisoner's Dilemma, while keeping silent and polluting is fine in the short term, in the longer term, the best scenario is 'tit for tat' – if I cooperate with you – stop polluting, you will cooperate with me – stop polluting too and the world will be a cleaner place – we both gain. If we keep betraying each other, we will both be losers at the end. And that is where we are with pollution. If we pollute with NPS pollutants, we are unlikely to be found out and everyone pays for the clean up. An individual, company or country can gain from non-compliance in the short term if the others comply.

But what will happen in the long term?

Think of two particular types of pollution (one in the atmosphere and one in water) that could be examples of NPS pollution.

What does this mean for international agreements on pollution?

Detection and monitoring of pollution

Pollution can be measured directly or indirectly.

Direct measurements record the amount of a pollutant in water, the air or soil.

Direct measurements of air pollution include measuring:

- the acidity of rainwater
- amount of a gas, for example carbon dioxide, carbon monoxide, nitrogen oxides in the atmosphere
- amount of particles emitted by a diesel engine
- amount of lead in the atmosphere.

Direct measurements of water or soil pollution include testing for:

- nitrates and phosphates
- amount of organic matter or bacteria
- heavy metal concentrations.

Indirect measurements record changes in an abiotic or biotic factor which are the result of the pollutants. Indirect measurements of pollution include:

- measuring abiotic factors that change as a result of the pollutant (eg oxygen content of water)
- recording the presence or absence of indicator species – species that are only found if the conditions are either polluted (eg rat-tailed maggot in water) or unpolluted (eg leafy lichens on trees).

Pollution management strategies

Pollution can be managed in three main ways:

- by changing the human activity which produces it
- by regulating or preventing the release of the pollutant or
- by working to clean up or restore damaged ecosystems.

The pollution management model in Figure 1.5.6 lists the actions available in each category of management and will be referred to throughout the book when specific pollutants are considered.

Process of pollution	Level of pollution management
HUMAN ACTIVITY PRODUCING POLLUTANT ↓	**Altering human activity** The most fundamental level of pollution management is to change the human activity that leads to the production of the pollutant in the first place, by promoting alternative technologies, lifestyles and values through: • campaigns • education • community groups • governmental legislation • economic incentives/disincentives.

RELEASE OF POLLUTANT INTO ENVIRONMENT	Controlling release of pollutant
	Where the activity/production is not completely stopped, strategies can be applied at the level of regulating or preventing the release of pollutants by: • legislating and regulating standards of emission • developing/applying technologies for extracting pollutant from emissions.
IMPACT OF POLLUTANT ON ECOSYSTEMS	Clean-up and restoration of damaged systems
	Where both the above levels of management have failed, strategies may be introduced to recover damaged ecosystems by: • extracting and removing pollutant from ecosystem • replanting/restocking lost or depleted populations and communities

▲ **Figure 1.5.6** Pollution management targeted at three different levels

DDT and malarial mosquitoes

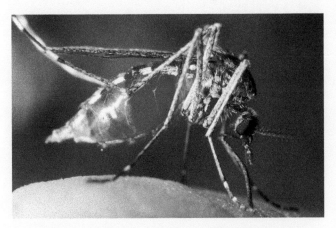

▲ **Figure 1.5.7** Malarial mosquito sucking blood from a human

In 1970, the WHO (World Health Organization) banned the use of DDT, a persistent organochlorine insecticide. It is still used in some countries in the tropics but in small quantities for spraying inside houses to kill the malarial mosquito, *Anopheles*, which is the vector for malarial parasites.

The question is whether banning DDT did more harm than good.

It is believed that malaria kills 2.7 million people a year, mostly children under the age of five, and infects 300–500 million a year. It is also thought that DDT prevented millions of deaths due to malaria. So why the ban? In her book, *Silent Spring*, Rachel Carson discusses the effect of DDT on birds of prey in thinning their eggshells and reducing their population numbers. But some say that evidence was slight for bird egg shell thinning and DDT is an effective insecticide against the malarial mosquito.

The manufacture and use of DDT was banned in the US in 1972, on the advice of the US Environmental Protection Agency. The use of DDT has since been banned in most other MEDCs, but it is not banned for public health use in most areas of the world where malaria is endemic. DDT was recently exempted from a proposed worldwide ban on organophosphate chemicals. DDT for malarial control involves spraying the walls and backs of furniture, so as to kill and repel adult mosquitoes that may carry the malarial parasite. Although other chemicals could be used, DDT is cheap and persistent and good at the job. Outside DDT is not used because of its persistence and toxicity. Also, its persistence means that mosquitoes become resistant (the ones that survive, breed and develop a population of resistant mosquitoes).

Malaria incidence is increasing, partly due to resistance, partly to changes in land use and migration of people to areas where malaria is endemic. In treating the cause, DDT use is just one tool along with other chemicals, mosquito nets and removal of stagnant water where mosquitoes breed.

There is hyperbole, bias and misinformation in the debate on DDT but malaria probably does not receive enough funding for research as it is mostly a disease of the poor.

To do

Do your own research on DDT. What evidence can you find for both sides of the argument?

Be careful in looking at sources. Are they biased? Can they substantiate their claims?

Do you now think that DDT should have been banned or should still be used?

CASE STUDY

State the environmental value system that you identify in your choice of lifestyle and how sustainable it is.

Explain how your own environmental value system compares with others.

BIG QUESTIONS

Foundations of environmental systems and societies

Examine in what ways might the solutions explored in the pollution management model alter your predictions for our future.

Discuss what are the strengths and weaknesses of using models to assess sustainability.

Reflective questions

→ Environmental value systems shape the way we perceive the environment. What other value systems shape the way we view the world?

→ Models are a simplified construction of reality. In the construction of a model, how can we know which aspects of the world to include, and which to ignore?

→ The laws of thermodynamics are examples of scientific laws. In what ways do scientific laws differ to the laws of human science subjects, such as economics?

→ EIAs incorporate baseline studies before a development project is undertaken. To what extent should environmental concerns limit our pursuit of knowledge?

→ On what basis might we decide between the judgements of the experts if they disagree?

→ What influences your EVS?

→ Human impact crosses national boundaries. How can agreement on international environmental issues be reached?

→ Can models facilitate international collaboration on environmental issues?

Quick review

Each question is worth 1 mark

1. A *system* may best be defined as

    A. a set of components that function predictably.

    B. an assemblage of parts and their relationships forming a whole.

    C. a set of components that function unpredictably.

    D. an assemblage of functioning parts without inputs or outputs.

2. Inputs to a closed system may be

    A. matter only   C. matter and energy

    B. energy only   D. heat only

3. What do outputs from an open system consist of?

    A. Energy only.

    B. Matter only.

    C. Energy and matter.

    D. Neither energy nor matter.

4. A lake with a stream flowing into it, but with water lost only by evaporation, is an example of a system which is

    A. isolated   C. unstable and closed

    B. stable and closed   D. open

5. Which of the following factors would prevent the ecosphere being classified as a *closed system*?

    A. The input of solar energy.

    B. The re-radiation to space of heat energy.

    C. The arrival of rocks as meteorites from space.

    D. The unstable state of its equilibrium.

6. Which statement is correct?

    A. A lake is an example of an isolated system.

    B. An open system exchanges energy but not matter with its surroundings.

    C. The most common systems found on Earth are closed systems.

    D. A closed system exchanges energy but not matter with its surroundings.

7. The carrying capacity of an environment for a given species

    A. can never be exceeded.

    B. is greater for a population with a slow reproductive rate.

    C. is achieved when birth rates equal death rates.

    D. can only be exceeded with unsustainable use of resources.

8. Which of the following conditions would lead to *unsustainable* harvesting of timber from a forest?

    I. Harvesting trees before they are fully mature.

    II. Regularly harvesting the full natural income from the forest.

    III. Reducing mineral content of soil through harvesting.

    A. I and III only   C. I and II only

    B. III only   D. I, II and III

9. *Sustainable yield* can be defined as

    I. annual growth and recruitment – annual death and emigration.

    II. (total biomass at time $t + 1$) – (total biomass at time t).

    III. the highest rate at which natural capital can be exploited without reducing its original stock.

    A. I and II only   C. II and III only

    B. I and III only   D. I, II and III

10. Which of the following populations are most likely to be sustainable?

	Population density	Mean individual consumption	High dependence on
A.	high	low	renewable resources
B.	high	high	renewable resources
C.	high	high	non-renewable resources
D.	low	low	non-renewable resources

2.1 Species and populations

Significant ideas:

→ A species interacts with its abiotic and biotic environment, and its niche is described by these interactions.

→ Populations change and respond to interactions with the environment.

→ All systems have a carrying capacity for a given species.

Applications and skills:

→ **Interpret** graphical representations or models of factors that affect an organism's niche. Examples include predator–prey relationships, competition, and organism abundance over time.

→ **Explain** population growth curves in terms of numbers and rates.

Knowledge and understanding:

→ A **species** is a group of organisms sharing common characteristics that interbreed and produce fertile offspring.

→ A **habitat** is the environment in which a species normally lives.

→ A **niche** describes the particular set of abiotic and biotic conditions and resources to which an organism or population responds.

→ The **fundamental niche** describes the full range of conditions and resources in which a species could survive and reproduce. The **realized niche** describes the actual conditions and resources in which a species exists due to biotic interactions.

→ The non-living, physical factors that influence the organisms and ecosystem, eg temperature, sunlight, pH, salinity, precipitation are termed **abiotic factors**.

→ The **interactions** between the organisms, eg predation, herbivory, parasitism, mutualism, disease, competition are termed **biotic factors**.

→ Interactions should be understood in terms of the influences each species has on the population dynamics of others, and upon the carrying capacity of the others' environment.

→ A **population** is a group of organisms of the same species living in the same area at the same time, and which are capable of interbreeding.

→ **S** and **J population curves** describe a generalized response of populations to a particular set of conditions (abiotic and biotic factors).

→ **Limiting factors** will slow population growth as it approaches the carrying capacity of the system.

This topic covers the ecology of the ESS course.

> ## Tips
>
> When you study this:
>
> - Use named **examples** to illustrate concepts or your arguments.
>
> - Always give the full name of the animal or plant you are mentioning, for example not 'fish' but 'Atlantic salmon', not 'tree' but 'common oak tree'.
>
> - When you use a habitat or local ecosystem as an example, give as much detail as possible about it, for example not 'beach' but 'rocky shore, north-facing, at Robin Hood's Bay, North Yorkshire, UK'.

'And this, our life, exempt from public haunt, finds tongues in trees, books in the running brooks, sermons in stones, and good in everything.'

William Shakespeare

What is what in ecology

Ecosystems are made up of the organisms and physical environment and the interactions between the living and non-living components within them.

Examples of species are humans, giraffes and pine trees. Each species is given a scientific name composed of two parts: the genus name and then the species name. Scientific names are always underlined or in italics and the genus name is given first with a capital letter:

Common name	Scientific or binomial name
Human	*Homo sapiens*
Giraffe	*Giraffa camelopardalis*
Scots pine	*Pinus sylvestris*
Aardvark	*Orycteropus afer*

Snails of one species in a pond form a population but the snails in another pond are a different population. A road or river may separate two populations from each other and stop them interbreeding. And this may cause speciation (see 3.2).

Population density is the average number of individuals in a stated area, for example gazelles km^{-2}, or bacteria cm^{-3}.

Three factors affect population size:

- natality(birth rate),

- mortality (death rate), and

- migration:

  - immigration (moving into the area)

  - emigration (moving out of the area).

This natural environment includes the physical (abiotic) environment. Many populations of different species (a community) may share the same habitat.

A **niche** is how an organism makes a living. This includes:

Biotic factors:

- every relationship that organism may have

- where it lives

> ### Key term
>
> A **species** is a group of organisms (living things) sharing common characteristics that interbreed and produce fertile offspring.

> ### Key term
>
> A **population** is a group of organisms of the same species living in the same area at the same time, and which are capable of interbreeding.

> ### Key term
>
> A **habitat** is the environment in which a species normally lives.

Abiotic factors are the non-living, physical factors that influence the organisms and ecosystem, eg temperature, sunlight, pH, salinity, pollutants.

Biotic factors are the living components of an ecosystem – organisms, their interactions or their waste – that directly or indirectly affect another organism.

A **niche** describes the particular set of abiotic and biotic conditions and resources to which an organism or population responds.

- **Fundamental niche** describes the full range of conditions and resources in which a species could survive and reproduce.

- **Realized niche** describes the actual conditions and resources in which a species exists due to biotic interactions.

- how it responds to resources available, to predators, to competitors
- how it alters these biotic factors.

Abiotic factors:

- how much space there is
- availability of light, water etc.

No two species can inhabit the same ecological niche in the same place at the same time: if many species live together they must have slightly different needs and responses so are not in the same niche.

For example, lions and cheetahs both live in the same area of the African savanna but they hunt different prey. Lions typically take down bigger herbivores such as zebra and Cape Buffalo whereas cheetahs will focus on the smaller antelopes such as the Thompson's gazelle and impalas.

▲ **Figure 2.1.1** Lion preying on a zebra

Limiting factors prevent a community, population or organism growing larger. There are many limiting factors which restrict the growth of populations in nature. Examples of this are phosphate being in limited supply (limiting) in most aquatic systems, and low temperature in the tundra which freezes the soil and limits water availability to plants. Limiting factors will slow population growth as it approaches the carrying capacity of the system.

Limiting factors are factors which slow down growth of a population as it reaches its carrying capacity.

Carrying capacity is the maximum number of a species or 'load' that can be sustainably supported by a given area.

International-mindedness

The **butterfly effect** is a term from chaos theory and refers to small changes that happen in a complex system that lead to seemingly unrelated results that are impossible to predict. It was first used in meteorology by Edward Lorenz in 1972 in a talk entitled 'Does the flap of a butterfly's wings in Brazil set off a tornado in Texas?' Since then, it has been applied to systems other than the weather eg asteroid travel paths, human behaviour.

One risk in applying the butterfly effect to complex environmental issues is that we might then think there is nothing to be done to improve things. But there is order in systems however complex and there is no evidence as yet to show that even many butterflies flapping their wings affect weather patterns.

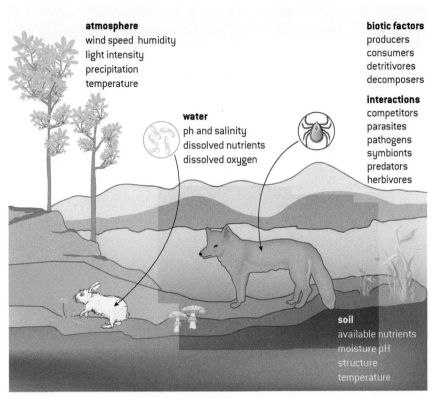

atmosphere
wind speed humidity
light intensity
precipitation
temperature

water
ph and salinity
dissolved nutrients
dissolved oxygen

biotic factors
producers
consumers
detritivores
decomposers

interactions
competitors
parasites
pathogens
symbionts
predators
herbivores

soil
available nutrients
moisture pH
structure
temperature

▲ **Figure 2.1.2** Biotic and abiotic factors and interactions within an ecosystem

Population interactions

No organism can stay the same: it grows, eats, ages and dies. All habitats change too. Animals enter and leave it, plants grow and shade the ground, water flows into and out of it. Animal migration may change a habitat greatly. Plagues of locusts can devastate all vegetation in their path including our crops. In 2013, a severe locust plague hit Madagascar with many swarms, each with over 1 billion locusts. The rice crop, livestock and rare wild animals were at risk and only aerial spraying of insecticide stopped some of the damage. Fire, natural disasters and human activities all change ecosystems. Interactions between individuals, populations and communities change ecosystems too. Each species influences the population sizes of others and the carrying capacity of the environment for that species.

Interactions between the organisms, eg predation, herbivory, parasitism, mutualism, disease, competition are termed biotic factors. All interactions result in one species having an effect on the population dynamics of the others and on the carrying capacity of the others' environment.

Competition

All the organisms in any ecosystem have some effect on every other organism in that ecosystem. Also any resource in any ecosystem exists only in a limited supply. When these two conditions apply jointly, competition takes place.

Intraspecific competition is between members of the same species. When the numbers of a population are small, there is little real

> **Key term**
>
> **Population dynamics** is the study of the factors that cause changes to population sizes.

Practical Work

* Investigate the impact of light intensity on the rate of photosynthesis in aquatic plants.

* Investigate insect herbivory on a plant.

competition between individuals for resources. Provided the numbers are not too small for individuals to find mates, population growth will be high.

Take, for example, a seagull colony on an oceanic outcrop (figure 2.1.3).

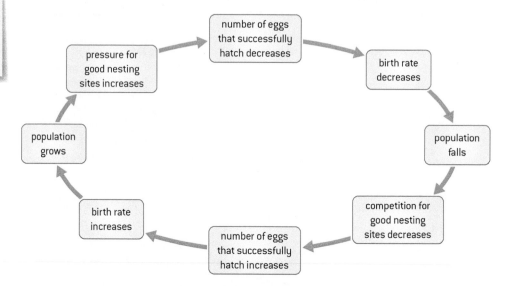

▲ **Figure 2.1.3** Competition within a seagull colony

As the population grows, so does the competition between individuals for the resources until eventually the carrying capacity of the ecosystem is reached. In this situation, often the stronger individuals claim the larger share of the resources.

Some species deal with intraspecific competition by being territorial, eg deer. An individual or pair holds an area and fends off rivals. Individuals that are the most successful reproductively will hold the biggest territory and hence have access to more resources, and will be more successful at breeding.

Intraspecific competition tends to stabilize population numbers. It produces something called a sigmoid or logistic growth curve which is S-shaped (see population changes section, p62).

Interspecific competition: Individuals of different species could be competing for the same resource. Interspecific competition may result in a balance, in which both species share the resource. The other outcome is that one species may totally out-compete the other: this is the principle of **competitive exclusion**. An example of both of these outcomes can be seen in a garden that has become overrun by weeds. A number of weed species coexist together, but often the original domestic plants have been totally excluded.

In a temperate deciduous woodland light is a limiting resource. Plant species that cannot get enough light will die out in a woodland. This is especially true of small flowering plants on the woodland floor that are not only shaded out by trees but by shrubs and bushes as well. Beech trees have very closely overlapping leaves, resulting in an almost bare woodland floor.

▲ **Figure 2.1.4** Snowdrops flowering in a temperate woodland in spring

But even in woods shaded by trees, flowers manage to grow. Carpets of snowdrops, primroses and bluebells are an integral part of all Northern European deciduous woodlands in the spring. The key to the success of these species is that they grow, flower and reproduce before the shrub and tree species burst into leaf. They avoid competing directly with species that would out-compete them for light by completing the stages of their yearly cycle that require the most energy and therefore the greatest photosynthesis when competition is less.

Competition reduces the carrying capacity for each of the competing species, as both species use the same resource(s).

Predation

Predation is when one animal, the predator, eats another animal, the prey. Examples are plenty, like lions eating zebras and wolves eating moose. The predator kills the prey. Be aware that not only animals eat other animals, some plants (insectivorous plants) consume insects and other small animals. Look at the example of the Canadian lynx and the snowshoe hare in sub-topic 1.3 as an example of negative feedback control.

Sometimes, however, a wider definition of predation is used: predation is the consumption of one organism by another. This broad definition includes not only predation in the narrower sense of the word but also herbivory and parasitism.

Herbivory

Herbivory is defined as an animal (herbivore) eating a green plant. Some plants have defence mechanisms against this, for example thorns or spines (some cacti), a stinging mechanism (stinging nettles), or toxic chemicals (poison ivy). Herbivores may be large (eg elephants, cattle) or small (eg larvae of leaf miner insects that eat the inside of leaves) or in between (eg rabbits).

Parasitism

Parasitism is a relationship between two species in which one species (the parasite) lives in or on another (the host), gaining its food from it. Normally parasites do not kill the host, unlike in predation. However, high parasite population densities can lead to the host's death. Examples of parasites are vampire bats and intestinal worms.

Mutualism

Mutualism is a relation between two or more species in which all benefit and none suffer. It is a form of symbiosis (living together). The other types of symbiosis are parasitism (above) and commensalism (when one partner is helped and the other is not significantly harmed, eg an epiphyte such as an orchid or fern growing half-way up a tree trunk).

Most people think of lichens as examples of mutualism. A **lichen** is a close association of a fungus underneath and a green alga on top. The fungus benefits by obtaining sugars from the photosynthetic alga. The alga benefits from minerals and water that the fungus absorbs and passes on to the alga.

▲ **Figure 2.1.5** Poplar sawfly larva (*Trichiocampus viminalis*) eating an aspen leaf (*Populus tremula*) in Glen Affric, Scotland, UK

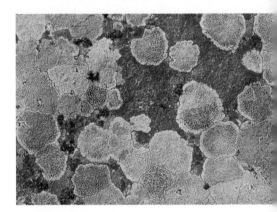

▲ **Figure 2.1.6** A lichen

Figure 2.1.7 Nitrogen-fixing root nodules on a legume root

Figure 2.1.8 Clownfish and sea anemone

Another example is the relationship between leguminous plants (beans, clover, vetch, peas) and nitrogen-fixing bacteria – *Rhizobium*. The bacteria live inside root nodules in the legumes. They absorb nitrogen from the soil and make it available to the plant in the form of ammonium compounds. The plants in turn supply the bacteria with sugar from photosynthesis. This mutualistic relationship enables legumes to live on very poor soils. As a consequence, leguminous plants are among the earliest pioneer species during succession on poor soil. Clover is also often used to increase the nutrient content of agricultural soil.

Mycorrhizal fungi and tree roots are another example. The fungi form a sheath around the feeding roots of many trees. They provide the tree with phosphates that they take up from the soil. The tree provides the fungi with glucose that it produces from photosynthesis. Both grow better than they do without the other one.

Sea anemones and clownfish are also mutualistic. The clownfish provide food for the sea anemone in the form of their feces. The anemone's stinging tentacles protect the clownfish from predators, but do not affect the clownfish.

To do

Summary of interactions between species

1. Copy and fill in the last column.

Type of Interaction		Species A	Species B	Example
Competition	Neither species benefits, both species suffer	−	−	
Predation	One species kills the other species for food	+	−	
Parasitism	The parasite benefits at the cost of the host	+	−	
Mutualism	Both species benefit	+	+	

2. What four things do all organisms need to survive? (Clue: think back to your first biology lesson.)

3. What is the difference between interspecific and intraspecific competition?

4. What effect does intraspecific competition have on the individuals of a species?

5. What is the link between competition and species diversity?

6. Why is species diversity believed to be beneficial for a community?

Population changes

Over time the numbers within a population change. If we were to collect a few bacterial cells, place them in a suitable supply of nutrients and then, under a microscope, count the number of cells every hour, we would find that there would be many more bacteria at the end of

a 24-hour period than at the start. Bacteria can reproduce asexually by splitting in two (binary fission) so, if you start with one bacterium, there will be 2,4,8,16,32,64 etc. if there are no **limiting factors** slowing growth. This is called **exponential or geometric growth** (figure 2.1.9).

S-curves

S-curves start with exponential growth. No limiting factors affect the growth at first. However, above a certain population size, the growth rate slows down gradually, finally resulting in a population of constant size.

The graph in figure 2.1.10 illustrates this for a colony of yeast grown in a constant but limited supply of nutrient. During the first few days the colony grows slowly as it starts to multiply (lag phase) then it starts to grow very rapidly as the multiplying colony has a plentiful nutrient supply (exponential phase). Eventually the population size stabilizes as only a set number of yeast cells can exploit the limited resources (stationary phase). Any more yeast cells and there is not enough food to go around. The numbers stabilize at the **carrying capacity** of the environment which is the maximum number or load of individuals that an environment can carry or support.

The maximum population size is called the **carrying capacity** (K) of the ecosystem. The area between the exponential growth curve and the S-curve is called **environmental resistance**.

J-curves

J-curves (see figure 2.1.11) show a 'boom and bust' pattern. The population grows exponentially at first and then, suddenly, collapses. These collapses are called **diebacks**. Often the population exceeds the carrying capacity on a long-term or continuing basis before the collapse occurs (**overshoot**). It is important to note 'long-term basis' as the carrying capacity can be exceeded in the short term. It seems likely that the human race is overshooting its carrying capacity at the moment.

The J-curve does not show the gradual slowdown of population growth with increasing population size.

A J-shaped population growth curve is typical of microbes, invertebrates, fish and small mammals.

S- and J-curves are idealized curves. In practice, many limiting factors act on the same population and the resulting population growth curve normally looks like a combination of an S- and a J-curve.

> *'It is well to remember that the entire universe, with one trifling exception, is composed of others.'*
>
> *John Andrew Holmes Junior*

Key terms

S and **J population curves** describe a generalized response of populations to a particular set of conditions (abiotic and biotic factors).

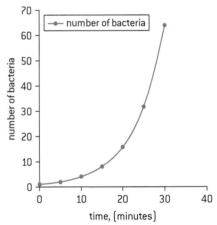

▲ **Figure 2.1.9** Exponential growth in a bacterial population over time

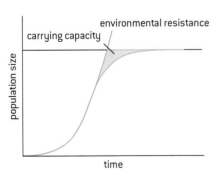

▲ **Figure 2.1.10** S-shaped growth curve of a population

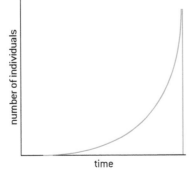

▲ **Figure 2.1.11** J-shaped growth curve of a population

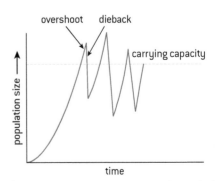

▲ **Figure 2.1.12** Fluctuation of population size around the carrying capacity

2.2 Communities and ecosystems

Significant ideas:

→ The interactions of species with their environment result in energy and nutrient flow.

→ Photosynthesis and respiration play a significant role in the flow of energy in communities.

→ The feeding relationships in a system can be modelled using food chains, food webs and ecological pyramids.

Applications and skills:

→ **Construct** models of feeding relationships, eg food chains, food webs and ecological pyramids, from given data.

→ **Explain** the transfer and transformation of energy as it flows through an ecosystem.

→ **Analyse** the efficiency of energy transfers through a system.

→ **Construct** system diagrams representing photosynthesis and respiration.

→ **Explain** the relevance of the laws of thermodynamics to the flow of energy through ecosystems.

→ **Explain** the impact of a persistent/non-biodegradable pollutant in an ecosystem.

Knowledge and understanding:

→ A **community** is a group of populations living and interacting with each other in a common habitat.

→ An **ecosystem** is a community and the physical environment it interacts with.

→ Respiration and photosynthesis can be described as processes with inputs, outputs and transformations of energy and matter.

→ **Respiration** is the conversion of organic matter into carbon dioxide and water in all living organisms, releasing energy. Aerobic respiration can simply be described as

glucose + oxygen → carbon dioxide + water

→ During respiration large amounts of energy are dissipated as heat, increasing the entropy in the ecosystem while enabling the organisms to maintain relatively low entropy/high organization.

→ Primary producers in the majority of ecosystems convert light energy into chemical energy in the process of photosynthesis.

→ The photosynthesis reaction is:

carbon dioxide + water → glucose + oxygen

→ **Photosynthesis** produces the raw material for producing biomass.

→ The trophic level is the position that an organism occupies in a food chain, or a group of organisms in a community that occupy the same position in food chains.

→ **Producers** (autotrophs) are typically plants or algae and produce their own food using photosynthesis and form the first trophic level in a food chain. Exceptions include **chemosynthetic** organisms which produce food without sunlight.

→ Feeding relationships involve **producers**, **consumers** and **decomposers**. These can be modelled using **food chains**, **food webs** and using **ecological pyramids**.

→ **Ecological pyramids** include pyramids of numbers, biomass and productivity and are quantitative models and are usually measured for a given area and time.

→ In accordance with the second law of thermodynamics, there is a tendency for

numbers and quantities of biomass and energy to decrease along food chains; therefore the pyramids become narrower towards the apex.

→ **Bioaccumulation** is the build-up of persistent/non-biodegradable pollutants within an organism or trophic level because they cannot be broken down.

→ **Biomagnification** is the increase in concentration of persistent or non-biodegradable pollutants along a food chain.

→ Toxins such as **DDT** and mercury accumulate along food chains due to the decrease of biomass and energy.

→ **Pyramids of numbers** can sometimes display different patterns, for example, when individuals at lower trophic levels are relatively large (inverted pyramids).

→ A **pyramid of biomass** represents the standing stock/storage of each trophic level measured in units such as grams of biomass per square metre ($g\,m^{-2}$) or Joules per square metre ($J\,m^{-2}$) (units of biomass or energy).

→ Pyramids of biomass can show greater quantities at higher trophic levels because they represent the biomass present at a given time, but there may be marked seasonal variations.

→ **Pyramids of productivity** refer to the **flow** of energy through a trophic level, indicating the **rate** at which that stock/storage is being generated.

→ Pyramids of productivity for entire ecosystems over a year always show a decrease along the food chain.

A community contains all the biotic (living) components of a habitat. A tropical rainforest is a community of plants and animals, bacteria and fungi. An aquarium is a community as well.

The term was first used in 1930 and modified by Arthur Tansley, a British ecologist, to describe the complex relationships between organisms and their abiotic environment. Ecosystems may be of varying sizes from a drop of rain water to a forest. Human ecosystems include a household or a school or a nation state. Ecosystems do not exist independently but interact to make up the biosphere. As virtually all parts of the Earth have been impacted by humans, all ecosystems may be considered as examples of human-affected ecosystems.

'In all things of nature, there is something of the marvellous.'

Aristotle

> **Key term**
>
> A **community** is a group of populations living and interacting with each other in a common habitat (the same place).

> **Key term**
>
> An **ecosystem** is a community and the physical environment it interacts with.

Respiration and photosynthesis

There are three key ecological concepts that are vital to your understanding of how everything else works. These are: **photosynthesis, respiration and productivity**. If you have a grasp of the basics of these, everything else makes more sense.

Respiration

All living things must respire to get energy to stay alive. If they do not do this, they die. Respiration involves breaking down food, often in the form of glucose, to release energy which is used in living processes. These processes are: Movement, Respiration, Sensitivity, Growth, Reproduction, Excretion, Nutrition and some people remember these by their first letters which spell MRS GREN. Respiration can use oxygen (aerobic) or not (anaerobic).

Key term

Respiration is the conversion of organic matter into carbon dioxide and water in all living organisms, releasing energy.

In aerobic respiration, energy is released and used and the waste products are carbon dioxide and water. Whether plants or animals, bacteria or fungi, all living things respire all the time, in the light and dark, when asleep or awake. Aerobic respiration can be summarized as:

$$\text{Glucose} + \text{oxygen} \longrightarrow \text{Energy} + \text{water} + \text{carbon dioxide}$$

$$C_6H_{12}O_6 + 6\,O_2 \longrightarrow \text{Energy} + 12\,H_2O + 6\,CO_2$$

Much of the energy produced in respiration is heat energy and is released (dissipated) into the environment. This increases the entropy (see 1.3) of the system while the organism maintains a relatively high level of organization (low entropy).

Photosynthesis

Green plants convert light energy into chemical energy in photosynthesis. This is a transformation of energy from one state to another.

Key term

Photosynthesis is the process by which green plants make their own food from water and carbon dioxide using energy from sunlight.

The leaves of plants contain chloroplasts with the green pigment chlorophyll. In the chloroplasts the energy of sunlight is used to split water and combine it with carbon dioxide to make food in the form of the glucose. Glucose is then used as the starting point for the plant to make every other molecule that it needs. In complex chemical pathways in cells, plants:

- add nitrogen and sulphur to make amino acids and then proteins,
- rearrange carbon, hydrogen and oxygen and add phosphorus to make fatty acids and lipoproteins which make up cell membranes.

Photosynthesis produces the raw material for producing biomass. Animals are totally dependent on the chemicals produced by plants. Although we can make most of the ones we need, we can only obtain essential amino acids from plants.

The waste product of photosynthesis is oxygen. This is really useful as oxygen is used in respiration.

Photosynthesis can be summarized as:

$$\text{carbon dioxide} + \text{water} \xrightarrow[\text{chlorophyll}]{\text{light energy}} \text{glucose} + \text{oxygen}$$

$$6CO_2 + 12H_2O \xrightarrow[\text{chlorophyll}]{\text{light energy}} C_6H_{12}O_6 + 6O_2$$

Green plants respire in the dark and photosynthesize and respire in the light. Water reaches the leaves from the roots by transpiration.

When all carbon dioxide that plants produce in respiration is used up in photosynthesis, the rates of the two processes are equal and there is no net release of either oxygen or carbon dioxide. This usually occurs at dawn and dusk when light intensity is not too high. This point is called the **compensation point** of a plant and it is neither adding biomass nor using it up to stay alive at this point. It is just maintaining itself. This is important to remember when we come to think about succession and biomes.

To do

Both respiration and photosynthesis are systems – biochemical ones – and we can draw systems diagrams for them with inputs, output, storages and flows which will be transformations or transfers of energy and matter.

Draw a systems diagram for each of respiration and photosynthesis.

Where can you link the two?

Food chains and trophic levels

All energy on Earth comes from the Sun so solar energy (solar radiation) is the start of every food chain. (Well, very nearly all, as some deep ocean vents give out heat from the Earth's mantle and some organisms get their energy from this through a process known as chemosynthesis. But most of us get ours from the Sun's energy.)

A **food chain** is the flow of energy from one organism to the next. A food chain shows the feeding relationships between species in an ecosystem. Arrows connect the species, usually pointing towards the species that consumes the other: so in the direction of transfer of biomass (and energy).

Organisms are grouped into trophic (or feeding) levels (Greek for food is *trophe*). Trophic levels usually start with a primary producer (plant) and end with a carnivore at the top of the chain – a **top carnivore.**

It is possible to classify the way organisms obtain energy into two categories.

1. **Producers**

    a. **Autotrophs** (green plants) which make their own food from carbon dioxide and water using energy from sunlight.

    b. **Chemosynthetic organisms** which make their own food from other simple compounds eg ammonia, hydrogen sulphide or methane, do not require sunlight and are often bacteria found in deep oceans.

2. **Consumers** (also called **heterotrophs**) which feed on autotrophs or other heterotrophs to obtain energy (herbivores, carnivores, omnivores, detritivores and decomposers).

But within the consumers there is a hierarchy of feeding.

Name of group	Trophic level	Nutrition: source of energy	Function
Primary producers (PP) **Green plants**	1st	Autotrophs: Make their own food from solar energy, CO_2 and H_2O	• Provide the energy requirements of all the other trophic levels • Habitat for other organisms • Supply nutrients to the soil • Bind the soil/stop soil erosion
Primary consumers (PC) **Herbivores**	2nd	Heterotrophs: Consume PP	These consumers keep each other in check through negative feedback loops (see 1.3). They also: • Disperse seeds

> **Key term**
>
> A **trophic level** is the position that an organism occupies in a food chain, or a group of organisms in a community that occupy the same position in food chains.

▲ **Figure 2.2.1**

Name of group	Trophic level	Nutrition: source of energy	Function
Secondary consumers (SC) **Carnivores and omnivores**	3rd	Heterotrophs: Consume herbivores and other carnivores, sometimes PP	• Pollinate flowers • Remove old and diseased animals from the population
Tertiary consumers (TC) **Carnivores and omnivores**	4th	Heterotrophs: Consume herbivores and other carnivores, sometimes PP	
Decomposers Bacteria and fungi		Obtain their energy from dead organisms by secreting enzymes that break down the organic matter	This group of organisms provide a crucial service for the ecosystem: • Break down dead organisms • Release the nutrients back into the cycle • Control the spread of disease
Detritivores Snails, slugs, blow fly maggots, vultures		Derive their energy from detritus or decomposing organic material – dead organisms or feces or parts of an organism, eg shed skin from a snake, a crab carapace	

▲ **Figure 2.2.2** Decomposer fungi in a woodland

Food webs

It would be very unusual to find an ecosystem with only a simple food chain. There are many more organisms involved and one may eat several other species.

It is possible to construct food chains for an entire ecosystem, but this starts to create a problem.

The food chains below are from a European oak woodland. In fact they are based on real food chains at Wytham Wood in Oxford, UK where some pioneer ecologists worked in the 1920s.

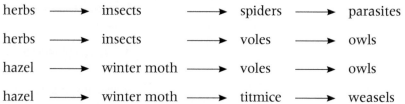

herbs ⟶ insects ⟶ spiders ⟶ parasites

herbs ⟶ insects ⟶ voles ⟶ owls

hazel ⟶ winter moth ⟶ voles ⟶ owls

hazel ⟶ winter moth ⟶ titmice ⟶ weasels

In the four different food chains, only ten species are listed and some of them are in more than one food chain. If we continued to list all the species in the wood and their interactions in every food chain, the list would run for many pages.

Food chains only illustrate a direct feeding relationship between one organism and another in a single hierarchy. The reality is very different. The diet of almost all consumers is not limited to a single food species. So a single species can appear in more than one food chain.

A further limitation of representing feeding relationships by food chains is when a species feeds at more than one trophic level. Voles are **omnivores** and as well as eating insects, they also eat plants. Humans eat plants and animals and the animals may be herbivores and carnivores. We would then have to list all the food chains again that contained voles or humans but move them to the second trophic level rather than the third in a shorter food chain.

The reality is that there is a complex network of interrelated food chains which create a **food web**.

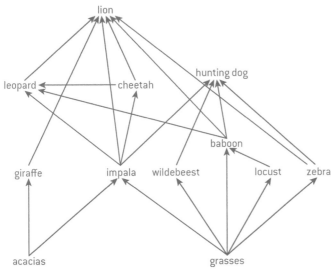

▲ **Figure 2.2.3** Food web on the African savannah

The earliest food webs were published in the 1920s by Elton (on Bear Island, Norway) and Hardy (on plankton and herring in the North Sea). Elton's food web is in figure 2.2.4.

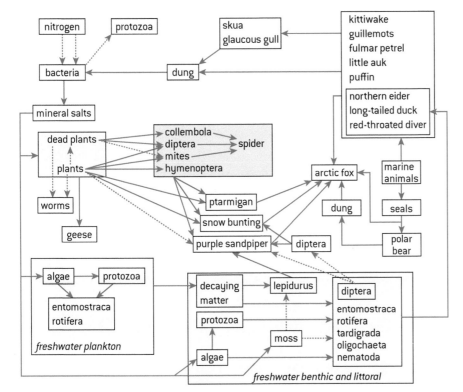

▲ **Figure 2.2.4** One of the first food webs observed by Elton on Bear Island, Norway

TOK

Feeding relationships can be represented by different models. How can we decide when one model is better than another?

To do

Tabulate the differences between a food chain and a food web.

69

To do

Carnivores in the tundra ecosystem

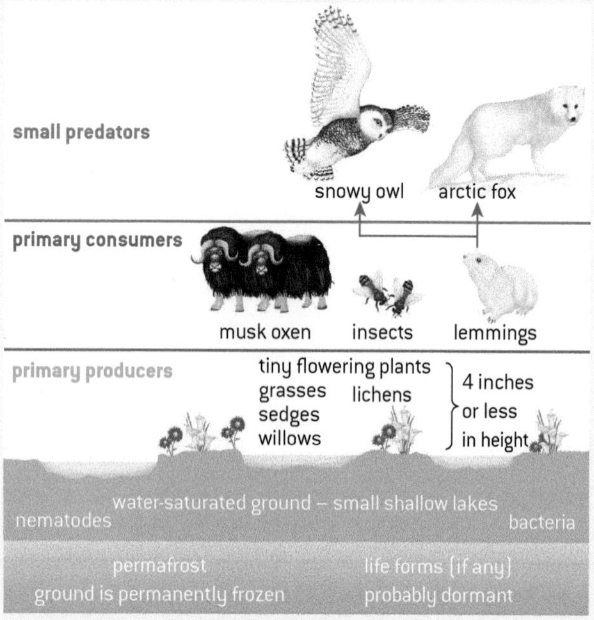

small predators
snowy owl
arctic fox
primary consumers
musk oxen
insects
lemmings
primary producers
tiny flowering plants
grasses
sedges
willows
lichens
4 inches or less in height
water-saturated ground – small shallow lakes
nematodes
bacteria
permafrost
ground is permanently frozen
life forms (if any) probably dormant

Figure 2.2.5 A food web in the tundra; source Dave Harrison, used with permission

There are several species of bear in the tundra. Polar bears live further north, but are also found in the tundra searching for food. The Kodiak is the largest bear in the Alaskan tundra. It is usually a brown colour. Brown bears are not as fierce as their reputation makes them out to be. They seldom eat meat. Wolves are the top predators of the tundra. They travel in small families (packs) and prey on caribou and other large herbivores that are too slow to stay with their groups. Some wolves change to a bright white colour in the winter. Otters live near rivers and lakes so they can feed on fish. Shrews are the smallest carnivores of the tundra. Even bats are found in the tundra during the summer. They feed on the swarms of insects that fill the air.

The primary production is not sufficient to support animal life if only small areas of tundra are considered. The large herbivores and carnivores are dependent on the productivity of vast areas of tundra and have adopted a migratory way of life. Small herbivores feed and live in the vegetation mat, eating the roots, rhizomes and bulbs. The populations of small herbivores like lemmings show interesting fluctuations that also affect the carnivores dependent on them, such as the arctic fox and snowy owl.

The blue squares represent the appearance and frequency of snowy owls after almost exponential population increases of lemmings. There is then a lag period of about two years before lemming numbers increase again.

1. Draw a food web for the tundra with **only** the animals mentioned.
2. Why do you think the snowy owls only appear when lemming numbers have fallen? (Hint: climate and decomposers.)

lemmings per hectare
40
27
13
0
1929
1930
1931
1932
1933
1934
1935
1936
1937
1938
1939
1940
1941
1942
1943
year

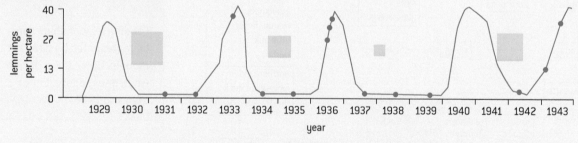

Figure 2.2.6 Snowy owl and lemming numbers in the tundra from 1929 to 1943

To do

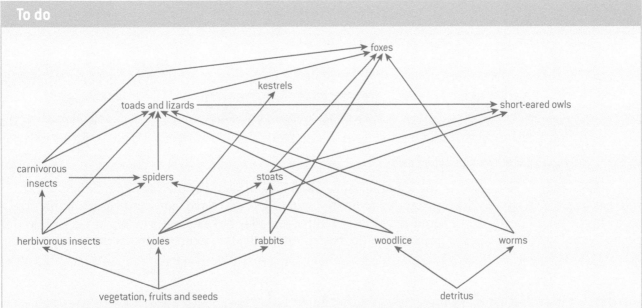

▲ **Figure 2.2.7** A simplified food web from the acid heathland at Studland, Dorset, UK

1. What is the longest food chain in this food web?

2. Name two species that are found at two trophic levels.

3. If all kestrels die, what may happen to (a) voles and (b) short-eared owls?

4. If there is a great increase in the rabbit population, what happens to (a) rabbit predators and (b) the vegetation?

5. If a pesticide is added to kill spiders, what may happen to the foxes?

Ecological pyramids

Pyramids are graphical models of the quantitative differences between amounts of living material stored at each trophic level of a food chain.

- They allow easy examination of energy transfers and losses.

- They give an idea of what feeds on what and what organisms exist at the different trophic levels.

- They also help to demonstrate that ecosystems are systems that are in balance.

All pyramids may be represented as in figure 2.2.8.

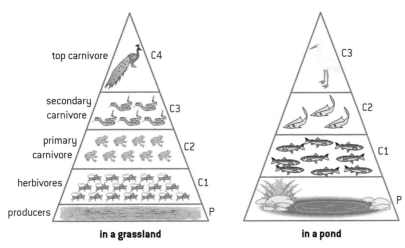

▲ **Figure 2.2.8**

A **pyramid of numbers** shows the number of organisms at each trophic level in a food chain at one time – the **standing crop**. The units are number per unit area.

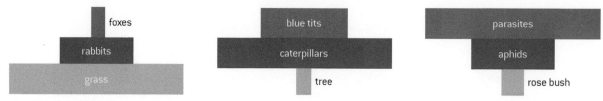

▲ **Figure 2.2.9** Pyramids of numbers

The length of each bar gives a measure of the relative numbers. Most pyramids are broad at their base and have many individuals in the producer (P) level. But some may have a large single plant, a tree, as the producer so the base is one individual which supports many consumers.

▲ **Figure 2.2.10** Pyramid of numbers for a grazing ecosystem   ▲ **Figure 2.2.11** Pyramid of numbers for an oak wood

Advantage

- This is a simple, easy method of giving an overview and is good at comparing changes in population numbers with time or season.

Disadvantages

- All organisms are included regardless of their size, therefore a pyramid based on an oak tree would be inverted (have a small bottom and get larger as it goes up the trophic levels).

- Does not allow for juveniles or immature forms.

- Numbers can be too great to represent accurately.

A **pyramid of biomass** contains the biomass (mass of each individual × number of individuals) at each trophic level. Biomass is the quantity of (dry) organic material in an organism, a population, a particular trophic level or an ecosystem.

The units of a pyramid of biomass are in units of mass per unit area, often grams per square metre ($g\,m^{-2}$) or kilograms per water volume (eg, $kg\,km^{-3}$). A pyramid of biomass is more likely to be a pyramid shape but there are some exceptions, particularly in oceanic ecosystems where the producers are phytoplankton (unicellular green algae). Phytoplankton reproduce fast but are present only in small amounts at any one time. As a pyramid represents biomass at one time only, eg in winter, the phytoplankton bar may be far less than that of the zooplankton which are the primary consumers.

▲ **Figure 2.2.12** Pyramids of biomass (units gm^{-2})

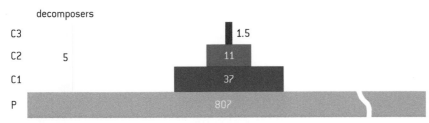

▲ **Figure 2.2.13** Pyramid of biomass for a lake

Advantage

- Overcomes some of the problems of pyramids of numbers.

Disadvantages

- Only uses samples from populations, so it is impossible to measure biomass exactly.

- Organisms must be killed to measure dry mass.

- The time of the year that biomass is measured affects the result. In the case of algae, their biomass changes by large amounts during the year therefore the shape of the pyramid would depend on the season. The giant redwood trees of California have accumulated their biomass over many years yet algae in a lake at the equivalent trophic level may only have needed a few days to accumulate the same biomass. This pyramid will not show these differences.

- Pyramids of total biomass accumulated per year by organisms at a trophic level would usually be pyramidal in shape. But two organisms with the same mass do not have to have the same energy content. A dormouse stores a large amount of fat, around 37 kJ g^{-1} of potential chemical energy yet a carnivore of equivalent mass would contain larger amounts of carbohydrates and proteins, around 17 kJ g^{-1} potential energy. Some organisms contain a high proportion of non-digestible parts such as in the exoskeletons of marine crustaceans.

Pyramids of numbers and biomass are snapshots at one time and place. Depending on when the pyramid was investigated, for the same food web in the same ecosystem, the pyramid can vary with season and year. In the spring, there will be more producers growing, in autumn, perhaps more consumers living on the producers. Pyramids of numbers may sometimes be inverted (figure 2.2.9).

A **pyramid of productivity** shows the rate of flow of energy or biomass through each trophic level. It shows the energy or biomass being generated and available as food to the next trophic level during a fixed period of time. So, unlike pyramids of numbers and biomass, which are snapshots at one time, these pyramids show the flow of energy over time. They are always pyramid-shaped in healthy ecosystems as they must follow the second law of thermodynamics (1.3). They are measured in units of energy or mass per unit area per period of time, often Joules per square metre per year (J m^{-2} yr^{-1}). Productivity values are rates of flow, whereas biomass values are stores existing at one particular time.

Supermarket analogy

The turnover of two supermarkets cannot be compared by just looking at the goods displayed on the shelves; the rate at which goods are being stocked and sold needs to be known. Both shops may have well stocked shelves but the rate of removal of goods from a city centre shop may be considerably more than a village shop. In the same way, pyramids of biomass simply represent the stock on the shelves, whereas pyramids of productivity show the rate at which that stock is being removed by customers and restocked by shop assistants.

The bars are drawn in proportion to the total energy utilized at each trophic level. As only about 10% of the energy in one level is passed on to the next, in pyramids of productivity, each bar will be about 10% of the lower one. Sometimes the term pyramid of energy is used which can be either the standing stock (biomass) or productivity. We shall avoid it here as it is confusing.

Pyramid	Units
Numbers (standing crop)	$N\,m^{-2}$
Biomass (standing crop)	$g\,m^{-2}$
Productivity (flow of biomass/energy)	$g\,m^{-2}yr^{-1}$ $J\,m^{-2}yr^{-1}$

▲ **Figure 2.2.14** Pyramid units. Note the notation: N = numbers, g = grams, J = joules, the negative indices replace / , eg N/m²

Advantages

- Most accurate system, shows the actual energy transferred and allows for rate of production.
- Allows comparison of ecosystems based on relative energy flows.
- Pyramids are not inverted.
- Energy from solar radiation can be added.

Disadvantages

- It is very difficult and complex to collect energy data as the rate of biomass production over time is required.
- There is still the problem (as in the other pyramids) of assigning a species to a particular trophic level when they may be omnivorous.

To do

On graph paper, draw and label pyramids from the data in the table. Comment on these.

	Number pyramid	Biomass pyramid / kJ m⁻²	Productivity pyramid / 000 kJ m⁻² yr⁻¹
Primary producers	100,000	2,500	500
Primary consumers	10,000	200	50
Secondary consumers	2,000	15	5
Top consumers	500	1	–

To think about

Consequences of pyramids and ecosystem function

1. The concentration of toxic substances in food chains.
2. The limited length of food chains.
3. The vulnerability of top carnivores.

Bioaccumulation and biomagnification

If a chemical in the environment (eg a pesticide or a heavy metal) breaks down slowly or does not break down at all, plants may take it up and animals may take it in as they eat or breathe. If they do not excrete or

egest it, it accumulates in their bodies over time. If the chemical stays in the ecosystem for a prolonged period of time the concentration builds up. Eventually, the concentration may be high enough to cause disease or death. This is **bioaccumulation**.

If a herbivore eats a plant that has the chemical in its tissues, the amount of the chemical that is taken in by the herbivore is greater than that in the plant that is eaten – because the herbivore grazes many plants over time. If a carnivore eats the herbivores, it too will take in more of the chemical than each herbivore contained as it eats several herbivores over time. In this way the chemical's concentration is magnified from trophic level to trophic level. While the concentration of the chemical may not affect organisms lower in the food chain, the top trophic levels may take in so much of the chemical that it causes disease or their death. This is **biomagnification**.

A serious problem with pesticides is how long they last in the environment once they are sprayed. Some decompose into harmless chemicals as soon as they touch the soil. Glyphosate (first sold by Monsanto as Roundup) is one of these: once it touches the soil, it is inactivated. Others are persistent and do not break down in this way. They enter the food web and move through it from trophic level to trophic level as they do not break down even inside the bodies of organisms. They are non-biodegradable (POPs, see sub-topic 1.5). Many early insecticides such as DDT, dieldrin and aldrin fall into this group and they are stored in the fat of animals. Seals and penguins in Antarctica and polar bears in the Arctic have been found with pesticides in their tissues. The nearest land where the pesticides have been used is thousands of kilometres away. How may the pesticides have reached them?

To do

In this food web, the smaller fish (minnows) eat plankton (microscopic plants and animals) in the water. The minnow is eaten by the larger fish called pickerel. These are eaten by herons, ospreys and cormorants and herons eat the minnows as well. The numbers give the percentage concentration of DDT.

1. How many trophic levels are in this food web?
2. How many times more concentrated is the DDT in the body of the cormorant than the water? Explain how this happens.
3. In which species does bioaccumulation occur?
4. In which species does biomagnification occur?

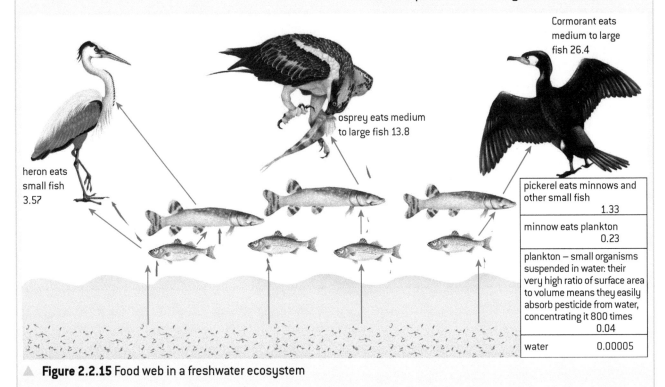

▲ **Figure 2.2.15** Food web in a freshwater ecosystem

To do

1. An ecosystem consists of one oak tree on which 10,000 herbivores are feeding. These herbivores are prey to 500 spiders and carnivorous insects. Three birds are feasting on these spiders and carnivorous insects. The oak tree has a mass of 4,000 kg, the herbivores have an average mass of 0.05 g, the spiders and carnivorous insects have an average mass of 0.2 g and the three birds have an average mass of 10 g.

   a. Construct a pyramid of numbers.

   b. Construct a pyramid of biomass.

   c. Explain the differences between these two pyramids.

2. Explain whether the energy 'loss' between two subsequent trophic levels is in contradiction with the first law of thermodynamics (see 1.3).

3. Assuming an ecological efficiency of 10%, 5% and 20% respectively (see figure 2.2.16), what will be the energy available at the tertiary consumer level (4th trophic level), given a net primary productivity of 90,000 kJ m^{-2} yr^{-1}? What percentage is this figure of the original energy value at the primary producer level?

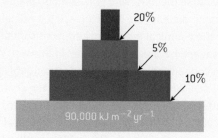

20%
5%
10%
90,000 kJ m^{-2} yr^{-1}

▲ **Figure 2.2.16**

Story of Minamata Bay

Minamata is a small factory town in Japan, dominated by one factory, the Chisso factory. Chisso make petrochemical-based substances from fertilizer to plastics. Waste water containing methylmercury from this process was released into Minamata Bay. Between 1932 and 1968 Chisso released an estimated 24 tonnes of mercury and methylmercury into Minamata Bay. Beginning in the 1950s, several thousand people living locally started to suffer from mercury poisoning.

What had happened? Waste water containing elemental mercury and methylmercury from this process was released into Minamata Bay. Also, some bacteria can change elemental mercury to the modified form called methylmercury. Methylmercury is easily absorbed into the bodies of small organisms such as shrimp. When the shrimp are eaten by fish, the methylmercury enters the fish. The methylmercury does not break down

easily and can stay in the fish bodies for a long time. As the fish eat more and more shrimp, the amount of methylmercury increases. The same increase in concentration happens when people then eat the fish. Mercury bioaccumulated in the food chain. People of Minamata ate a lot of shellfish and were poisoned by mercury. It took over 30 years to recognize the cause of their illnesses and compensation is still being given by the Chisso Corporation although the mercury release stopped in 1968.

There is a slow orders-of-magnitude build-up along the food chain: very many bacteria absorb very small amounts of mercury – many shrimp eat a lot of bacteria building up the mercury concentration – lots of fish eat lots of shrimp again building up the concentration and finally a small number of humans at the top of the food chain eventually eat a lot of fish and absorb high levels of methylmercury.

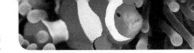

Why top carnivores are in trouble

It is often the highest trophic level in a food chain (the top carnivore) that is the most susceptible to alterations in the environment. In the UK, the population of the peregrine falcon (a bird of prey) crashed in the late 1950s probably due to agricultural chemicals such as DDT accumulating and then magnifying in the food chain. This appeared to cause egg-shell thinning and reduced breeding success. These chemicals were banned and from the mid 1960s, the peregrine population began to slowly recover despite persecution and the threat from egg collectors.

The top of the food chain is always vulnerable to the effects of changes further down the chain. Top carnivores often have a limited diet so a change in their food prey has a knock-on effect. Their population numbers are low because of the fall in efficiency along a food chain, therefore their ability to withstand negative influences is more limited than species lower in the food chain with larger populations.

Practical Work

* Construct a pyramid of numbers of for a local ecosystem.

* Build up a food chain for local ecosystem.

To think about

Polar bears and the new DDT

▲ **Figure 2.2.17** Polar bears

The new DDT could be polybrominated diphenyl ether (PBDE). It is manufactured in the United States and was widely used in the 1990s as a flame retardant to coat electrical appliances, sofas, carpets and car seats. The problem is that this chemical was designed to last the lifetime of the product, but in fact it lasts much longer. When sofas, carpets and car seats were thrown away, PBDE entered the rivers, the oceans and the atmosphere. The Arctic, where all the world's polar bears live, is one of the great sinks of the planet. Chemical pollutants such as PBDE are carried towards the Arctic Ocean by the great rivers of Russia and Canada. PBDE already in the sea is taken north by ocean currents and carried by the wind. As it moves through the food chain from plankton to predator, PBDE bioaccumulates and is biomagnified so that long-lived top carnivores such as the polar bear accumulate the most concentrated amounts of them. High amounts of PBDE have now been found in the body tissue of polar bears and killer whales. The long-term environmental effect of PBDE is unknown, but it will probably damage immune systems, brain functions and bone strength. It also messes up the polar bear's sex hormones. One female bear on Spitzbergen had both male and female organs, a condition called imposex and often linked to chemical pollution.

The length of food chains

As a rule of thumb, only 10% of the energy in one trophic level is transferred to the next – the **trophic efficiency** is 10%. A major part of the energy is used in respiration to keep the organism alive and is finally lost as heat to the environment. This is a result of the second law of thermodynamics (1.3) which states that energy is degraded to lower quality and finally to heat. More is lost because herbivores destroy more plant material than they actually eat – by trampling on it, or they reject it because it is too tough, old or spiky. Some material is not eaten at all and some dies and decomposes before it can be eaten. The 90% loss of energy in going from one trophic level to the next means there is very little energy available after about four trophic levels in terrestrial ecosystems and five in aquatic ecosystems.

Top carnivores are vulnerable because of the loss of energy from each trophic level. There is only so much energy available and that is why big, fierce animals are rare. It is hard for them to accumulate enough energy to grow to a large size and to maintain their bodies.

To do

Model of the structure of an ecosystem

A model is a simplified diagram that shows the structure and workings (functions) of a system.

Copy and label the model (right) to show the relationship between trophic levels.

Add arrows and names of the various processes.

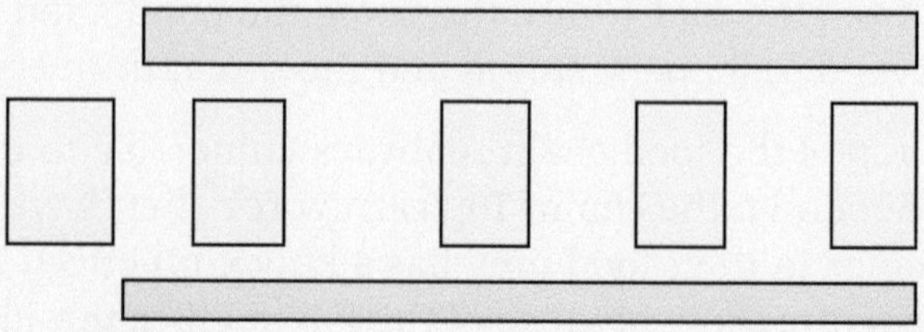

▲ **Figure 2.2.18**

To do

Constructing a food web from information

Make a list of all the organisms described in the description below and construct a food web diagram to show all feeding relationships.

The Aigamo paddy farming system is a self-sustaining agro-ecosystem based on rice, ducks (aigamo) and fish. The ducks eat up insect pests and the golden snails, which attack rice plants. They also eat the seeds and seedlings of weeds, using their feet to dig up the weed seedlings, thereby oxygenating the water and encouraging the roots of the rice plants to grow more strongly. The 'pests' and 'weeds' are food sources for rearing the ducks. The ducks are left in the fields 24 hours a day and are completely free-range until the rice plants form ears of grain in the field. At that point, the ducks have to be rounded up (otherwise they will eat the rice grains) and are fed exclusively on waste grain. There they mature, lay eggs, and fatten up for the market.

The ducks are not the only inhabitants of this system. The aquatic fern, *Azolla*, or duckweed, which harbours a mutualistic blue-green bacterium that can fix atmospheric nitrogen, is also grown on the surface of

the water. The *Azolla* is an efficient nitrogen fixer, and is readily eaten by the ducks, as well as attracting insects to be similarly enjoyed by the ducks. The plant is very prolific, doubling itself every three days, so it can be harvested for cattle-feed as well. In addition, the plants spread out to cover the surface of the water, providing hiding places for another inhabitant, the roach (a fish), and protecting them from the ducks. The roach feed on duck feces, on *Daphnia* (a crustacean) and various worms, which in turn feed on the plankton. Both fish and ducks provide manure to fertilize the rice plants throughout the growing season, and the rice plants in turn provide shelter for the ducks.

The Aigamo paddy field, then, is a complex, well-balanced, self-maintaining, self-propagating ecosystem. The only external input is the small amount of waste grain fed to the ducks, and the output is a delicious, nutritious harvest of organic rice, duck and roach. It is amazingly productive. A two hectare farm of which 1.5 ha are paddy fields can yield annually seven tonnes of rice, 300 ducks, 4,000 ducklings and enough vegetables to supply 100 people.

2.3 Flows of energy and matter

Significant ideas:

→ Ecosystems are linked together by energy and matter flows.

→ The Sun's energy drives these flows and humans are impacting the flows of energy and matter both locally and globally.

Applications and skills:

→ **Analyse** quantitative models of flows of energy and matter.

→ **Construct** quantitative model of flows of energy or matter for given data.

→ **Analyse** the efficiency of energy transfers through a system.

→ **Calculate** the values of both gross primary productivity (GPP) and net primary productivity (NPP) from given data.

→ **Calculate** the values of both gross secondary productivity (GSP) and net secondary productivity (NSP) from given data.

→ **Discuss** human impacts on energy flows, the carbon and nitrogen cycles.

🌐 Knowledge and understanding:

→ As **solar radiation** (insolation) enters the Earth's atmosphere some energy becomes unavailable for ecosystems as the energy is absorbed by inorganic matter or reflected back into the atmosphere.

→ Pathways of radiation through the atmosphere involve a loss of radiation through **reflection and absorption** as shown in figure 2.3.1.

→ Pathways of energy through an ecosystem include:

- conversion of light energy to chemical energy

- transfer of chemical energy from one trophic level to another with varying efficiencies

- overall conversion of ultraviolet and visible light to heat energy by an ecosystem

- re-radiation of heat energy to the atmosphere.

→ The conversion of energy into biomass for a given period of time is measured as **productivity**.

→ **Net primary productivity (NPP)** is calculated by subtracting respiratory losses (R) from gross **primary productivity (GPP)**. **NPP = GPP − R**

→ **Gross secondary productivity (GSP)** is the total energy / biomass assimilated by consumers and is calculated by subtracting the mass of fecal loss from the mass of food eaten. **GSP = food eaten − fecal loss**

→ **Net secondary productivity (NSP)** is calculated by subtracting respiratory losses (R) from GSP. **NSP = GSP − R**

→ **Maximum sustainable yields** are equivalent to the net primary or net secondary productivity of a system.

→ Matter also flows through ecosystems linking them together. This flow of matter involves **transfers** and **transformations**.

→ The **carbon and nitrogen cycles** are used to illustrate this flow of matter using flow diagrams. These cycles contain storages (sometimes referred to as sinks) and flows which move matter between storages.

→ **Storages in the carbon cycle** include organisms and forests (both organic), or the atmosphere, soil, fossil fuels and oceans (all inorganic).

→ **Flows in the carbon cycle** include consumption (feeding), death and decomposition, photosynthesis, respiration, dissolving and fossilisation.

→ **Storages in the nitrogen cycle** include organisms (organic), soil, fossil fuels, atmosphere and water bodies (all inorganic).

→ **Flows in the nitrogen cycle** include nitrogen fixation by bacteria and lightning, absorption, assimilation, consumption (feeding), excretion, death and decomposition, and denitrification by bacteria in waterlogged soils.

→ **Human activities** such as burning fossil fuels, deforestation, urbanization and agriculture impact energy flows as well as the carbon and nitrogen cycles.

Key points

- Almost all energy that drives processes on Earth comes from the Sun.
- This is called **solar radiation** and is made up of visible wavelengths (light) and those wavelengths that humans cannot see (ultraviolet and infrared).
- Some 60% of this is intercepted by atmospheric gases and dust particles. Nearly all the ultraviolet light is absorbed by ozone.
- Most of the infrared light (heat) is absorbed by carbon dioxide, clouds and water vapour in the atmosphere.
- Both ultraviolet and visible light energy (short wave) are converted to heat energy (long wave) (following the laws of thermodynamics).
- The systems of the biosphere are dependent on the amount of energy reaching the ground, not the amount reaching the outer atmosphere. This amount varies according to the time of day, the season, the amount of cloud cover and other factors.
- Most of this energy is not used to power living systems, it is reflected from soil, water or vegetation or absorbed and re-radiated as heat.
- Of the energy reaching the Earth's surface, about 35% is reflected back into space by ice, snow, water and land.
- Some energy is absorbed and heats up the land and seas.
- Of all the energy coming in, only about 1–4% of it is available to plants on the surface of the Earth.
- This energy is captured by green plants which convert light to chemical energy.
- Then the chemical energy is transferred from one trophic level to the next.

The fate of solar radiation reaching the Earth

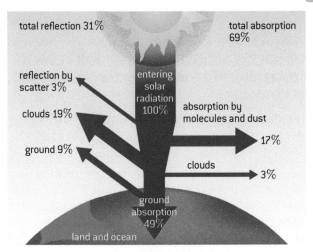

▲ **Figure 2.3.1** The fate of solar radiation hitting the Earth

Our Sun is about 4.5 billion years old and halfway through its lifespan. It has burned up about half of its hydrogen in nuclear fusion to make helium and release energy. This energy is in packets called photons and it takes eight minutes for a photon leaving the Sun to reach the Earth. The energy leaving the Sun is about 63 million joules per second per square metre ($J\ s^{-1}\ m^{-2}$). The solar energy reaching the top of the atmosphere of Earth is 1,400 $J\ s^{-1}\ m^{-2}$ (or 1,400 watts per second). This is the Earth's **solar constant**.

The only way in which life can turn solar energy into food is through photosynthesis by green plants. For a crop plant, such as wheat, which is an efficient converter, the figures are as follows. The plant can only

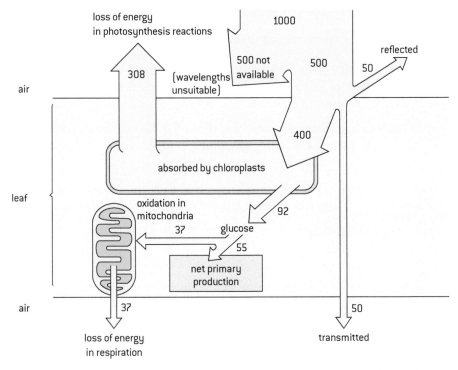

▲ **Figure 2.3.2** Photosynthetic efficiency of a crop plant. This is based on the input of 1,000 units of solar radiation

Key term

Productivity is the conversion of energy into biomass over a given period of time. It is the rate of growth or biomass increase in plants and animals. It is measured per unit area per unit time, eg per metre^2 per year. $(m^{-2} yr^{-1})$.

absorb about 40% of the energy that hits a leaf. About 5% is reflected, 50% lost and 5% passes straight through the leaf. But plants only use the red and blue wavelengths of light in photosynthesis and reflect the other colours (which is why plants look green). So of the 40%, just over 9% can be used. This is the GPP of the plant. Just under half of this is required in respiration to stay alive so 5.5% of the energy hitting a leaf becomes NPP (new plant material).

Of all the solar radiation falling on the Earth, plants only capture 0.06% of it (GPP) and use some of that to stay alive. What is left over (NPP) is the amount of food available to all the animals including humans. In general the efficiency of conversion of energy to food is low at about 2–3% in terrestrial systems but even lower at about 1% in many aquatic systems as water absorbs more of the light before it reaches the plants, though it is variable and there are exceptions (such as marine zooplankton feeding on phytoplankton).

Productivity

Gross refers to the total amount of something made as a result of an activity, eg profit from a business or salary from a job.

Net refers to the amount left after deductions are made, eg costs of production or deductions of tax and insurance from a salary. It is what you have left and is always lower than the gross amount.

Primary in ecology means to do with plants. **Secondary** is to do with animals.

Biomass is the living mass of an organism or organisms but sometimes refers to dry mass.

Now we have these clear, we can put some together.

Key terms

Gross productivity (GP) is the total gain in energy or biomass per unit area per unit time. It is the biomass that could be gained by an organism before any deductions.

Net productivity (NP) is the gain in energy or biomass per unit area per unit time that remains after deductions due to respiration.

Gross primary productivity (GPP) is the total gain in energy or biomass per unit area per unit time by green plants. It is the energy fixed (or converted from light to chemical energy) by green plants by photosynthesis. But, some of this is used in respiration so...

Net productivity (NP) results from the fact that all organisms have to respire to stay alive so some of this energy is used up in staying alive instead of being used to grow.

We usually talk about productivity and not production in ecology – that way we know the area or volume and the time period to which we refer.

Primary productivity: autotrophs are the base unit of all stored energy in any ecosystem. Light energy is converted into chemical energy by photosynthesis using chlorophyll within the cells of plants.

Gross primary productivity (GPP): plants are the first organisms in the production chain. They fix light energy and convert it to sugars so it is theoretically possible to calculate a plant's energy uptake by measuring the amount of sugar produced (GPP).

However measuring the sugar produced is extremely difficult as much of it is used up by plants in respiration almost as soon as it is produced. A

more useful way of looking at production of plants is the measurement of **net primary productivity (NPP)**.

An ecosystem's NPP is the rate at which plants accumulate dry mass (actual plant material) usually measured in g m^{-2}. This glucose produced in photosynthesis has two main fates.

- Some provides for growth, maintenance and reproduction (life processes) with energy being lost as heat during processes of respiration.

- The remainder is deposited in and around cells as new material and represents the stored dry mass – this store of energy is potential food for consumers within the ecosystem.

So, NPP represents the difference between the rate at which plants photosynthesize, GPP, and the rate at which they respire. This accumulation of dry mass is usually termed biomass and provides a useful measure of both the production and the utilization of resources.

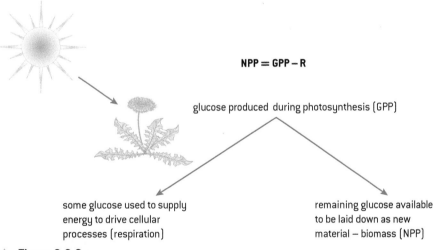

NPP = GPP – R

glucose produced during photosynthesis (GPP)

some glucose used to supply energy to drive cellular processes (respiration)

remaining glucose available to be laid down as new material – biomass (NPP)

▲ **Figure 2.3.3**

The total amount of plant material is the theoretical maximum amount of energy that is available to all the animals, both the herbivores and the carnivores that feed on them. It has two fates:

- lost from food chains as it dies and decays OR

- eaten by herbivores which means it is removed from primary productivity.

The amount of biomass produced varies.

- Spatially: some biomes have much higher NPP rates than others – eg tropical rainforest vs tundra.

- Temporally: Many plants have seasonal patterns of productivity linked to changing availability of basic resources – light, water and warmth (see succession 2.4).

Net secondary productivity (NSP)

As with plants, not all energy that goes into the herbivore is available to make new biomass, it has different fates.

- Only food that crosses the wall of the alimentary canal (gut wall) of animals is absorbed and is used to power life processes (**assimilated food energy**):

  - Some of the assimilated food energy is used in cellular respiration to provide energy for life processes.

  - Some is removed as nitrogenous waste, in most animals as urine.

  - The rest is stored in the dry mass of new body tissue.

- Some of the ingested plant material will pass straight through the herbivore and be released as feces (egestion). This is not absorbed and provides animals with no energy.

  > net productivity of herbivores (net secondary productivity) = energy in the food ingested − the energy lost in egestion − energy used in respiration

Total food ingested including the food that is egested is the measure of **gross secondary productivity (GSP)**. Therefore net secondary productivity can be thought of in the same way as net primary productivity.

Only a very small percentage of the original NPP of plants is turned into secondary productivity by herbivores and it is this secondary productivity, which is available to consumers at the next trophic level. This change of primary productivity to secondary productivity follows the general conditions of energy transfer up the trophic levels.

Carnivores, animals that eat other animals, are the next up the trophic ladder. Secondary consumers are those that eat herbivores and tertiary consumers are those whose main source of energy is other carnivores. The ability of carnivores to assimilate energy follows the same basic path as that of herbivores, though secondary and tertiary consumers have higher protein diets, meat, which is more easily digested and assimilated.

Carnivores

- On average they assimilate 80% of the energy in their diets.

- They egest less than 20%.

- Usually they have to chase moving animals so higher energy intake is offset by increased respiration during hunting.

- Biomass is locked up in the prey foods – non-digestible skeletal parts, such as bone, horn and antler – so they have to assimilate the maximum amount of energy that they can from any digestible food.

Key terms

Gross secondary productivity (GSP) is the total energy / biomass assimilated (taken up) by consumers and is calculated by subtracting the mass of fecal loss from the mass of food eaten.

> GSP = food eaten − fecal loss

Animals are known as heterotrophs or heterotrophic organisms to distinguish them from plants (autotrophs). Troph is derived from the Ancient Greek word for food, so plants are auto-feeding and animals other-feeding (hetero = other) or feed on others.

Net secondary productivity (NSP) is the total gain in energy or biomass per unit area per unit time by consumers after allowing for losses to respiration. There are other losses in animals as well as to respiration but respiration is the main one.

NSP is calculated by subtracting respiratory losses (R) from GSP.

Herbivores

- Assimilate about 40% of the energy in their diet.
- They egest 60%.
- They graze static plants.

Flows of energy and matter

Flow	Energy flows through systems	Matter also flows through ecosystems (nutrients, oxygen, carbon dioxide, water)
How much?	Infinite (the Sun is always shining somewhere)	Finite
When?	Once	Cycles and recycles repeatedly
Outputs	All organisms give out energy all the time as respiration releases heat	All organisms release waste nutrients, carbon dioxide and water
Quality	Degrades from higher to lower quality energy (light to heat) so entropy increases	May change form but does not degrade
Storages	Temporarily stored as chemical energy	Is stored long and short term in chemical forms

▲ **Figure 2.3.4** Comparing energy and matter flows

Transfers and transformations

Both matter and energy move or flow through ecosystems (see 1.2).

Both types of flow use energy – transfers, being simpler, use less energy and are therefore more efficient than transformations.

Cycles and flows

Energy flows through an ecosystem in one direction, starting as solar radiation and finally leaving as heat released through the respiration of decomposers. On the other hand chemical nutrients in the biosphere cycle: nutrients are absorbed by organisms from the soil and atmosphere and circulate through the trophic levels and are finally released back to the ecosystem, usually via the detritus food chain. These are the **biogeochemical cycles**.

Nutrient cycles

There are around 40 elements that cycle through ecosystems, though some exist only in trace amounts. All the biogeochemical cycles have both organic (when the element is in a living organism) and inorganic (when the element is in a simpler form outside living organisms) phases. Both are vital: the efficiency of movement through the organic phase determines how much is available to living organisms. Yet the major reservoir for all the main elements tends to be outside of the food chain as inorganic molecules in rock and soils. Flow in this inorganic phase tends to be much slower than the movement

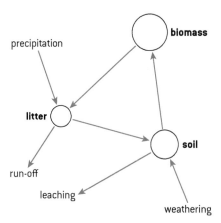

▲ **Figure 2.3.5** Gersmehl's nutrient model

of these nutrients through organisms, the organic phase. The major biogeochemical cycles are those of water, carbon, nitrogen, sulphur and phosphorus, all of which follow partially similar routes and all of which have similar characteristics:

- Movement of matter, such as nutrients, through an ecosystem is very different from the movement of energy.

- Energy travels from the Sun, through food webs and is eventually lost to space as heat.

- Nutrients and matter are finite and are recycled and reused (via the decomposer food chain).

- Organisms die and are decomposed and nutrients are released, eventually becoming parts of living things again, when they are taken up by plants. These are the biogeochemical cycles.

The carbon cycle

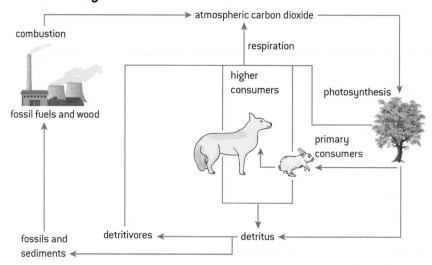

▲ **Figure 2.3.6** The carbon cycle

Where is the carbon stored?

In carbon or carbon dioxide sinks. These may be:

Organic (with complex carbon molecules):

- Organisms (biomass) in the biosphere – living plants and animals.

- Fossilized life forms, eg fossil fuels.

Inorganic (simple carbon molecules):

- Locked up or fixed into solid forms and stored as sedimentary rocks and fossil fuels. Most carbon is stored here and locked up for millions of years.

- The oceans where carbon is dissolved or locked up as carbonates in the shells of marine organisms.

- Soil.

- A small proportion is carbon dioxide in the atmosphere (0.37%).

Carbon flows

The carbon cycle, in which carbon circulates through living and non-living systems occurs in the ecosphere. Here carbon is found in four main storages: the soil, living things (biomass), the oceans and the atmosphere. Carbon not in the atmosphere is stored in carbon dioxide sinks (soil, biomass and oceans), as complex organic molecules or dissolved in seawater.

Carbon cycles between living (biotic) and non-living (abiotic) chemical cycles: it is fixed by photosynthesis and released back to the atmosphere through respiration. Carbon is also released back to the atmosphere through combustion of fossil fuels and biomass. When dead organisms decompose, when they respire and when fossil fuels are burned, the carbon is oxidized to carbon dioxide and this, water vapour and heat are released. By photosynthesis, plants recapture this carbon – **carbon fixation** – and lock it up in their bodies for a time as glucose or other large molecules.

When plants are harvested and cut down for food, firewood or processing, the carbon is also released again to the atmosphere. As we burn fossil fuels and cut down trees, we are increasing the amount of carbon in the atmosphere and changing the balance of the carbon cycle. Carbon can remain locked in either cycle for long periods of time, ie in the wood of trees or as coal and oil.

Human activity has disrupted the balance of the global carbon cycle (carbon budget) through increased combustion, land use changes and deforestation.

The carbon budget

The amount of carbon on Earth is a finite amount and we have a rough idea of where it goes. The diagram of the carbon cycle in figure 2.3.7 shows carbon sinks (storages) and flows in gigatonnes of carbon (GtC). A gigatonne is one billion tonnes (10^9 tonnes).

Humans and the carbon cycle

Our annual current global emissions from burning fossil fuels are about 5.5 GtC. About 20% of this is from burning natural gas, 40% from burning coal and the other 40% from burning oil. Another 1.6 GtC are added through deforestation. So 7.1 GtC enter the atmosphere each year. Only about 2.4–3.2 GtC of this stay in the atmosphere. Some is taken up by living things. Diffusion of carbon dioxide into the oceans and uptake by oceanic phytoplankton accounts for 2.4 GtC. New growth in forests fixes about 0.5 GtC a year. But this still leaves between 1 and 1.8 GtC – a large amount – unaccounted for. We are not sure where it goes because of the complexity of the system. The amounts of carbon in GtC in other reservoirs are:

- atmosphere 750

- standing biomass 650

- soils 1,500

- oceans 1,720.

Since the pre-industrial period, we have added 200 GtC to the atmosphere.

To do

Draw your own diagram of the carbon cycle.

Include these flows and label them:

    photosynthesis

    respiration

    feeding

    death and decomposition

    fossilisation

    combustion

    dissolving

For each flow, draw:

    storages as boxes

    arrows to represent the sizes of the flow.

Label biotic and abiotic phases.

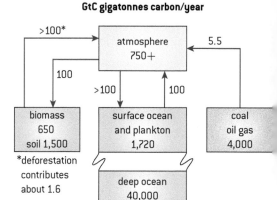

▲ **Figure 2.3.7** The carbon cycle with flow values

The nitrogen cycle

All living organisms need nitrogen as it is an essential element in proteins and DNA.

Nitrogen is the most abundant gas in the atmosphere but atmospheric nitrogen is unavailable to plants and animals, though some specialized microorganisms can fix atmospheric nitrogen.

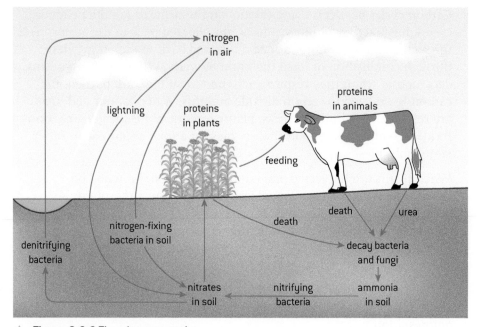

▲ **Figure 2.3.8** The nitrogen cycle

Nitrogen storages or sinks are:

- organisms
- soil
- fossil fuels
- the atmosphere
- in water.

Flows in the nitrogen cycle are:

- nitrogen fixation
- nitrification
- denitrification
- feeding
  - absorption
  - assimilation
  - consumption
- excretion
- death and decomposition.

For plants to take up nitrogen, it must be in the form of ammonium ions (NH_4^+) or nitrates (NO_3^-). Animals eat plants and so take in their nitrogen in the form of amino acids and nucleotides.

The nitrogen cycle can be thought of in three basic stages: nitrogen fixation, nitrification and denitrification.

Nitrogen fixation: when atmospheric nitrogen (N_2) is made available to plants through the fixation of atmospheric nitrogen. This conversion from gaseous nitrogen to ammonium ions can be carried out in one of five ways:

1. By nitrogen-fixing bacteria free-living in the soil (*Azotobacter*).

2. By nitrogen-fixing bacteria living symbiotically in root nodules of leguminous plants (*Rhizobium*). The plant provides the bacteria with sugars from photosynthesis, the bacteria provide the plant with nitrates.

3. By cyanobacteria (sometimes called blue-green algae) that live in soil or water. Cyanobacteria are the cause of the high productivity of Asian rice fields, many of which have been productive for hundreds or even thousands of years without nitrogen-containing fertilizers.

4. By lightning also causing the oxidation of nitrogen gas to nitrate which is washed into the soil.

5. The industrial **Haber** process is a nitrogen-fixing process used to make fertilizers. Nitrogen and hydrogen gases are combined under pressure in the presence of iron as a catalyst (speeds up the reaction) to form ammonia.

The last two processes are non-living nitrogen fixation.

Nitrification: some bacteria in the soil are called nitrifying bacteria and are able to convert ammonium to nitrites (*Nitrosomonas*) while other convert the nitrites (NO_2^-) to nitrates (*Nitrobacter*) which are then available to be absorbed by plant roots.

Denitrification: denitrifying bacteria (*Pseudomonas denitrificans*), in waterlogged and anaerobic (low oxygen level) conditions, reverse this process by converting ammonium, nitrate and nitrite ions to nitrogen gas which escapes to the atmosphere.

As well as nitrogen fixation, **decomposition** of dead organisms also provides nitrogen for uptake by plants. Decomposition of dead organisms supplies the soil with much more nitrogen than nitrogen fixation processes. Important organisms in decomposition are animals (insects, worms among others), fungi and bacteria. They break down proteins, producing different ions: ammonium ions, nitrite ions and finally nitrate ions. These ions can be taken up by plants which recycle the nitrogen.

Assimilation: Once living organisms have taken in nitrogen, they assimilate it or build it into more complex molecules. Protein synthesis in cells turns inorganic nitrogen compounds into more complex amino acids and then these join to form proteins. Nucleotides are the building blocks of DNA and these too contain nitrogen.

Humans and the nitrogen cycle

It is easy for humans to alter the cycle and upset the natural balance. When people remove animals and plants for food for humans, they extract nitrogen from the cycle. Much of this nitrogen is later lost to the sea in human sewage. But people can also add nitrogen to the cycle in the form of artificial fertilizers, made in the Haber process, or by planting leguminous crops with root nodules containing nitrogen-fixing bacteria. These plants enrich the soil with nitrogen when they decompose. The soil condition also affects the nitrogen cycle. If it becomes waterlogged near the surface, most bacteria are unable to break down detritus because of lack of oxygen but certain bacteria can. Unfortunately they release the nitrogen as gas back into the air. This is called denitrification. Excessive flow of rainwater through a porous soil, such as sandy soil, will wash away the nitrates into rivers, lakes and then the sea. This is called leaching and can lead to eutrophication.

To do

Copy the diagram of the nitrogen cycle and add these terms to it:

nitrogen fixation, nitrification, denitrification, decomposition, assimilation.

Copy and complete:

Nitrogen fixation is:

Nitrification is:

Denitrification is:

Assimilation (or protein formation) is:

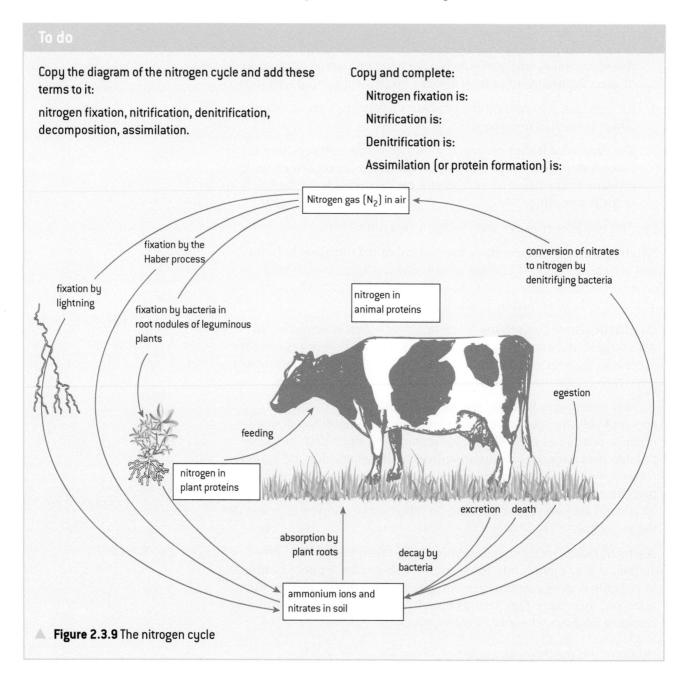

▲ **Figure 2.3.9** The nitrogen cycle

Energy flow diagrams

Energy flow diagrams allow easy comparison of various ecosystems. These show the energy entering and leaving each trophic level. Energy flow diagrams also show loss of energy through respiration and transfer of material as energy to the decomposer food chain.

flow of energy and material through an ecosystem

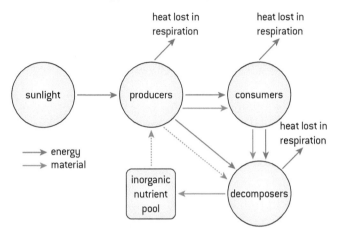

▲ **Figure 2.3.10** Generalized energy flow diagram through an ecosystem

---

To do

The diagram below shows the flow of energy through a food web, and should be used for the three questions (right).

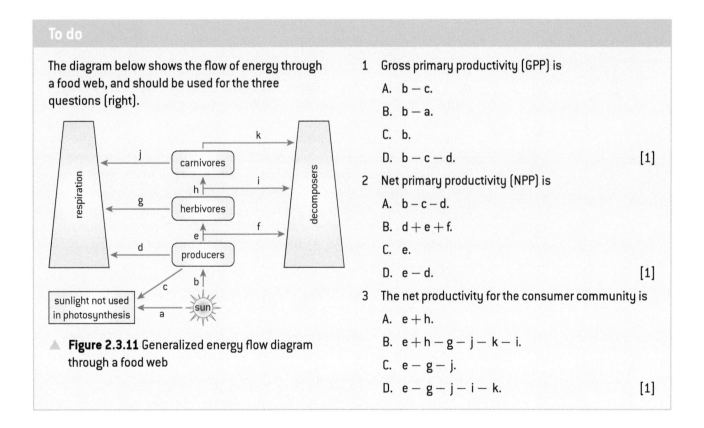

▲ **Figure 2.3.11** Generalized energy flow diagram through a food web

1 Gross primary productivity (GPP) is
   A. b − c.
   B. b − a.
   C. b.
   D. b − c − d.   [1]

2 Net primary productivity (NPP) is
   A. b − c − d.
   B. d + e + f.
   C. e.
   D. e − d.   [1]

3 The net productivity for the consumer community is
   A. e + h.
   B. e + h − g − j − k − i.
   C. e − g − j.
   D. e − g − j − i − k.   [1]

There are many different ways to draw energy flow diagrams and you need to be able to interpret these. Some examples are given in the next pages.

Assimilation and productivity efficiencies

There are two quantities that we need to know to establish these efficiencies.

1. What proportion of the NPP from one trophic level is assimilated by the next?

2. How much of this assimilated material is turned into the tissues of the organism and how much is respired?

For an animal raised for meat these questions are:

1. How much of the grass that an animal eats can it **assimilate** (absorb into its body)? This will determine how many animals the farmer can put in a field.

2. How much of what is assimilated is used for **productivity** (turned into meat)? On a commercial farm this will determine the profits.

$$\text{Efficiency of assimilation} = \frac{\text{gross productivity} \times 100}{\text{food eaten}}$$

$$\text{Efficiency of biomass productivity} = \frac{\text{net productivity} \times 100}{\text{gross productivity}}$$

Trophic efficiency

The efficiency of transfer from one trophic level to the next, eg the ratio of secondary productivity to primary productivity consumed, is considered, on average, to be about 10%. As always, things are not quite as straightforward as they at first appear. While the 10% rule is a generalization and a helpful aid to our understanding of energy flow, there are considerable variations. Trophic efficiencies generally range from 5% to 20%, ie only 5% to 20% of primary producer biomass consumed is converted into consumer biomass. A community of small mammals in a grassland ecosystem may only have a trophic efficiency of 0.1% as they are warm-blooded, have a high metabolic rate and large surface area compared to their volume, and so lose a great deal of energy in respiration and heat. In the oceans, zooplankton feeding on phytoplankton may have a trophic efficiency of 20% and consume most of the producer biomass. Cold-blooded animals (all except mammals and birds) have much slower assimilation rates than warm-blooded animals.

Trophic inefficiencies occur because:

- Not everything is eaten (if it were, the world would not be green as all plants would be consumed).

- Digestion is inefficient (food is lost in feces because the digestive system cannot extract all the energy from it).

- Heat is lost in respiration.

- Some energy assimilated is used in reproduction and other life processes.

Energy budgets

For an individual animal or population, we can measure the quantities of energy entering, staying within and leaving the animal or population. This is its **energy budget**. It can be measured in the laboratory for a population of silk worms or locusts and it is useful for farmers to know what stocking rate of animals per hectare they can use.

To do

1. Consider the assimilation efficiencies in the table on the right.

   a. Why do carnivores have a relatively high assimilation efficiency. (Think about the food they eat.)

   b. Do you think ruminant herbivores would be at the top or bottom of the range for herbivores? Why?

   c. Why does the giant panda have such a low assimilation efficiency? (Hint: its diet is mainly bamboo shoots.)

Organism	Assimilation efficiency
Carnivore	90%
Insectivore	70–80%
Herbivore	30–60%
Zooplankton feeding on phytoplankton	50–90%
Giant panda	20%

2. Copy figure 2.3.13 and add the energy storages and transfers in figure 2.3.12.

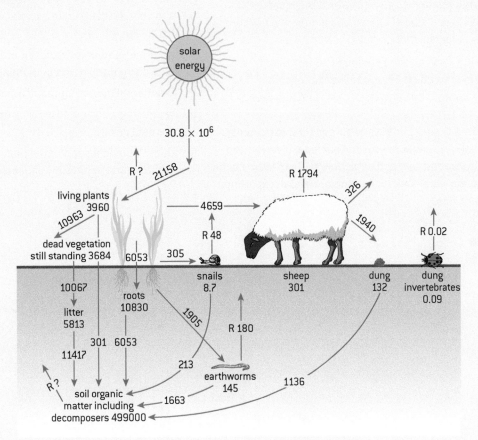

▲ **Figure 2.3.12** The energy budget in a sheep-grazed ecosystem

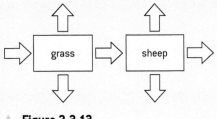

▲ **Figure 2.3.13**

To do

The classical energy flow example

Silver Springs, in central Florida is famous amongst ecologists as the place where Howard T. Odum researched energy flow in the ecosystem in the 1950s. Odum (1924–2002) was a pioneer ecologist working on ecological energetics. This was the first time an energy budget measurement was attempted when Odum measured primary productivity and losses by respiration. (Later, near the end of a long and illustrious career, he and David Scienceman developed the concept of **emergy** (embodied energy) which is a measure of the quality and type of energy and matter that go into making an organism.)

Figure 2.3.14 shows the energy flows and biomass stores measured by Odum at Silver Springs. This simple community consists of algae and duckweed (producers); tadpoles, shrimps and insect larvae (herbivores); water beetles and frogs (first carnivores); small fish (top consumers); and bacteria, bivalves and snails (decomposers and detritivores). Dead leaves also fall into the water and spring water flows out, exporting some detritus.

1. Why does the width of the energy flow bands become progressively narrower as energy flows through the ecosystem?

2. Suggest an explanation for the limit on the number of trophic levels to four or five at most in a community.

3. How is the energy transferred between each trophic level?

4. Insolation (light) striking leaves is 1,700,000 units but only 410,000 are absorbed. What happens to the unabsorbed light energy?

5. A further 389,190 units escapes from producers as heat. Why is this?

6. Account (mathematically) for the difference between gross and net primary productivity.

7. Draw a productivity pyramid from the data given.

8. Would it be possible to draw a biomass pyramid from the data given?

9. Does the model support the first law of thermodynamics? Show your calculations.

10. How does the diagram demonstrate the second law of thermodynamics?

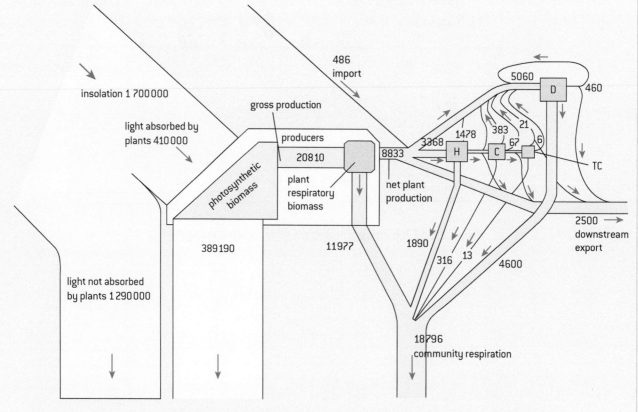

▲ **Figure 2.3.14** The energy flow values in Silver Springs community. Units kcal m^{-2} yr^{-1} (1kcal = 4.2 J)

Human activities and ecosystems

A process, effect or activity derived from humans is known as **anthropogenic ('anthro' meaning human)**. The enhanced greenhouse effect is anthropogenic. Do not confuse this with anthropomorphic which is giving human characteristics to other animals, plants or inanimate objects, eg your doll or your pets.

The concept of energy subsidy

Generally, when humans have an influence on an ecosystem, be it farming or living within it, we tend to simplify it and make it less diverse. Usually, this is on purpose. We cut down forest to grow crops and often this is just one species, eg wheat. So the complex food web that may have been there in a deciduous temperate forest becomes:

wheat ⟶ human

or

improved pasture grasses ⟶ cattle ⟶ human

Much of what we do in agriculture is also aimed at keeping things simple – killing pests and getting rid of weeds as these either eat or compete with the crops we want. Our aim is to maximize the NPP of the organisms we grow to maximize our profit. What happens is that we have to become ever more sophisticated in our farming practices – agribusiness – so we use artificial means to maintain the system. The Green Revolution which brought improved varieties of rice and other crops also brought the need to buy fertilizers for them or pesticides to kill the pests to which they were susceptible.

All farming practices require an **energy subsidy** which is the additional energy that we have to put into the system above that which comes from the Sun's energy. It may be the human labour, animal labour or machines using fuel to power the tractors and plows, pump water for cattle, make fertilizers and other chemicals, transport the crop. The result is that some agricultural systems are very productive with high NPP, particularly, eg, sugar cane.

As humans lived in larger groups and population density increased, they needed more food so farming methods became more sophisticated and used more energy. The advantage of an energy subsidy is that we can feed more people because food production seems more efficient but the energy has to come from somewhere (first law of thermodynamics). As communities become more complex, the energy subsidy increases. Hunter-gatherers have to add little energy to the system apart from their own work. Subsistence farming may involve draught animals, wind-power or water-power to irrigate or grind corn. All these are subsidized by human effort. Commercial farming now involves major use of fossil fuels to power machines, make chemicals to put on the crops or produce feedstuffs for animals. It is estimated that we use 50 times as much energy in MEDCs as a hunter-gatherer society and it is rising all the time.

Energy: yield ratio

In economic terms, we can look at a farming system as inputs and outputs or costs and profits. So we can look at energy in and energy out in the form of food. It seems that as agriculture has become more sophisticated, the ratio goes down. A simple slash and burn type agriculture (when land is cleared in the rainforest and then a variety of crops grown by a subsistence farmer) may have an energy:yield ratio of

Practical Work

* Measure the GPP and NPP in a local ecosystem.

* Investigate the biomass in a local food chain.

* Design an experiment to measure productivity in different ecosystems.

To do

1. What are the two main reasons why there has been an increase in the impact of human activities on the environment over time?

2. Write down the three trends that can be seen in relation to the impact of human activities over time on ecosystems.

1:30 or 40 (30–40 units of food energy for each one unit of input energy as work). With increasing input of energy, this could reduce to 10:1 for battery chicken or egg production, so far more energy is put in to the system than taken out. But the important thing is that the energy is in the form of high energy foods – concentrated energy such as protein and meat, not lower energy cereals. We are producing high energy foodstuffs.

The issue to remember is that energy has to keep flowing through ecosystems whether natural or influenced by humans. If it does not, the system alters rapidly. Blocking sunlight from reaching a plant stops photosynthesis and the plant dies. Stopping the energy subsidy to agriculture will result in chaos. In a natural ecosystem, the large number and variety of food chains and energy paths mean the system is complex and less likely to fail completely. If one species goes, others can take its role. The system is resilient. If there is only one species in an ecological niche, eg wheat, its failure can have a bigger impact.

To do

The data refer to carbon (in biomass) flows in a freshwater system at 40° N latitude:

	g C m^{-2} yr^{-1}
Gross productivity of phytoplankton	132
Respiratory loss by phytoplankton	35
Phytoplankton eaten by zooplankton	31
Fecal loss by zooplankton	6
Respiratory loss by zooplankton	12

From the data, write down word equations and calculate:

a. net productivity of phytoplankton
b. gross productivity of zooplankton
c. net productivity of zooplankton
d. % assimilation of zooplankton
e. % productivity of zooplankton.

Here are two more energy flow diagrams.

a. For ecosystem I, copy and draw a rectangle on the diagram to show the ecosystem boundary.
b. Explain why the storage boxes reduce in size as you go up the food chain.
c. Name three decomposers and explain how they lose heat.

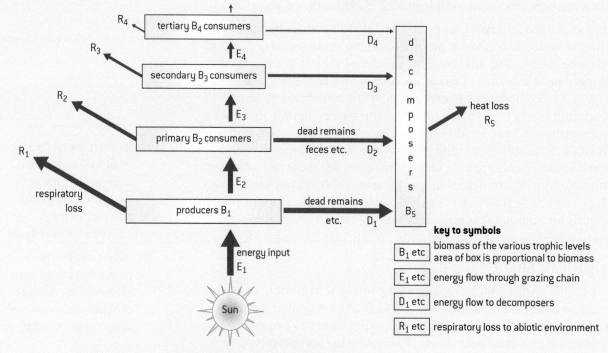

Figure 2.3.15 Energy flow diagram of an ecosystem I

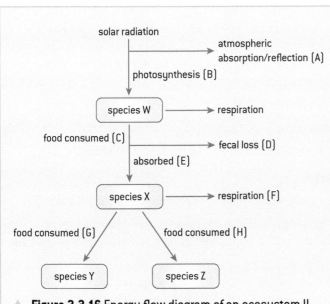

Figure 2.3.16 Energy flow diagram of an ecosystem II

d. For ecosystem II, identify from the diagram the letter(s) referring to the following energy flow processes and explain what happens to this energy at each stage as it passes through the ecosphere:

i. loss of radiation through reflection and absorption

ii. conversion of light to chemical energy in biomass

iii. loss of chemical energy from one trophic level to another

iv. efficiencies of transfer

v. overall conversion of light to heat energy by an ecosystem

vi. re-radiation of heat energy to atmosphere.

To do

Nutrient cycling in terrestrial ecosystems

Copy, fill the gaps and delete incorrect options in the paragraph below.

All living organisms need elements such as _____ and _____ . These are needed to produce worms/minerals/growth/organic material. The availability of such elements is finite – we cannot increase the amount. The plants take up the nutrients from the soils, and once they have been used are passed on to the carnivores/herbivores/photosynthesizers/producers and then the _____ which feed upon them. As organisms die, they _____ and nutrients are returned to the system. As for all systems, there are inputs, _____, storages and _____ . Nutrients are stored in _____ main compartments: the biomass (total mass of living organisms), the soil and the _____ (the surface layer of vegetation which may eventually become humus).

A model of the nutrient cycle is given below. Add the name of each of the transfers of nutrients to the boxes.

Figure 2.3.17 Gersmehl's model

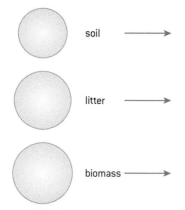

soil   →

litter   →

biomass   →

▲ **Figure 2.3.18** Making Gersmehl's model

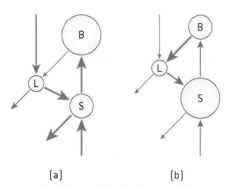

(a)   (b)

▲ **Figure 2.3.19** Gersmehl's models for two different ecosystems (see box to right)

The nutrient cycle varies according to the climate and type of vegetation. The size of each of the stores and size of the transfer can be different. Using the symbols given left (no re-sizing needed), copy and move them to make a nutrient cycle diagram for

1. a deciduous woodland and

2. a tropical rainforest.

Explain the size of the BIOMASS, SOIL and LITTER stores for each.

Grassland ecosystem

(Redraw and resize the boxes to make the correct nutrient cycle diagram)

1. In the ecosystem in figure 2.3.17, there are relatively large stores of nutrients in the litter and soil compared to living things. Give three reasons why this is the case.

2. What is the main nutrient flow from the soil? Why does this happen?

3. Is transport of minerals from one soil layer to another a transfer or a transformation process?

4. Look at the two nutrient models left, a tropical rainforest and a continental grassland (prairie) ecosystem. Label each with its respective ecosystem name.

5. Copy and complete the table of comparisons between the two ecosystems:

Comparison	Ecosystem	Explanation
Which ecosystem stores most nutrients in biomass?		
Which ecosystem has most undecomposed detritus?		
Which ecosystem has least humus?		
In which ecosystem is plant uptake of nutrients greater?		
In which ecosystem is decomposition slower?		
Which ecosystem loses nutrients from biomass quickest?		
In which ecosystem are most nutrients lost due to heavy rain?		
In which ecosystem does rainfall supply many nutrients?		

2.4 Biomes, zonation and succession

Significant ideas:

→ Climate determines the type of biome in a given area although individual ecosystems may vary due to many local abiotic and biotic factors.

→ Succession leads to climax communities that may vary due to random events and interactions over time. This leads to a pattern of alternative stable states for a given ecosystem.

→ Ecosystem stability, succession and biodiversity are intrinsically linked.

Applications and skills:

→ **Explain** the distributions, structure, biodiversity and relative productivity of contrasting biomes.

→ **Analyse** data for a range of biomes.

→ **Discuss** the impact of climate change on biomes.

→ **Describe** the process of succession in a named example.

→ **Explain** the general patterns of change in communities undergoing succession.

→ **Discuss** the factors which could lead to alternative stable states in an ecosystem.

→ **Discuss** the link between ecosystem stability, succession, diversity and human activity.

→ **Distinguish** the roles of r and K selected species in succession.

→ **Interpret** models or graphs related to succession and zonation.

Knowledge and understanding:

→ **Biomes** are collections of ecosystems sharing similar climatic conditions which can be grouped into five major classes – aquatic, forest, grassland, desert and tundra. Each of these classes will have characteristic limiting factors, productivity and biodiversity.

→ Insolation, precipitation and temperature are the main factors governing the distribution of biomes.

→ The **tricellular model** of atmospheric circulation explains the distribution of precipitation and temperature influencing structure and relative productivity of different terrestrial biomes.

→ Climate change is altering the distribution of biomes and causing biome shifts.

→ **Zonation** refers to changes in community along an environmental gradient due to factors such

as changes in altitude, latitude, tidal level or distance from shore (coverage by water).

→ **Succession** is the process of change over time in an ecosystem involving pioneer, intermediate and climax communities.

→ During succession the patterns of energy flow, gross and net productivity, diversity and mineral cycling change over time.

→ Greater habitat diversity leads to greater species and genetic diversity.

→ **r and K strategist species** have reproductive strategies that are better adapted to pioneer and climax communities respectively.

→ In early stages of succession, gross productivity is low due to the unfavourable initial conditions and low density of producers. The proportion of energy lost through community respiration is relatively low too, so net productivity is high, that is, the system is growing and biomass is accumulating.

→ In later stages of succession, with an increased consumer community, gross productivity may be high in a climax community. However, this is balanced by respiration, so net productivity approaches zero and the productivity:respiration (P:R) ratio approaches one.

→ In a complex ecosystem, the variety of nutrient and energy pathways contributes to its stability.

→ There is no one climax community but rather a set of alternative stable states for a given ecosystem. These depend on the climatic factors, the properties of the local soil and a range of random events which can occur over time.

→ Human activity is one factor which can divert the progression of succession to an alternative stable state, by modifying the ecosystem, for example the use of fire in an ecosystem, use of agriculture, grazing pressure, or resource use such as deforestation. This diversion may be more or less permanent depending upon the resilience of the ecosystem.

→ An ecosystem's capacity to survive change may depend on its diversity and resilience.

Key terms

A **biome** is a collection of ecosystems sharing similar climatic conditions.

The **biosphere** is that part of the Earth inhabited by organisms. It extends from the upper part of the atmosphere down to the deepest parts of the oceans which support life.

Biomes

How many biomes are there?

Opinion differs slightly on the number of biomes, which is because they are not a natural classification but one devised by humans, but it is possible to group biomes into five major types with sub-divisions in each type:

Aquatic – freshwater and marine

Freshwater – swamp forests, lakes and ponds, streams and rivers, bogs

Marine – rocky shore, mud flats, coral reef, mangrove swamp, continental shelf, deep ocean

Deserts – hot and cold

Forests – tropical, temperate and boreal (taiga)

Grassland – tropical or savanna and temperate

Tundra – Arctic and alpine.

Each of these biomes will have characteristic limiting factors, productivity and biodiversity. Insolation, precipitation and temperature are the main factors governing the distribution of biomes.

We shall look in more detail at these biomes: tropical rainforest, hot desert, tundra, temperate forest, deep ocean and temperate grassland.

To do

Copy the tables and fill in the gaps

Component	Local example	International/global example
Species		
Population		
Community		
Habitat		
Ecosystem		
Biome		

Biome	Named example
Tropical rainforest	
Hot desert	
Tundra	
Temperate forest	
Deep ocean	
Temperate grassland	

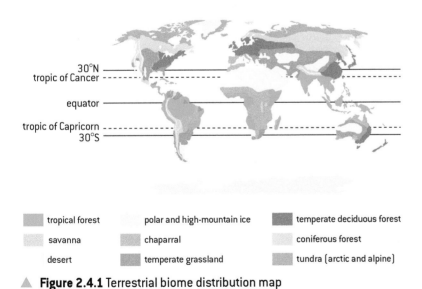

- tropical forest
- savanna
- desert
- polar and high-mountain ice
- chaparral
- temperate grassland
- temperate deciduous forest
- coniferous forest
- tundra (arctic and alpine)

▲ **Figure 2.4.1** Terrestrial biome distribution map

Why biomes are where they are

The climate is the major factor that determines what grows where and so what lives where. The other important factor is the terrain or geography – slope, aspect and altitude. Climate is made up of general weather patterns, seasons, extremes of weather and other factors but two factors are most important – temperature and precipitation (rain and snowfall).

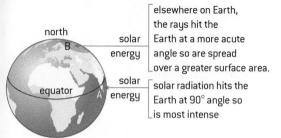

▲ **Figure 2.4.2** Solar radiation hitting the Earth

The temperature is hotter nearer the equator and generally gets cooler as we go towards the poles (increase latitude). This is due to the fact the suns rays hit the Earth at a more acute angle and so are spread over a greater surface area. You can see this effect if you shine a torch beam directly at an object which is flat in front of the torch or shining it at an angle.

Latitude (distance north or south from the equator) and **altitude** (height above sea level) both influence climate and biomes. It generally gets colder as you increase latitude or increase altitude. So there is snow on Mt Kilimanjaro and the Himalayas and Andes and they have alpine or polar biomes even though they are at lower latitudes.

Ocean currents and winds distribute surplus heat energy at the equator towards the poles. Air moving horizontally at the surface of the Earth is called wind. Winds blow from high to low pressure areas. Winds cause the ocean currents. It is water that is responsible for transferring the heat. Water can exist in three states – solid (ice and snow), liquid (water) and gas (water vapour). As it changes from state to state it either gives out or takes in heat. This is its **latent heat**. As water changes from solid to liquid (melts) to gas (evaporates), it takes in heat as more energy is needed to break the molecular bonds holding the molecules together. As water changes from gas to liquid (condenses) to solid (freezes), it gives out heat to its surroundings. It is this change that distributes heat around the Earth. Water is the only substance that occurs naturally in the atmosphere that can exist in the three states within the normal climatic conditions on Earth. Lucky for us, then!

As well as orbiting around the Sun, the Earth rotates and is tilted at 23.5 degrees on its axis. It takes 365 days (and a quarter) for the Earth to go once round the Sun and this gives us a year and our seasons.

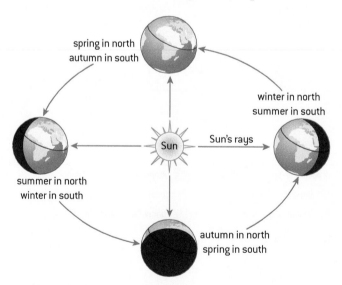

▲ **Figure 2.4.3** How the Earth's tilt causes seasons

Insolation, precipitation and temperature are the most important abiotic (physical) factors influencing biomes or what grows where. Increasing temperature causes increased evaporation so the relationship between precipitation and evaporation is also important. Plants may be short of water even if it rains or snows a lot if the water evaporates straight away (deserts) or is frozen as ice (tundra). So we must also consider the P/E

ratio (**precipitation to evaporation ratio**). This is easy to calculate. For example:

Tundra: Norway	Desert: Jordan
• 75 cm of snow falls/year	• 5 cm of rain falls/year
• 50 cm are lost by evaporation	• 50 cm are lost by evaporation
P/E ratio is 75/50 or 1.25	P/E ratio is 5/50 or 0.1
P/E ratio is much greater than 1	**P/E ratio is far less than 1**
• It rains or snows a lot and evaporation rates are low.	• Water moves upwards through the soil and then evaporates from the surface.
• Then there is leaching in the soil when soluble minerals are washed downwards.	• This leaves salts behind and the soil salinity increases to the point that plants cannot grow (salinization).

P/E ratio is approximately 1 when precipitation is about the same as evaporation; the soils tend to be rich and fertile.

Make sure you understand the section on productivity in 2.3. Different biomes have differing amounts of productivity due to limiting factors: raw materials or the energy source (light) for photosynthesis may be in short supply. Solar radiation and heat may be limited at the South Pole in winter, water in limited supply in a desert. All food webs depend on photosynthesis by green plants to provide the initial energy store so, if they cannot photosynthesize to their maximum capacity, other organisms will have a problem getting enough food.

Productivity is greater in low latitudes (nearer the equator), where temperatures are high all through the year, sunlight input is high and precipitation is also high. These conditions are ideal for photosynthesis. Moving towards the poles, where temperatures and amount of sunlight decline, the rate at which plants can photosynthesize is lower, and thus both GPP (gross primary productivity) and NPP values are lower. In the terrestrial areas of the Arctic, Antarctic and adjacent regions (ie in high latitudes), low temperatures, permanently frozen ground (permafrost), long periods in winter when there is perpetual darkness, and low precipitation (cold air cannot hold as much moisture as warm air) all tend to cause a reduction in photosynthesis and lower productivity values. Obviously, in desert areas (such as the Sahara, and much of Saudi Arabia), and semi-arid areas (eg central Australia, the southwest USA), the absence of moisture for long periods lowers productivity values severely, even though temperatures may be high and sunlight is abundant. Temperate deciduous forests would become temperate rainforest if precipitation were higher and temperate grassland if it were lower. However, these are generalizations and variations are considerable. In a few, sheltered, favourable places in Greenland and South Georgia – in the Arctic and sub-Antarctic, respectively – productivity values close to those of mid-latitude forest have been recorded.

To do

Review questions

1. What is latitude?

2. Does temperature increase or decrease with increasing (a) latitude, (b) altitude? Explain why.

3. Describe how the tilt of the Earth's rotational axis causes differences in the amount of heat received at the Earth's surface.

4. What are trade winds?

5. What causes high and low air pressure?

Whittaker, an American ecologist, first plotted biomes against temperature and precipitation.

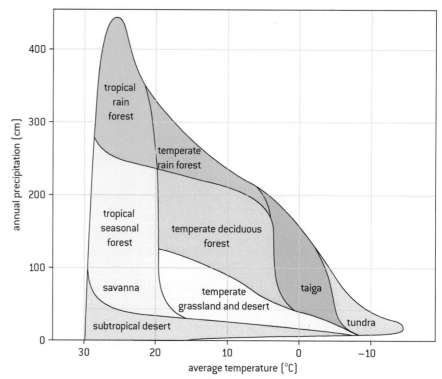

▲ **Figure 2.4.4** Diagram of annual precipitation and temperature showing biomes

Climate change and biome shift

With increase in mean global temperature and changes in precipitation, there is evidence that biomes are moving. There is general agreement that the climate is changing in these ways:

● Temperature increase of 1.5 to 4.5 °C by 2100 (according to the IPCC)

● Greater warming at higher latitudes

● More warming in winter than summer

● Some areas becoming drier, others wetter

● Stronger storms.

These changes are happening very fast, within decades, and organisms change slowly, over many generations through evolutionary adaptation. All they can do to adapt to fast change is to move and that is what they are doing.

These moves are:

● towards the poles where it is cooler

● higher up mountains where it is cooler – 500 m of altitude decreases temperature about 3 °C

● towards the equator where it is wetter.

Examples of biomes shifting are:

- in Africa in the Sahel region, woodlands are becoming savannas

- in the Arctic, tundra is becoming shrubland.

Plants can only migrate very slowly as seeds are dispersed by wind or animals. But animals can migrate longer distances, eg albatrosses, wildebeest, whales. But there are obstacles to migration – natural ones like mountain ranges and seas and ones caused by human activities such as roads, agricultural fields and cities. Animals may not be able to cross these and could become extinct.

There are hotspots – areas predicted to have a high turnover of species due to climate change. These are:

- The Himalayas – sometimes called the third pole – as species can move no higher than the land mass.

- Equatorial Eastern Africa – with a very drought-sensitive climate.

- The Mediterranean region.

- Madagascar.

- The North American Great Plains and Great Lakes.

Up to one billion people live in regions which are vulnerable to biome changing. But these changes can also bring new opportunities for exploitation of resources.

- Drilling for oil under the Arctic Ocean is becoming possible with the decrease in sea ice.

- The North-West Passage for ships between the North Pole and North America could become a trade route without icing up.

To think about

How the atmosphere circulates: The tricellular model of atmospheric circulation

The equator receives most insolation (solar radiation) per unit area on Earth. This heats up the air which rises (hot air rises because it is less dense). As it rises, it cools and the water vapour in the air condenses as rain. This causes the afternoon thunderstorms and low pressure areas of the tropical rainforests. There is so much energy in this air that it continues to rise until it is pushed away from the equator to north and south. As it moves away, it cools and then sinks (cooler air is more dense) at about 30° N and S of the equator forming high pressure areas (more air). This air is dry and this is where the desert biome lies. Some of the air then returns to the equator and some blows to higher latitudes. The air that blows towards the equator completes a circle or cell to end up where it started. This movement of air is given the name 'trade winds' which always blow towards the equator. At the equator, the north and south trade winds converge and rise again at the ITCZ (inter-tropical convergence zone). Here are the doldrums or area of little wind. The cell is called the Hadley cell.

The air that blows to higher latitudes at 30° N and S forms the winds known as the Westerlies and they collect water vapour from the oceans as they blow towards the poles. At about 60° N and S, these winds meet cold polar air and so rise as they are less dense. As they rise, the water in them condenses and falls as precipitation where the temperate forest and grassland biomes are found. This is another low pressure area and associated with depressions and heavy cyclonic rainfall.

The air then continues to flow, some to the poles and some back towards the equator. This air forms another cell known as the Ferrel cell. The air that continues towards the poles then descends as it gets cooler and more dense and forms a high pressure area at the poles, completing the polar cell. This then returns to lower latitudes as winds known as the Easterlies.

Three cells form between the equator and each pole, hence the name tricellular model.

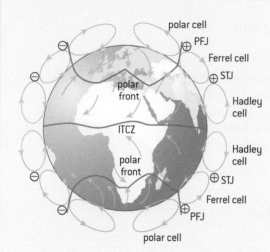

▲ **Figure 2.4.5** Idealized representation of the general circulation of the atmosphere showing the positions of Polar Front; ITCZ (Inter Tropical Convergence Zone); Subtropical Jets (STJ); Polar Front Jets (PFJ)

The winds do not blow directly north or south because the Earth is rotating towards the East, ie if you viewed it from above the North Pole, it would be turning counter-clockwise.

Because of this anything not fixed on the Earth (the oceans and the atmosphere) appear to veer to the right in the Northern hemisphere and left in the Southern. This apparent deflection is called the **Coriolis effect** and it means that the trade winds, westerlies and other winds are deflected to east or west.

North of the equator, the trade winds blow from the northeast and westerlies from the southwest: south of the equator, trade winds blow from the southeast, westerlies from the northwest. This allowed sailing ships to cross the major oceans on trade routes as they could find the prevailing winds by altering their latitude.

Some biomes in more detail.

Tropical rainforest

▲ **Figure 2.4.6** A tropical rainforest in Borneo

What	Hot and wet areas with broadleaved evergreen forest.
Where (distribution)	Within 5 degrees North and South of the equator.
Climate and limiting factors	High rainfall 2000–5000 mm yr^{-1}. High temperatures 26–28 °C and little seasonal variation. High insolation as near equator. P and E are not limiting but rain washes nutrients out of the soil (leaching) so nutrients may be limiting plant growth.

What's there (structure)	Amazingly high levels of biodiversity – many species and many individuals of each species. Plants compete for light and so grow tall to absorb it so there is a multi-storey profile to the forests with very tall emergent trees, a canopy of others, understorey of smaller trees and shrub layer under this – called stratification. Vines, climbers and orchids live on the larger trees and use them for support (epiphytes). In primary forest (not logged by humans), so little light reaches the forest floor that few plants can live here. Nearly all the sunlight has been intercepted before it can reach the ground. Because there are so many plant species and a stratification of them, there are many niches and habitats for animals and large mammals can get enough food. Plants have shallow roots as most nutrients are near the surface so they have buttress roots to support them.
Net productivity	Estimated to produce 40% of NPP of terrestrial ecosystems. Growing season all year round, fast rate of decomposition and respiration and photosynthesis. Plants grow faster. But respiration is also high and for a large mature tree in the rainforest, all the glucose made in photosynthesis is used in respiration so there is no net gain. However, when rainforest plants are immature, their growth rates are huge and biomass gain very high. Rapid recycling of nutrients.
Human activity	The problem is that more than 50% of the world's human population lives in the tropics and subtropics and one in eight of us live in or near a tropical rainforest. With fewer humans, the forest could provide enough resources for the population but there are now too many exploiting the forest and it does not have time to recover. This is not sustainable. In addition, commercial logging of valuable timber, eg mahogany, and clear felling to convert the land to grazing cattle all destroy the forest.
Issues	Logging, clear-felling, conversion to grazing. Tropical rainforests are mostly in LEDCs and have been exploited for economic development.
Examples	Amazon rainforest, Congo in Africa, Borneo rainforest.

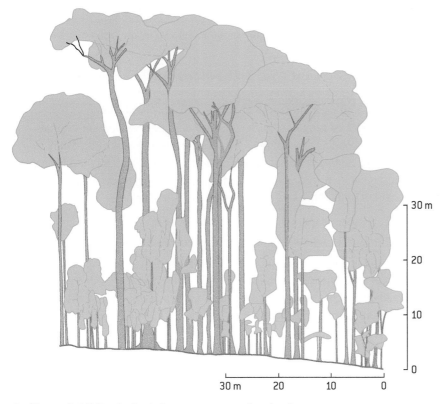

▲ **Figure 2.4.7** Tropical rainforest structure showing layers

Deserts

▲ **Figure 2.4.8** A desert in southwest North America

What	Dry areas which are usually hot in the day and cold at night as skies are clear and there is little vegetation to insulate the ground. There are tropical, temperate and cold deserts.
Where (distribution)	Cover 20–30% of the Earth's surface about 30 degrees North and South of the equator where dry air descends. Most are in the middle of continents. (Some deserts are cold deserts, eg the Gobi desert.) The Atacama desert in Chile can have no rain for 20 years or more. It is the driest place on Earth.
Climate and limiting factors	Water is limiting. Precipitation less than 250 mm per year. Usually evaporation exceeds precipitation – E>P.
What's there (structure)	Few species and low biodiversity but what can survive in deserts is well-adapted to the conditions. Soils are rich in nutrients as they are not washed away. Plants are drought-resistant and mostly cacti and succulents with adaptations to store water and reduce transpiration, eg leaves reduced to spines, thick cuticles to reduce transpiration. Animals too are adapted to drought conditions. Reptiles are dominant, eg snakes, lizards. Small mammals can survive by adapting to be nocturnal (come out at night and stay in a burrow in the heat of the day, eg kangaroo rat) or reduce water loss by having no sweat glands and absorbing water from their food. There are few large mammals in deserts.
Net productivity	Both primary (plants) and secondary (animals) are low because water is limiting and plant biomass cannot build up to large amounts. Food chains tend to be short because of this.
Human activity	Traditionally, nomadic tribes herd animals such as camels and goats in deserts as agriculture has not been possible except around oases or waterholes. Population density has been low as the environment cannot support large numbers. Oil has been found under deserts in the Gulf States and many deserts are rich in minerals including gold and silver. Irrigation is possible by tapping underground water stores or aquifers so, in some deserts, crops are grown. But there is a high rate of evaporation of this water and, as it evaporates, it leaves salts behind. Eventually these reach such high concentrations that crops will not grow (salinization).
Issues	Desertification – when an area becomes a desert either through overgrazing, overcultivation or drought or all of these, eg the Sahel.
Examples	Sahara and Namib in Africa, Gobi in China.

Temperate grasslands

▲ **Figure 2.4.9** Temperate grassland

What	Fairly flat areas dominated by grasses and herbaceous (non-woody) plants.
Where (distribution)	In centres of continents 40–60° North of equator.
Climate and limiting factors	P = E or P slightly > E. Temperature range high as not near the sea to moderate temperatures. Clear skies. Low rainfall, threat of drought.
What's there (structure)	Grasses, wide diversity. Probably not a climax community as arrested by grazing animals. Grasses die back in winter but roots survive. Decomposed vegetation forms a mat, high levels of nutrients in this. Burrowing animals (rabbits, gophers), kangaroo, bison, antelopes. Carnivores – wolves, coyotes. No trees.
Net productivity	600 g m^{-2} yr^{-1} so not very high.
Human activity	Used for cereal crops. Cereals are annual grasses. Black earth soils of the steppes rich in organic matter and deep so ideal for agriculture. Prairies in North America are less fertile soils so have to add fertilizers. Called world's bread baskets. Plus livestock – cattle and sheep that feed on the grasses.
Issues	Dust Bowl in 1930s in America when overcropping and drought led to soil being blown away on the Great Plains – ecological disaster. Overgrazing reduces them to desert or semi-desert.
Examples	North American prairies, Russian steppes in Northern hemisphere; pampas in Argentina, veld in South Africa (30–40° South).

Temperate forests

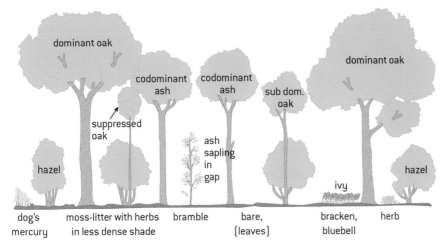

dominant oak

codominant ash

codominant ash

sub dom. oak

dominant oak

suppressed oak

ash sapling in gap

hazel

hazel

ivy

dog's mercury

moss-litter with herbs in less dense shade

bramble

bare, (leaves)

bracken, bluebell

herb

▲ **Figure 2.4.10** Temperate forest structure in Europe

▲ **Figure 2.4.11** Temperate forest

What	Mild climate, deciduous forest.
Where (distribution)	Between 40° and 60° North and South of the equator.
Climate and limiting factors	P > E. Rainfall is 500–1500 mm per year, colder in winter. Winters freezing in some (Eastern China and NE USA), milder in western Europe due to the Gulf Stream. Temp range −30 °C to + 30 °C. Summers cool.
What's there (structure)	Fewer species than tropical rainforests. For example in Britain, oaks, which can reach heights of 30–40 m, become the dominant species of the climax vegetation. Other trees, such as the elm, beech, sycamore, ash and chestnut, grow a little less high. Relatively few species and many woodlands are dominated by one species, eg beech. In USA there can be over thirty species per km². Trees have a growing season of 6–8 months, may only grow by about 50 cm a year. Woodlands show stratification. Beneath the canopy is a lower shrub layer varying between 5 m (holly, hazel and hawthorn) and 20 m (ash and birch). The forest floor, if the shrub layer is not too dense, is often covered in a thick undergrowth of brambles, grass, bracken and ferns. Many flowering plants (bluebells) bloom early in the year before the taller trees have developed their full foliage. Epiphytes, eg mistletoe, mosses, lichens and algae, grow on the branches. The forest floor has a reasonably thick leaf litter that is readily broken down. Rapid recycling of nutrients, although some are lost through leaching. The leaching of humus and nutrients and the mixing by biota produce a brown-coloured soil. Well-developed food chains in these forests with many autotrophs, herbivores (rabbits, deer and mice) and carnivores (foxes). Deciduous trees give way to coniferous towards polar latitudes and where there is an increase in either altitude or steepness of slope. P > E sufficiently to cause some leaching.
Net productivity	Second highest NPP after tropical rainforests but much lower than these because of leaf fall in winter so reduced photosynthesis and transpiration and frozen soils when water is limiting. Temperatures and insolation lower in winter too as further from the Sun.
Human activity	Much temperate forest has been cleared for agriculture or urban developments. Large predators (wolves, bears) virtually wiped out.
Issues	Most of Europe's natural primary deciduous woodland has been cleared for farming, for use as fuel and in building, and for urban development. Some that is left is under threat, eg US Pacific Northwest old-growth temperate and coniferous forests. Often mineral wealth under forests is mined.
Examples	US Pacific Northwest.

Arctic tundra

▲ **Figure 2.4.12** Arctic tundra

▲ **Figure 2.4.13** Distribution of Arctic tundra (shown in yellow)

What	Cold, low precipitation, long, dark winters. 10% of Earth's land surface. Youngest of all the biomes as it was formed after the retreat of the continental glaciers only 10,000 years ago. Permafrost (frozen soil) present and no trees.
Where (distribution)	Just south of the Arctic ice cap and small amounts in Southern hemisphere. (Alpine tundra is found as isolated patches on high mountains from the poles to the tropics.)
Climate and limiting factors	Cold, high winds, little precipitation. Frozen ground (permafrost). Permafrost reaches to the surface in winter but in summer the top layers of soil defrost and plants can grow. Low temperatures so rates of respiration, photosynthesis and decomposition are low. Slow growth and slow recycling of nutrients. Water, temperature, insolation and nutrients can be limiting.

In the winter, the Northern hemisphere, where the Arctic tundra is located, tilts away from the sun. After the spring equinox, the Northern hemisphere is in constant sunlight. For nearly three months, from late May to August, the sun never sets. This is because the Arctic regions of the Earth are tilted toward the Sun. With this continuous sun, the ice from the winter season begins to melt quickly.

During spring and summer, animals are active, and plants begin to grow rapidly. Sometimes temperatures reach 30 °C. Much of this energy is absorbed as the latent heat of melting of ice to water.

In Antarctica, where a small amount of tundra is also located, the seasons are reversed. |
| What's there (structure) | No trees but thick mat of low-growing plants – grasses, mosses, small shrubs. Adapted to withstand drying out with leathery leaves or underground storage organs. Growing season may only be 8 weeks in the summer. Animals also adapted with thick fur and small ears to reduce heat loss. Mostly small mammals, eg lemmings, hares, voles. Predators – Arctic fox, lynx, snowy owl. Most hibernate and make burrows. Simple ecosystems with few species.

Often bare areas of ground. Low biodiversity – 900 species of plants compared with 40,000 or more in the Amazon rainforest.

Soil poor, low inorganic matter and minerals. |
| Net productivity | Very low. Slow decomposition so many peat bogs where most of the carbon is stored. |
| Human activity | Few humans but mining and oil – see oil tars. Nomadic groups herding reindeer. |
| Issues | Fragile ecosystems that take a very long time to recover from disruption. May take decades to recover if you even walk across it. Mining and oil extraction in Siberia and Canada destroy tundra.

Many scientists feel that global warming caused by greenhouse gases may eliminate Arctic regions, including the tundra, forever. The global rise in temperature may damage the Arctic and Antarctic more than any other biome because the Arctic tundra's winter will be shortened, melting snow cover and parts of the permafrost, leading to flooding of some coastal areas. Plants will die, animal migrating patterns will change, and the tundra biome as we know it will be gone. The effect is uncertain but we do know the tundra, being the most fragile biome, will be the first to reflect any change in the Earth.

Very large amounts of methane are locked up in tundra ice in clathrates. If these are released into the atmosphere then huge increase in greenhouse gases (clathrates contain 3,000 times as much methane as is in the atmosphere now and methane is more than 20 times as strong a greenhouse gas as carbon dioxide). |
| Examples | Siberia, Alaska. |

▲ **Figure 2.4.14** Deep ocean animals –
fangtooth fish, tubeworms

Deep ocean

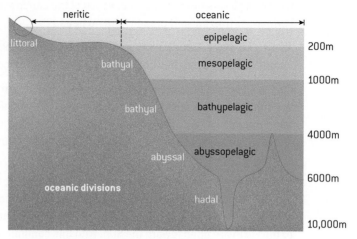

▲ **Figure 2.4.15** Deep ocean divisions

What	The ocean and seafloor beyond continental shelves.
Where (distribution)	65% of the Earth's surface. Most is abyssal plain of the ocean floor – averaging 3.5 miles deep.
Climate and limiting factors	Pressure increases with depth, temperature variation decreases to a constant −2°C at depth. Light limiting below 1,000 m – there is none. Nutrients – low levels and low primary productivity but some dead organic matter falls to deep ocean floors.
What's there (structure)	Top 200 m – some light for photosynthesis so phytoplankton and cyanobacteria live here and they and algae are the main producers. They are eaten by zooplankton, fish and invertebrates, eg squid, jellyfish. 200–1,000 m deep – as pressure increases with depth, fish here are muscular and strong to resist pressure. Very little light reaches here so large eyes, reflective sides and light-producing organs on their bodies. Many are red which absorbs shorter wavelengths of light that penetrate further. 1,000–4,000 m deep – higher diversity here, always dark. Fish are black with small eyes, bristles and bioluminescence – create own light to hunt or avoid predators. Very little muscle, large mouths. 4,000 m to bottom – huge pressures, constant cold. Mostly shrimps, some fish, jellyfish, tubeworms on bottom. Bottom surface – fine sediments made up of debris from above – plankton shells, dead organisms, whale and fish skeletons. Also mud and volcanic rocks in mid-ocean ridges. Where volcanoes erupt, there are hydrothermal vent communities high in sulphides where chemosynthetic bacteria gain their energy from the sulphur. These producers support communities of crabs, tubeworms, mussels, and even octopus and fish.
Net productivity	Low.
Human activity	Minimal but rocks rich in manganese and iron could be a resource.
Issues	Pollution from run-off from rivers, sewage, ocean warming due to climate change.
Examples	Arctic, Atlantic, Pacific Oceans.

Comparison of biomes

Biome	Net primary productivity g m^{-2} yr^{-1}	Annual precipitation mm yr^{-1}	Area 10^6 km^2	Plant biomass 10^9 t	Mean biomass kg m^{-2}	Animal biomass 10^6 t	Solar radiation W m^{-2} yr^{-1}
Tropical rainforests	2200	2000–5000	17.0	765	45	330	175
Temperate forests	1200	600–2500	12.0	385	32.5	160	125
Boreal forests	800	300–500	12.0	240	20	57	100
Tropical grasslands (savannas)	900	500–1300	15.0	60	4	220	225
Temperate grassland	600	250–1000	9.0	14	1.6	60	150
Tundra and alpine	140	<250	8.0	5	0.6	3.5	90
Desert (rock, sand, ice)	90	<250	24.0	0.5	0.02	0.02	75
Deep oceans	20–300	Variable	352	1000+	V low	800–2000	Variable

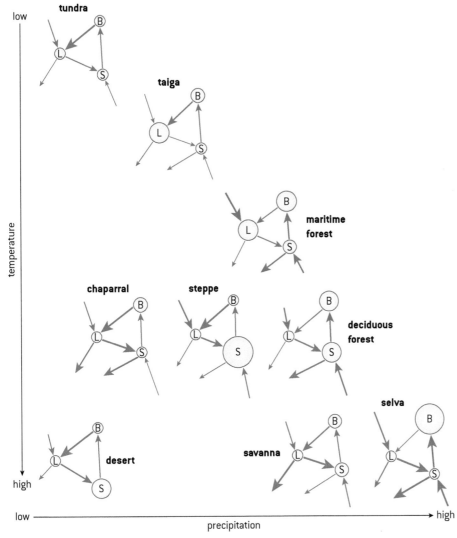

TOK

Controlled laboratory experiments are often seen as the hallmark of the scientific method. To what extent is the knowledge obtained by observational natural experiment less scientific than the manipulated laboratory experiment?

▲ **Figure 2.4.16** Nutrient cycling models for world biomes (after Gersmehl 1976)

To do

1. What does deciduous mean?
2. What is net primary productivity (NPP)?
3. Which biome has the highest NPP per m² per yr and why?
4. Which biome has the largest NPP and why?
5. Why is NPP low in tundra and deserts?
6. Why are there no large deciduous forests or tundra in the southern hemisphere?
7. Why are trees in temperate biomes deciduous?
8. How do the number of tree species and their distribution differ in temperate and tropical rainforests?
9. What are the main factors causing the distribution of biomes?
10. Which biome(s) are most threatened and why?
11. Which biome(s) have been most changed by human activity?
12. What effect may climate change have on biome distribution?

Key terms

Zonation is the change in community along an environmental gradient due to factors such as changes in altitude, latitude, tidal level or distance from shore/coverage by water.

Succession is the process of change over time in an ecosystem involving pioneer, intermediate and climax communities.

Succession and zonation

Do not confuse succession with zonation.

Succession is how an ecosystem changes in time.

Zonation is how an ecosystem is changing along an environmental gradient, eg altitude.

Zonation, eg rocky seashore, mountain slopes.	Succession, eg terrestrial.
Spatial and static.	Dynamic and temporal (takes place over long periods of time).
Caused by an abiotic gradient. Mountains – changes in temperature. Seashore – changes in time exposure to air / water.	Caused by progressive changes through time, eg as vegetation colonizes bare rock.

Sand dune colonization is unusual in that the succession is dynamic but one also observes vegetation zones during the various stages of the process.

Zonation

For each species, there is an ecological niche (2.1). That niche has boundary limits and outside these, the species cannot live. There are many abiotic and biotic factors that influence these limits. The most important ones on mountains are:

- Temperature – which decreases with increasing altitude and latitude.
- Precipitation – on mountains, most rainfall is at middle altitudes so deciduous forest grows. Higher up, the air is too dry and cold for trees.
- Solar insolation – more intense at higher altitudes and plants have to adapt – often with red pigment in their leaves to protect themselves against too much insolation.
- Soil type – in warmer zones, decomposition is faster so soils are deeper and more fertile. Higher up, decomposition is slow and soils tend to be acidic.
- Interactions between species – competition may crowd out some species and grazing may alter plant composition. Mycorrhizal fungi (2.1) may be very important in allowing trees to grow in some zones.

Human activities alter zonation. Road building on mountains may allow tourism into previously inaccessible areas or deforestation or agriculture.

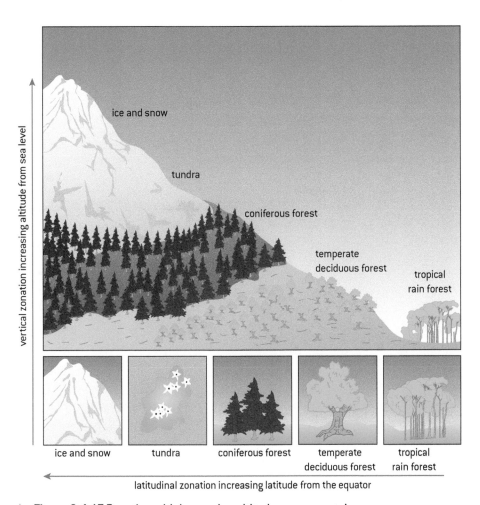

▲ **Figure 2.4.17** Zonation with increasing altitude on a mountain

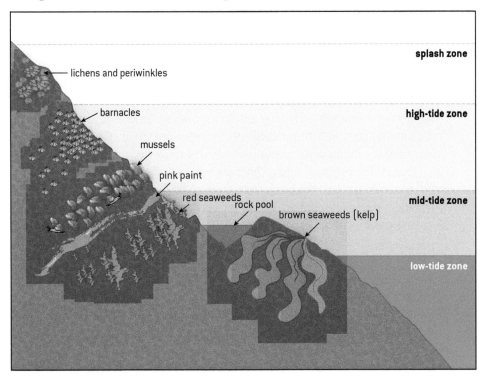

▲ **Figure 2.4.18** Zonation of species on a rocky shore due to increasing exposure to air higher up the shore

Graphical representation of zonation is often by a kite diagram where the width of the 'kites' corresponds to the number of that species (see sub-topic 2.5).

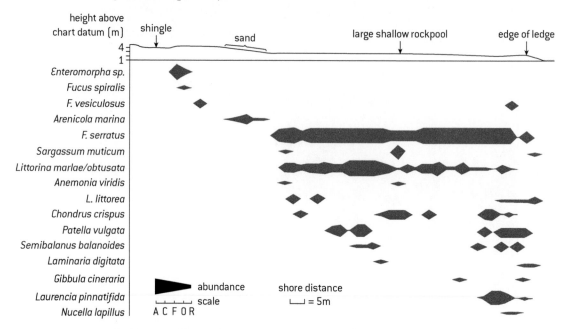

▲ **Figure 2.4.19** Kite diagram showing zonation of species on a rocky shore. (For ACFOR key see 2.5)

Succession

Key points

- Succession is the change in species composition in an ecosystem over time.
- It may occur on bare ground (**primary succession**) where soil formation starts the process or where soil already formed but the vegetation has been removed (**secondary succession**).
- Early in succession, gross primary productivity (GPP) and respiration are low and so net primary productivity (NPP) is high as biomass accumulates.
- In later stages, while GPP may remain high, respiration increases so NPP may approach zero and the productivity:respiration ratio (P:R) approaches one.
- A climax community is reached at the end of a succession when species composition stops changing. But there may be several states of a climax community depending on abiotic factors.
- The more complex the ecosystem (higher biodiversity, increasing age), the more stable it tends to be.
- In agricultural systems, humans often deliberately stop succession when NPP is high and crops are harvested.
- Humans also interrupt succession by deforestation, grazing with animals or controlled burning.
- Sometimes the ecosystem recovers from this interruption and succession continues, sometimes the interruption is too great and the system is less resilient and so succession is stopped.
- Species biodiversity is low in early stages and increases as succession continues, falling a little in a climax community.
- The higher the diversity, the higher the resilience.
- Mineral cycling also changes over the succession, increasing with time.

Primary succession occurs on a bare inorganic surface. It involves the colonization of newly created land by organisms. It occurs as new land is either created or uncovered such as river deltas, after volcanic eruptions, on sand dunes.

Bare land almost anywhere on the planet does not stay bare for very long. Plants very quickly start to colonize the bare land and over time an entire plant community develops. This change is directional as one community is replaced by another. This process is succession. Succession results in a natural increase in complexity to the structure and species composition of a community over time.

Stages in primary succession

Figure 2.4.20 shows the stages of succession. Stages 2, 3 and 4 are intermediate.

Bare, inorganic surface ↓	A lifeless abiotic environment becomes available for colonization by pioneer plant and animal species. Soil is little more than mineral particles, nutrient poor and with an erratic water supply.
Stage 1 Colonization ↓	First species to colonize an area are called **pioneers** adapted to extreme conditions. Pioneers are typically **r-selected species** showing small size, short life cycles, rapid growth and production of many offspring or seeds. Simple soil starts from windblown dust and mineral particles.
Stage 2 Establishment ↓	Species diversity increases. Invertebrate species begin to visit and live in the soil increasing humus (organic material) content and water-holding capacity. Weathering enriches soil with nutrients.
Stage 3 Competition ↓	Microclimate continues to change as new species colonize. Larger plants increase cover and provide shelter, enabling **K-selected species** to become established. Temperatures, sun and wind are less extreme. Earlier pioneer r-species are unable to compete with K-species for space, nutrients or light and are lost from the community.
Stage 4 Stabilization ↓	Fewer new species colonize as late colonizers become established shading out early colonizers. Complex food webs develop. K-selected species are specialists with narrower niches. They are generally larger and less productive (slower growing) with longer life cycles and delayed reproduction.
Climax community	The final stage or **climax community** is stable and self-perpetuating. It exists in a steady-state dynamic equilibrium. The climax represents the maximum possible development that a community can reach under the prevailing environmental conditions of temperature, light and rainfall.

▲ **Figure 2.4.20**

A **hydrosere** is a succession in water.

deep freshwater, no rooted plants because of lack of light in deep water

community only microorganisms and phytoplankton

sediments get carried into the pond allowing rooted submerged and floating plants to start to grow

sediments continue to build up

reeds and grasses develop around pond margin, trapping more sediment

a marsh community builds up around the pond margins

reeds take over more of the ponds as more silt builds up

as the soil around the edge dries from waterlogged to damp, tree species such as willow and alder become established

▲ **Figure 2.4.21** Succession in a lake

Ponds and lakes get continuous inputs of sediment from streams and rivers that open into them. Some of this sediment passes through but a lot sinks to the pond bottom. As plant communities develop they add dead organic material to these sediments. Over time these sediments build up allowing rooted plants to invade the pond margins as the pond slowly fills in. This eventually leads to the establishment of climax communities around the pond margins and in smaller ponds the eventual disappearance of the pond.

Secondary succession

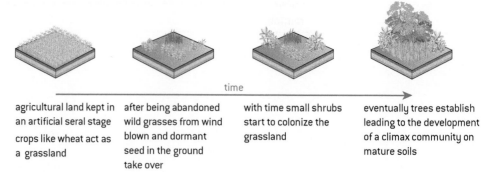

time

agricultural land kept in an artificial seral stage crops like wheat act as a grassland

after being abandoned wild grasses from wind blown and dormant seed in the ground take over

with time small shrubs start to colonize the grassland

eventually trees establish leading to the development of a climax community on mature soils

▲ **Figure 2.4.22** Stages of secondary succession in abandoned agricultural land

Where an already established community is suddenly destroyed, such as following fire or flood or even human activity (plowing) an abridged version of succession occurs.

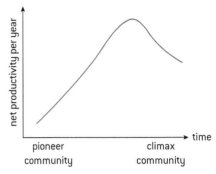

This secondary succession occurs on soils that are already developed and ready to accept seeds carried in by the wind. Also there are often dormant seeds left in the soil from the previous community. This shortens the number of stages the community goes through.

Changes occurring during a succession

During a succession the following changes occur:

- The size of organisms increases with trees creating a more hospitable environment.

- Energy flow becomes more complex as simple food chains become complex food webs.

- Soil depth, humus, water-holding capacity, mineral content and cycling all increase.

- Biodiversity increases because more niches (lifestyle opportunities) appear and then falls as the climax community is reached.

- NPP and GPP rise and then fall.

- Productivity : respiration ratio falls.

Primary productivity varies with time. When plants first colonize bare ground, it is low as there are not many plants and they are starting from a seed. It rises quickly as more plants germinate and the biomass accumulates. When a climax community is reached (stable community of plant and animal species), productivity levels off as energy being fixed by the producers is approximately equal to the rate at which energy is being used in respiration, and emitted as heat.

In the early stages, **gross primary productivity** is low due to the initial conditions and low density of producers. The proportion of energy lost through community respiration is relatively low too, so **net productivity** is high, ie the system is growing and biomass is accumulating.

In later stages, with an increased producer, consumer and decomposer community, gross productivity continues to rise to a maximum in the climax community. However, this is balanced by equally high rates of respiration particularly by decomposers, so net productivity approaches zero and the productivity : respiration (P:R) ratio approaches 1.

During succession, gross primary productivity tends to increase through the pioneer and early stages and then decreases as the climax community reaches maturity. This increase in productivity is linked to growth and biomass.

Early stages are usually marked by rapid growth and biomass accumulation – grasses, herbs and small shrubs. Gross primary productivity is low but net primary productivity tends to be a large proportion of GPP as with little biomass in the early stages, respiration is low. As the community develops towards woodland and biomass increases so does productivity. But NPP as a percentage of GPP can fall as respiration rates increase with more biomass.

Studies have shown that standing crop (biomass) in succession to deciduous woodland reaches a peak within the first few centuries. Following the establishment of mature climax forest, biomass tends to fall as trees age, growth slows and an extended canopy crowds out

Figure 2.4.23 Productivity changes in a succession

net productivity per year

time

pioneer community

climax community

ground cover. Also older trees become less photosynthetically efficient and more NPP is allocated to non-photosynthetic structural biomass such as root systems.

Early stage	Middle stage	Late stage
Low GPP but high percentage NPP	GPP high	Trees reach their maximum size
Little increase in biomass	Increased photosynthesis Increases in biomass as plant forms become bigger	Ratio of NPP to R is roughly equal

▲ **Figure 2.4.24** Biomass accumulation and successional stage

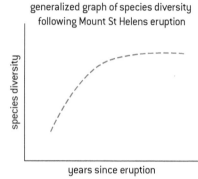

generalized graph of species diversity following Mount St Helens eruption

species diversity

years since eruption

▲ **Figure 2.4.25** Species diversity changes in a succession

Species diversity in successions

In early stages of succession, there are only a few species within the community. As the community passes through subsequent stages so the number of species found increases. Very few pioneer species are ever totally replaced as succession continues. The result is increasing diversity – more species. This increase tends to continue until a balance is reached between possibilities for new species to establish, existing species to expand their range and local extinction.

Evidence following the eruption of the Mount St Helens volcano in 1980 has provided ecologists with a natural laboratory to study succession. In the first 10 years after the eruption species diversity increased dramatically but after 20 years very little additional increase in the diversity occurred.[4]

Disturbance

Communities are affected by periods of disturbance to a greater or lesser extent. Even in large forests trees eventually age, die and fall over leaving a gap. Other communities are affected by flood, fire, landslides, earthquakes, hurricanes and other natural hazards. All of these have an effect of making gaps available that can be colonized by pioneer species within the surrounding community. This adds to both the productivity and diversity of the community.

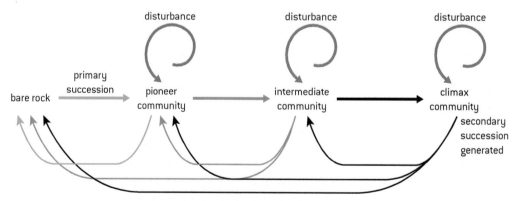

disturbance can send any seral stage back to an earlier seral stage or create gaps in a later community that then regenerate increasing both productivity and diversity of the whole community

▲ **Figure 2.4.26** Effects of disturbance in a succession

4   Carey, S., J. Harte and R. del Moral. 2006. Effect of community assembly and primary succession on the species–area relationship in disturbed systems. *Ecography* 29, pp866–872

Succession on sand dunes

On the southern coast of England in Dorset is Studland Bay where sand dunes have continued to be formed since the 16th century.

▲ **Figure 2.4.27** Sand dunes at Studland Bay, Dorset, UK.

This begins with a bare surface of sand. Vegetation colonizes the sand. The pioneer plants tend to be low growing – why? They have fat fleshy leaves with a waxy coating and are able to survive being submersed – temporarily.

Later, the predominant plant species is marram grass on the seaward side due to its ability to cope with the environmental conditions. It, like the other grasses, have leaves which are able to fold to reduce its surface area. Leaves are waxy to reduce transpiration and can be aligned to the wind direction. It incorporates silica into its cell structure to give the leaves extra strength and flexibility.

As a result of the humus from the previous stages, a sandy soil has now developed. This is now able to support 'pasture' grasses and bushes. Species such as hawthorn, elder, brambles and sea buckthorn (which has nitrogen-fixing root nodules so can thrive in nutrient-poor soil) are present. As the scrub develops, shorter species will be shaded out.

The oldest dunes will have forest – first pine and finally oak and ash woodland growing on them; the **climatic climax vegetation** for the area. Here the species diversity declines due to competition – for what?

In every case, vegetation colonizes in a series of stages. The final one is in dynamic equilibrium with its climatic environment and hence is known as climatic climax vegetation. In the UK this is temperate deciduous forest.

As succession develops, there are increases in vegetation cover, soil depth and humus content, soil acidity, moisture content and sand stability.

foredunes: *an object such as a plant or rock causes sand to build up on the lee side.*

sandy beach, tidal litter

mobile dunes

plants begin to bind sand together

semi-fixed dunes: *the dunes could be 20m high here*

dune slack: *once a hollow is formed, perhaps by blowout, sand is removed by the wind until the damp sand near the water table can not be transported.*

marsh plants

Scrub, health & woodland: *climax vegetation in the absence of management/interference.*

increasing soil depth and quality

site number/description	strand line (1)	embryo dunes (2)	fore dunes (3)	white/yellow dunes (4)	fixed dunes (5)	dune slack (6)	dune scrub (7)	dune heath (8)	woodland (9)
distance from the sea (m)	0–20	20–80	80–150	150–300	300–500	500–700	(400) variable	700–2500	2500+
approximate age (years)	–	0–50	50–100	100–125	125–150	150–250	–	250–400	>400
soil colour	–	yellow			yellow/grey		grey	brown	
soil surface pH	8.5	8.0	7.5	7.0	6.5		6.0		4.5
% calcium carbonate	10	8	8	5	1		<0.1		
% humus	<1			2.5	5	10		20	>40
dominant plant species	sea rocket, saltwort	sand, couch, lyme grass	marram, grass, sea holly	marram, sand sedge	harebell, willow	creeping willow, common sallow	birch, brambles	heather, gorse	pines, oaks

▲ **Figure 2.4.28** Diagram of idealized sand dunes

Arrested and deflected successions

Succession may be stopped or 'arrested' at a stage by an abiotic factor, eg soil conditions such as waterlogging, or a biotic factor such as heavy grazing. This results in an arrested or **sub-climax** community which will only continue its development if the limiting factor is removed.

Under other circumstances a climax community may be affected by either a natural event, eg fire or landslide, or human activity such as agriculture, regular use of fire or habitat destruction. This will lead to a deflected or **plagioclimax** community such as pasture, arable farmland or plantations with reduced biodiversity. Again if the human activity ceases the plagioclimax community will develop into the climatic climax community.

Why do farmers want to maintain their crops at a plagioclimax? Clue – think about the productivity : respiration ratios.

Significance of changes during succession

During succession energy flow, gross and net productivity, diversity and mineral cycling change. In early stages gross productivity is low due to the initial conditions and low density of producers. The proportion of energy lost through community respiration is relatively low too, so net productivity is high, ie the system is growing and biomass is accumulating. In later stages, with an increased consumer community, gross productivity may be high in a climax community. However, this is balanced by respiration, so net productivity approaches zero and the productivity : respiration (P:R) ratio approaches 1. **Biodiversity** increases during succession as more species arrive and then decreases slightly if a stable climax community is reached. **Mineral cycling** tends to be slow at the early stages of succession but increases strongly during the succession process.

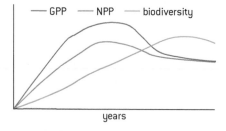

▲ **Figure 2.4.29** Changes during succession

There is a basic conflict between natural succession and human requirements in agriculture. We want to achieve high rates of productivity with no standing crop left. The natural system leads to increasing complexity, longer food chains, higher biodiversity, more biomass and a well-organized stratified ecosystem. Food production aims for a simple system where weed plants are controlled and monoculture maximizes yield that does not reach a climax community. But we have not placed a value on other services – natural income that natural systems provide – a balance in the carbon cycle, nutrient cycling, climate buffer of forests and oceans, clean water provision, aesthetic services that natural systems provide. Less productive places are as necessary as productive fields. We need waste places, a mixture of habitats, quality as well as quantity. While we may be able to grow crops on more land, should we? What is the balance between human rights and environmental rights that we need to find?

K- and r-strategists' reproductive strategies

Species can be roughly divided into **K-** and **r-strategists or K- and r-selected species**. K and r are two variables that determine the shape of the population growth curve. K is the carrying capacity and r describes the shape of the exponential part of the growth curve.

K- and r-strategies describe the approach different species 'take' to getting their genes into the next generation and ensuring the survival of the species.

Different species vary in the amount of time and energy they use to raise their offspring. There are two extremes.

K-strategists, eg humans and other large mammals:

- Have small numbers of offspring.

- Invest large amounts of time and energy in parental care.

- Most offspring survive.

- They are good competitors.

- Population sizes are usually close to the carrying capacity, hence their name.

- In stable, climax ecosystems, K-strategists out compete r-strategists.

r-strategists, eg invertebrates and fish:

- Use lots of energy in the production of vast numbers of eggs.

- No energy is used in raising them after hatching.

- They lay their eggs and leave them forever.

- They reproduce quickly.

- Are able to colonize new habitats rapidly.

- Make opportunistic use of short-lived resources.

- Because of their fast reproductive and growth rates, they may exceed the carrying capacity, with a population crash as a result. They predominate in unstable ecosystems.

Typical characteristics of r- and K- strategists

r-strategist	K-strategist
Short life	Long life
Rapid growth	Slower growth
Early maturity	Late maturity
Many small offspring	Fewer large offspring
Little parental care or protection	High parental care and protection
Little investment in individual offspring	High investment in individual offspring
Adapted to unstable environment	Adapted to stable environment
Pioneers, colonizers	Later stages of succession
Niche generalists	Niche specialists
Prey	Predators
Regulated mainly by external factors	Regulated mainly by internal factors
Lower trophic level	Higher trophic level
Examples: annual plants, flour beetles, bacteria	Examples: trees, albatrosses, humans

It is important to appreciate that K- and r-strategists are the extremes of a continuum of reproductive strategies and many species show a mixture of these characteristics.

<aside>
To do

In 1980, Mt St Helens volcano erupted in Washington State, USA. Research what has happened to the vegetation in the area since then. Start here http://vulcan.wr.usgs.gov/Volcanoes/MSH/Recovery/framework.html

Sketch curves for gross productivity, net productivity and respiration as a function of time in one graph for Mt St Helens since 1980. Indicate the different succession stages in the graph.
</aside>

Survivorship curves

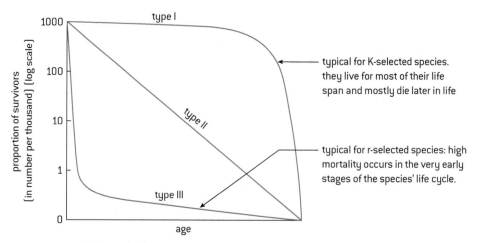

▲ **Figure 2.4.30** Survivorship curves

A survivorship curve shows the fate of a group of individuals of a species. Three hypothetical survivorship curves are shown in figure 2.4.30. Note that the vertical axis is logarithmic.

Curve II is rather rare. It represents species that have an equal chance of dying at any age. It occurs for example in the hydrozoan *Hydra* and some species of birds.

To do

Look at the diagram of a theoretical survival model of a small bird population.

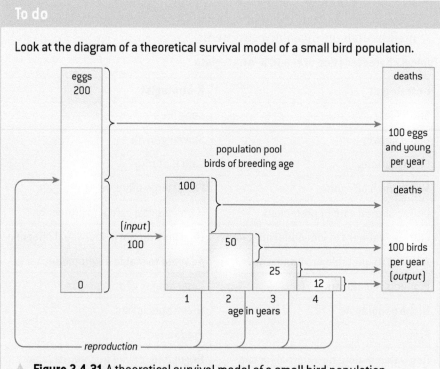

▲ **Figure 2.4.31** A theoretical survival model of a small bird population

1. What is the life span of these birds?
2. What is the potential natality?
3. How many survivors after the first year?
4. What is the percentage mortality at end of year 1?
5. What is the percentage mortality at end of year 2?
6. What is the percentage mortality at end of year 3?
7. What would the survivorship curve for these birds look like?

To do

1. Which kind of organisms are r-strategists?

2. List characteristics of r-strategists shown by the shape of the curve.

3. Which species are most likely to be regulated by density-independent factors, eg weather?

4. Describe and explain the shape of the survivorship curve for K-selected species.

2.5 Investigating ecosystems – Practical work

Significant ideas:

→ The description and investigation of ecosystems allows for comparisons to be made between different ecosystems and for them to be monitored, modelled and evaluated over time, measuring both natural change and human impacts.

→ Ecosystems can be better understood through the investigation and quantification of their components.

Applications and skills:

→ **Design** and carry out ecological investigations.

→ **Construct** simple identification keys for up to eight species.

→ **Evaluate** sampling strategies.

→ **Evaluate** methods to measure at least three abiotic factors in an ecosystem.

→ **Evaluate** methods to investigate the change along an environmental gradient and the effect of a human impact in an ecosystem.

→ **Evaluate** methods for estimating biomass at different trophic levels in an ecosystem.

→ **Evaluate** methods for measuring/estimating populations of motile and non-motile organisms.

→ **Calculate** and **interpret** data for species richness and diversity.

→ **Draw** graphs to illustrate species diversity in a community over time or between communities.

Knowledge and understanding:

→ The study of an ecosystem requires that it be named and located eg Deinikerwald, Baar, Switzerland, a mixed deciduous–coniferous managed woodland.

→ Organisms in an ecosystem can be identified using a variety of tools including **keys**, comparison to herbarium / specimen collections, technologies and scientific expertise.

→ **Sampling strategies** may be used to measure biotic and abiotic factors and their change in space, along an environmental gradient, over time, through succession or before and after a human impact, for example as part of an EIA.

→ Measurements should be repeated to increase reliability of data. The number of repetitions required depends on the factor being measured.

→ Methods for estimating the **biomass** and energy of trophic levels in a community

include measurement of dry mass, controlled combustion and extrapolation from samples. Data from these methods can be used to construct ecological pyramids.

→ Methods for estimating the **abundance of non-motile organisms** include the use of quadrats for making actual counts, measuring population density, percentage cover and percentage frequency.

→ Direct and indirect methods for estimating the **abundance of motile organisms** can be described and evaluated. Direct methods include actual counts and sampling. Indirect methods include the use of **capture-mark-recapture** with the application of the Lincoln Index.

→ **Species richness** is the number of species in a community and is a useful comparative measure.

> → **Species diversity** is a function of the number of species and their relative abundance and can be compared using an index. There are many versions of diversity indices but students are only expected to be able to apply and evaluate the result of the Simpson diversity index. Using its formula, the higher the result, the greater the species diversity. This indication of diversity is only useful when comparing two similar habitats or the same habitat over time.

Studying ecosystems

All ecosystem investigations should follow the guidelines in the **IB animal experimentation policy**. This may be more stringent than your local, national standards so check it carefully before designing an experiment.

Consider if you could:

- Replace the animal by using cells, plants or simulations.
- Refine the experiment to alleviate harm or distress.
- Reduce the number of animals involved.

The IB policy states that you may not carry out an animal experiment if it involves:

- pain, undue stress or damage to health of the animal
- death of the animal
- drug intake or dietary change beyond those easily tolerated by the animal.

If humans are involved, you must also have their written permission and not carry out experiments that involve the possibility of transfer of blood-borne pathogens.

Techniques for data collection

This topic lends itself to a lot of traditional environmental studies and a number of basic techniques underpin many of the investigations. Once you understand these various methods of data collection you can combine them to collect the relevant data for a wide range of investigations.

- Where to collect the data
    - Quadrats
    - Transects
- What to measure
    - Measuring abiotic factors
        - Marine
        - Freshwater
        - Terrestrial

- Measuring biotics
  - Biomass and productivity
  - Catching small motile animals
    - ▸ Terrestrial
    - ▸ Aquatic
  - Keys
- Measuring abundance
  - Lincoln Index
  - Simpson diversity index

Where to collect the data

Quadrats

How many quadrat samples, and of what size?

▲ **Figure 2.5.1** Quadrats

The size of the quadrat chosen is dependent on the size of the organisms being sampled.

Quadrat size	Quadrat area	Organism
10 × 10 cm	0.01 m²	Very small organisms such as lichens on tree trunks or walls, or algae.
0.5 × 0.5 m	0.25 m²	Small plants: grasses, herbs, small shrubs. Slow moving or sessile animals: mussels, limpets.
1.0 × 1.0 m	1 m²	Medium size plants: large bushes.
5.0 × 5.0 m	25 m²	Mature trees.

▲ **Figure 2.5.2**

There is a balance to strike between increasing accuracy with increasing size and time available and the number of times a quadrat is placed.

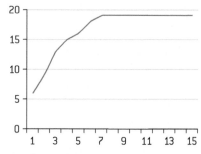

▲ **Figure 2.5.3** Number of species and quadrat size

These will vary depending on the ecosystem, size of organisms and their distribution. But you can work out how many samples to take and what size the quadrats should be quite simply.

As you increase the number of samples, plot the number of species found. When this number is stable, you have found all species in the area, so in figure 2.5.3, eight samples are enough.

If you increase the size of the quadrat (eg from side length 10 cm to 15 cm, 20 cm and so on) and plot the number of species found, when this number reaches a constant, that is the quadrat size to use.

How to place quadrats

Quadrats can be placed randomly or continuously or systematically (according to a pattern).

1. Random quadrats may be placed by throwing the quadrat over your shoulder but we do not recommend this as it could be both dangerous and not random – you may decide where to throw.

   The conventional method (figure 2.5.4) is to use random number tables:

   - Map out your study area.
   - Draw a grid over the study area.
   - Number each square.
   - Use a random number table to identify which squares you need to sample.

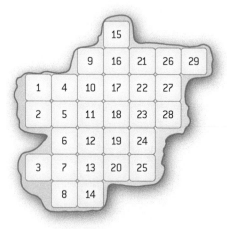

▲ **Figure 2.5.4**

2. Stratified random sampling is used when there is an obvious difference within an area to be sampled and two sets of samples are taken.

   This study area (figure 2.5.5) has two distinctly different vegetation types and three separate areas to be studied. Samples need to be taken in each area.

   - Deal with each area separately.
   - Draw a grid for each area.
   - Number the squares in each area (they can be the same numbers or different).
   - Use a random number table to identify which squares you need to sample in each area.

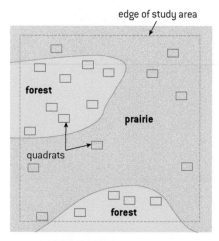

▲ **Figure 2.5.5**

Transects

Continuous and systematic sampling is along a transect line.

You might use this to look at changes in organisms as a result of changes along an environmental gradient, eg zonation along a slope, a rocky shore or grassland to woodland, or to measure the change in species composition with increasing distance from a source of pollution. Transects are quick and relatively simple to conduct.

NOTE
Many line transects (at least 3, preferably 5) need to be combined to obtain sufficient reliable data.

Key term

A **transect** is a sample path/line/strip along which you record the occurrence and/or distribution of plants and animals in a particular study area.

There are 2 main types of transect that could be useful to you.

1. **Line transect:** consists of a string or measuring tape which is laid out in the direction of the environmental gradient and species touching the string or tape are recorded.

2. **Belt transect:** this is a strip of chosen width through the ecosystem. It is made by laying two parallel line transects, usually 0.5 or 1 metre apart, between which individuals are sampled.

Transect lines may be continuous or interrupted.

1. In a **continuous transect** (line or belt transect) the whole line or belt is sampled.

2. In an **interrupted transect** (line or belt) samples are taken at points along the line or belt. These points are usually taken at regular horizontal or vertical intervals. This is a form of systematic sampling. Quadrats are placed at intervals along the belt.

What to measure

Measuring abiotic components of the system

Ecosystems can be roughly divided into marine, freshwater and terrestrial ecosystems. Each of these ecosystem types has a different set of physical (abiotic) factors and you should be able to:

1. Describe and evaluate methods for measuring these.

2. Describe and evaluate how to measure spatial and temporal variations in abiotic factors.

Marine ecosystems

Abiotic factors: salinity, pH, temperature, dissolved oxygen, wave action.

Normal seawater	35‰
Brackish water (Baltic Sea)	Between 1‰ and 10‰
Freshwater	0.5 ‰

The salinity can be determined by measuring the electrical conductivity or the density of the water.

Seawater usually has a pH of above 7 (basic). The pH can be measured using a pH meter.

Many meters have interchangeable probes and so can be used to measure a number of abiotic factors.

Temperature

Temperature affects the metabolic rates of marine organisms: this is due to the fact that many are ectothermic (their body temperature is about the same as the surrounding water). Lower temperatures = low metabolic rates. So changes in temperature caused by thermal pollution may have a significant impact on some organisms.

> **Key term**
>
> **Salinity** is the concentration of salts expressed in ‰ (parts of salt per thousand parts of water).

Dissolved oxygen

Solubility of oxygen in water is affected by:

- Temperature: higher temperatures = lower concentrations of dissolved oxygen. Many marine organisms rely on dissolved oxygen for respiration hence changes in temperature will impact the marine ecosystem.

- Water pollution: this can cause low oxygen concentrations and thus problems for marine organisms.

Dissolved oxygen can be measured using an oxygen-selective electrode connected to an electronic meter, datalogging, or by a Winkler titration. (A series of chemicals is added to the water sample and dissolved oxygen in the water reacts with iodide ions to form a golden-brown precipitate. Acid is then added to release iodine which can be measured, and is proportional to the amount of dissolved oxygen, which can then be calculated.) Oxygen-selective electrodes give quick results, but need to be well maintained and calibrated in order to give accurate results. The Winkler titration is more labour intensive.

Wave action

Wave action is important in coastal zones where organisms live close to the water surface. Areas with high wave activity usually have high concentrations of dissolved oxygen. Typical examples are coral reefs and rocky coasts.

Freshwater ecosystems

Abiotic factors: turbidity, flow velocity, pH, temperature, dissolved oxygen.

Turbidity

High turbidity = cloudy water

Low turbidity = clear water

The turbidity is important because it limits the penetration of sunlight and thereby the depth at which photosynthesis can occur. Turbidity can be measured with optical instruments or by using a Secchi disc.

A Secchi disc is a white or black-and-white disc attached to a graduated rope. The disc is heavy to ensure that the rope goes vertically down. The procedure is:

1. Slowly lower the disc until it disappears from view.

2. Read the depth from the graduated rope.

3. Slowly raise the disc until it is just visible again.

4. Read the depth from the graduated rope.

5. Calculate the average depth. This depth is known as the Secchi depth.

For reliable results a standard procedure should be followed:

- Always stand or always sit in the boat.

- Always wear your glasses or always work without them.

- Always work on the shady side of the boat.

This should be repeated in the same spot 3–5 times.

> **Key term**
>
> **Turbidity** is the cloudiness of a body of fresh water.

▲ **Figure 2.5.6** The Secchi disc

Flow velocity

This is the speed at which the water is moving and it determines which species can live in a certain area. Flow velocity varies with:

1. Time: melt water in the spring gives high flow rates, summer drought low flow rates.

2. Depth: Surface water may flow more slowly than that in the middle of the water column.

3. Position in the river: Inside bend has shallow slower-moving water, outside bend has deeper fast-moving water.

There are three basic methods for measuring flow velocity:

1. Flow meter: These are generally expensive and can be unreliable as mixing water with electricity has its problems.

2. Impellers: a simple mechanical device as shown in figure 2.5.7:

   a. The impeller is mounted on a graduated stick and the base placed on the floor of the river / stream. The height of the impeller can be adjusted and the velocity measured at different depths, BUT it can only be used in clear shallow water, as you must be able to see the impeller.

   b. The impeller is held at the end of the side arm and lowered into the water facing upstream.

   c. The impeller is released and the time it takes to travel the distance of the side arm is measured.

   d. Repeat 3–5 times for accurate results.

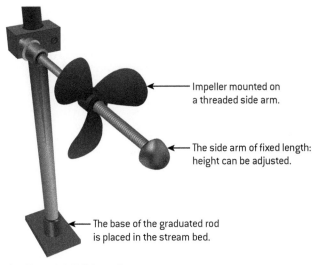

Impeller mounted on a threaded side arm.

The side arm of fixed length: height can be adjusted.

The base of the graduated rod is placed in the stream bed.

▲ **Figure 2.5.7** Impeller

3. Floats

The easiest way to measure flow velocity is to measure the time a floating object takes to travel a certain distance. The floating object should preferably be partly submerged to reduce the effect of the wind. Oranges and grapefruits make suitable floats. This method gives the surface flow velocity only. The average flow velocity of a river can be estimated from the surface flow velocity by dividing the surface velocity by 1.25.

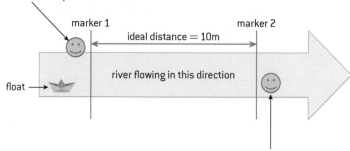

person 1 drops the float above the first marker
and shouts start as it passes the marker

marker 1   marker 2

ideal distance = 10m

river flowing in this direction

float →

person 2 starts the stopwatch on command from
person 1 and stops it as the float passes marker 2
and catches the float

▲ **Figure 2.5.8**

This should be repeated 3–5 times for accuracy.

WARNING: This method gives seconds / metre NOT metres / second

pH

pH values of freshwater range from moderately acidic to slightly basic, depending on surrounding soil, rock and vegetation. It can be measured with a pH meter or datalogging pH probe.

For temperature and dissolved oxygen see marine ecosystems (p129).

Terrestrial ecosystems

Abiotic factors: temperature, light intensity, wind speed, soil texture, slope, soil moisture, drainage and mineral content.

Air temperature

Temperature varies temporally and spatially and can be measured using simple liquid thermometers, min–max thermometers, or more complex (electronic) thermometers. The latter equipment can be used to measure temperature continuously during a longer time as can a data-logging temperature probe.

Light intensity

This can be measured with electronic meters. The fact that light intensity varies with time (sunny period, clouds, time of the day, season) should be taken into account.

Wind speed

There are a variety of techniques used to measure wind speed:

- A revolving cup anemometer consists of three cups that rotate in the wind. The number of rotations per time period is counted and converted to a wind speed. Revolving cup anemometers can be mounted permanently or hand-held.

- A ventimeter is a calibrated tube over which the wind passes. This reduces the pressure in the tube, which makes a pointer move. It is easy to use and inexpensive.

- By observation of the effect of the wind on objects. The observations are then related to the Beaufort Scale (a scale of wind speed from 0 to 12).

Rainfall

Rainfall can be collected using a rain gauge. Some schools have an established weather station – in which case collecting rainfall data is easy. Many schools will not have a weather station but rain gauges are very easy to make and there are plenty of websites that can give you advice on how to make your own. Once you have made your rain gauge:

1. Place your rain gauge in a suitable spot in the study area – somewhere away from the influence of buildings, trees and other obstacles that may affect rainfall.

2. Check rain gauge every 24 hours – at the same time every day. Pour rain into a graduated cylinder and record daily amount of rainfall.

 Figure 2.5.9 Rain gauge

Soil

Soil has a significant impact on plant growth and there are a variety of aspects of the soil that can be measured.

Texture (Particle size)

Soil is made up of particles (gravel, sand, silt, clay) and the average size and distribution of them affects a soil's drainage and water-holding capacity.

Particle	How to measure
Gravel: very coarse, coarse and medium	Measure individually – simple, but time-consuming procedure
Gravel: fine and very fine	Sieved through a series of sieves with different mesh sizes.
Sand: All sizes	
Silt and clay	Sedimentation or optical techniques. Sedimentation techniques are based on the fact that large particles sink faster than small particles. Optical techniques use light scattering by the particles (light scattering is what makes suspensions of soil particles in water look cloudy). Both sedimentation and light scattering can nowadays be done using automated instruments but are expensive for secondary school use.

Soil moisture

This is the amount of water in the soil. It can be measured by drying soil samples.

1. Place a sample of the soil in a crucible.

2. Weigh it and record the weight.

3. Dry the sample.

Drying can be done in a conventional drying oven or a microwave oven. In a conventional oven:

- Set the oven to 105 °C; hot enough to dry the soil but not so hot as to burn off organic matter.

- Leave for 24 hours and weigh the sample, repeat this until its mass becomes constant. This takes several days.

In a microwave oven:

- Place the sample in the microwave for 10 minutes.
- Weigh the sample, and return to the oven for 5 minutes – repeat until its mass becomes constant.

A minimum of 3–5 samples should be tested.

Organic content

The organic content of a soil is plant and animal residues in various stages of decay and it has several functions.

- Supplies nutrients to the soil.
- Holds water (like a sponge).
- Helps reduce compaction and crusting.
- Increases infiltration.

Organic content can be determined by the loss on ignition (LOI) method.

1. Dry the sample as above.
2. Heat the soil at high temperatures of 500 to 1,000 °C for several hours.
3. Weigh the sample and repeat this until its mass becomes constant.

Mineral content and pH

There is a wide range of soil nutrients essential for a fertile soil. These are easy to measure through traditional soil testing kits or the ones available in many gardening centres.

Soil pH can also be measured using a soil testing kit or a pH probe.

Measuring biotic components of the system

To measure biotic components we need to observe and question.

- Why is it as it is?
- What has changed recently?
- Why does this grow here and not there?
- What impact do more people walking here have?

So walk around your institution's grounds or the local area.

- Is there a playing field?
- Is there a footpath on soil rather than concrete?
- Does the ground slope?
- Is it more shady or more moist in one area than another and what difference does that make to the type and number of species living there?

Measuring biomass and productivity

Plant biomass

Measuring plant biomass is simple but destructive. Generally speaking it is best to take above-ground biomass as trying to get roots etc. can be very difficult.

For low vegetation / grasses:

1. Place a suitably sized quadrat (see figure 2.5.2).

2. Harvest all the above-ground vegetation in that area.

3. Wash it to remove any insects.

4. Dry it at about 60–70°C until it reaches a constant weight. Water content can vary enormously so all the water should be removed and the mass given as dry weight.

5. For accurate results this should be repeated 3–5 times so that a mean per unit area can be obtained.

6. The result can then be extrapolated to the total biomass of that species in the ecosystem.

For trees and bushes:

1. Select the tree or bush you which to test.

2. Harvest the leaves from 3–5 branches.

3. And repeat steps 3–6 in the above method.

Primary productivity

In aquatic ecosystems (both marine and freshwater ecosystems) the light and dark bottle technique can be used to measure both the gross and net productivity of aquatic plants (including phytoplankton). This is simple but has given us a good idea of the productivity of the oceans and of many lakes.

The productivity is usually calculated from the oxygen concentrations in the bottles. The procedure is:

1. Take two bottles filled with water from the ecosystem.
   a. One of the bottles is made of clear glass.
   b. The other is of dark glass or is covered to exclude light.

2. Measure the oxygen concentration of the water by chemical titration (Winkler method) or an oxygen probe, and record it as mg oxygen per litre of water.

3. Place equal amounts of plants of the same species into each of the bottles.

4. Both bottles must be completely filled with water and capped. (No air should be present.)

5. Allow to stand and incubate for several hours.

6. Measure the oxygen levels in both bottles and compare with the original oxygen level of the water. The incubation can take place in the laboratory or outdoors in the ecosystem of investigation.

In the light bottle, photosynthesis and respiration have been occurring. In the dark, only respiration occurs.

In terrestrial ecosystems, you can do a similar experiment with square 'patches':

1. Select three equally sized patches with similar vegetation (eg grass).

2. The first patch (A) is harvested immediately and the biomass measured (see above).

3. The second patch (B) is covered with black plastic (no photosynthesis, just respiration).

4. The third patch (C) is just left as it is.

5. After a suitable time period (depends on the season), patches B and C are harvested and the biomass measured (as above).

6. Now GPP, NPP and R can be calculated (usually per m^2).

Secondary productivity

In a typical experiment, a herbivore is fed with a known amount of food. The procedure is that the food and the herbivore(s) are weighed. After a suitable time period, the remaining food, the herbivore(s) and the feces are weighed.

Catching small motile animals

These are more problematic as they move around, so how do we count small animals? Obviously they have to be caught first. Make sure you can identify the insects you are likely to catch – have a key handy to help you.

WARNING: Under no circumstances should any animal be stressed or killed during any investigation – there are humane ways to catch and count small animals.

Terrestrial ecosystems

There is a range of safe harmless techniques that can be used to catch insects:

1. Pitfall traps.

2. Sweep nets.

3. Tree beating.

WARNING:

- Make sure there are no venomous insects in your local area.

- DO NOT handle the insects directly – move the insects with tweezers or a pooter.

The pitfall trap is ideal for catching insects and other small crawling animals that cannot fly away (see figure 2.5.10). Insects can be attracted by decaying meat or sweet sugar solution (this must be covered so the insects do not fall in it and drown) and will fall into the trap.

Several of these traps can be placed around the study area. They should be checked at regular intervals (every 6 hours) and the species and number of that species recorded.

▲ **Figure 2.5.11** Sweep netting

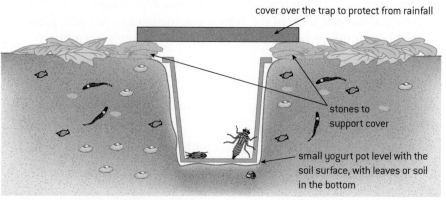

▲ **Figure 2.5.10** A pitfall trap

WARNING

- DO NOT put any fluid in the bottom of the trap – you do not want to kill the insect.

- DO NOT leave the traps unchecked for more than 24 hours.

Sweep nets

Sweep nets of various sizes can be swept through grasses at various heights in order to catch many insects.

These can then be emptied into a large clear container and the species and numbers recorded.

Tree beating

This method can find insects in tree branches. Simply place a catching tray beneath a tree branch and gently tap the branch. The tray will catch anything that falls from the tree and you can log the species and their numbers.

Night-flying moths will be attracted to a light behind which a white sheet is hung and the moths settle on this for you to observe.

Small insects and invertebrates can be caught with a pooter – a small jar with two tubes attached (see figure 2.5.13). You suck gently on one tube and the animal is pulled into the jar. You cannot swallow it as there is gauze at the end of the mouthpiece tube!

Aquatic ecosystems

The organisms of most interest will be the stream invertebrates and the most efficient way to catch them is through kick samples.

Kick sampling is another simple technique:

- Place the sweep net downstream from you.

- Shuffle your feet into the streambed for 30 seconds.

- Empty the contents of the net into a tray filled with stream water.

- Use a pipette to sort the various insects into small plastic cups and record your results.

- Repeat three times to ensure good results.

In aquatic systems, nets of various mesh sizes and net sizes can be used to catch plankton, small invertebrates or larger fish. These can be towed behind boats or held in running water. Simple plastic sieves are effective. Kick sampling loosens invertebrates, which drift into the net. Turning stones over is also effective. Some of these methods are destructive and kill the organisms. You may think this is not acceptable but most do not harm the organisms caught and you should always return them to their habitats if at all possible.

Keys

Once you have collected the organisms, you may want to find out what they are called. A (biological) key is used to identify species. Ecologists make keys to specific groups of organisms, eg soil invertebrates in specific ecosystems, to help other interested people identify species. Keys come in two formats, a diagrammatic, dichotomous or 'spider' key and a paired statement key.

Look at examples of published keys. Diagram keys are useful but professionals use paired statement keys because printed descriptions are more exact than pictures. Both keys are used by starting at the top each time and following the lines or 'go to' numbers.

▲ **Figure 2.5.12** Tree beating

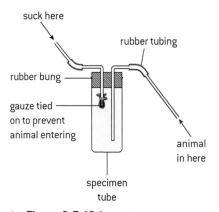

suck here

rubber tubing

rubber bung

gauze tied on to prevent animal entering

animal in here

specimen tube

▲ **Figure 2.5.13** A pooter

current carries material into net

stream flow

▲ **Figure 2.5.14** Kick sampling

A list of pond organisms in a temperate ecosystem is shown in figure 2.5.15. It is from the UK Environmental Education Centre in Canterbury, Kent.

A paired statement key has no pictures but is more accurate .

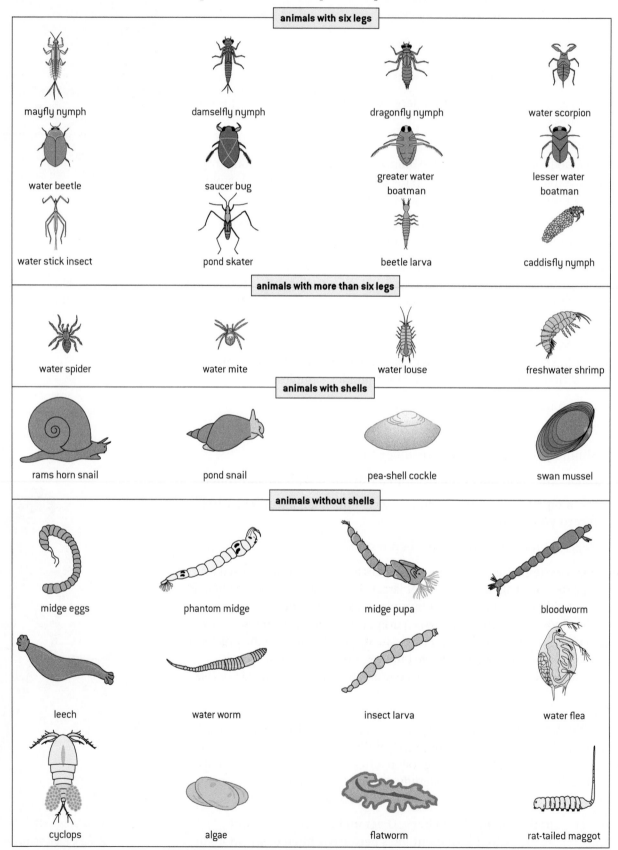

Figure 2.5.15 A picture key of pond animals

Measuring abundance

Having established what organisms are where, what use is it?

Plants

There are a number of ways of assessing plant species abundance:

- Density: mean number of plants per m².

- Frequency: the percentage of the total quadrat number that the species was present in, may also be measured within the quadrat.

- Percentage cover: because plants spread out and grow percentage cover is often measured instead of individual numbers. This is an estimate of the coverage by each species and it sometimes helps if the quadrat is divided up for this. Species may overlap or lie in different storeys in a forest, so the percentage cover within a quadrat may be well over 100% or much less if there is bare ground. The percentage cover can be estimated either by comparing the sample area with figure 2.5.16 and then it can be graded on a scale from 0 to 5, or on the ACFOR scale by using figure 2.5.17.

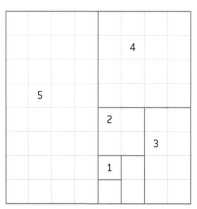

▲ **Figure 2.5.16**

Percentage cover (%)	ACFOR scale	Score
50	Abundant	5
25–50	Common	4
12–25	Frequent	3
6–12	Occasional	2
<6, or single individual	Rare	1
absent		0

▲ **Figure 2.5.17** Percentage cover scales

Lincoln Index (capture, mark, release and recapture)

Sessile or slow moving animals can be counted as individuals, e.g. limpets and barnacles. More mobile animals are harder to assess and the Lincoln Index is used to estimate the population size of animals which move about or do not appear during the day.

Method

1. Establish the study area.

2. Capture a sample of the population. The actual method of capture will depend on the size of animals; you can take your pick from the methods discussed earlier.

3. Mark each of the organisms captured and record how many you have marked: this must be done in a non-harmful way that does not expose them to higher predation levels than non-marked individuals. For example, dog whelks on a rocky shore or woodlice in a woodland can be marked with a spot of non-toxic subtle coloured paint (nothing bright).

4. Release the captured individuals back into the environment and allow sufficient time to remix with the population.

5. Take a second sample in the same way as the first: count the number of organisms captured in this sample and count how many of them are marked. At least 10% of the marked sample should be recaptured if this estimate is going to be fairly accurate.

Assumptions made are that:

- Mixing is complete, ie the marked individuals have spread throughout the population.

- Marks do not disappear.

- Marks are not harmful nor increase predation by making the individual more easily seen.

- It is equally easy to catch every individual.

- There are no immigration, emigration, births or deaths in the population between the times of sampling.

- Trapping the organisms does not affect their chances of being trapped a second time.

Lincoln Index formulae

$$\frac{m_2}{n_2} = \frac{n_1}{N}$$

OR

$$N = \frac{n_1 \times n_2}{m_2}$$

Where

n_1 = number of animals first marked and released

n_2 = number of animals captured in the second sample

m_2 = number of marked animals in the second sample

N = Lincoln Index or total population (the figure you are after)

Simpson diversity index

Species diversity is the number of different species and the relative numbers of individuals of each species, ie species richness and species evenness.

Ecologists try to express diversity in a number. The higher the number, the greater the species diversity. This makes it possible to compare similar ecosystems or to see whether ecosystems are changing in time. The most common way to turn diversity into a number is by the Simpson diversity index.

But be careful. The name 'Simpson diversity index' actually describes three related indices (Simpson's Index, Simpson's index of diversity and Simpson's reciprocal index). Here we are using Simpson's reciprocal index in which 1 is the lowest value (when there would be just one species) and a higher value means more diversity. The highest value is equal to the number of species in the sample. In the other indices, the value ranges from 0 to 1.

$$D = \frac{N(N-1)}{\Sigma n(n-1)}$$

Where

D = Simpson diversity index

N = total number of organisms of all species found

n = number of individuals of a particular species

Example

	Number of individuals of species		
	A	B	C
Ecosystem 1	25	24	21
Ecosystem 2	65	3	4

The diversity index of ecosystem 1 can be calculated like this:

$$N = 25 + 24 + 21 = 70$$

$$D = \frac{70 \times 69}{(25 \times 24) + (24 \times 23) + (21 \times 20)} = 3.07$$

In ecosystem 2, the diversity index is 1.22.

Both ecosystems have the same species richness (3) but in system 1 these are more evenly distributed so species diversity, by this measure, is higher.

A high value for D indicates a highly diverse ecosystem and often a stable and ancient site. In contrast, low values for D are found in disturbed ecosystems like logged forests. Pollution also results in low values for D. Agricultural land has extremely low values for D, as farmers try to prevent competition between their crops and other species (weeds). Be careful though as low values for D in Arctic tundra may represent ancient and stable sites as growth is so slow there and diversity is low.

The Simpson diversity index is most often used for vegetation, but it can also be applied to animal populations.

Note: you do not need to memorize the Lincoln Index or Simpson diversity index formulae.

REVIEW

Explain what are the main impacts of humans on flows of energy and matter in the biosphere.

Evaluate the model of succession described here.

BIG QUESTIONS

Ecosystems and ecology

Reflective questions

→ Feeding relationships can be represented by different models. How can we decide when one model is better than another?

→ What role does indigenous knowledge play in passing on scientific knowledge?

→ When is quantitative data superior to qualitative data in giving us knowledge about the world?

→ Controlled laboratory experiments are often seen as the hallmark of the scientific method, To what extent is the knowledge obtained by observational natural experiment less scientific than the manipulated laboratory experiment?

→ The tropical rainforest biome is vital to the ecological equilibrium of the Earth, therefore who should decide how humans use it?

→ Why do we use internationally standardized methods of ecological study when making comparisons across international boundaries?

Quick review

All questions are worth 1 mark

The diagram below shows a complete food web and should be used for questions 1 and 2.

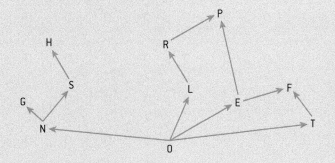

1. Which are producers?

   A. P and H

   B. P, H, R and F

   C. O, N and T

   D. O

2. Which are secondary consumers?

   A. G, S, R, P and F

   B. H and P

   C. N, L, E and T

   D. O

Below is a diagram of a food web. The letters represent different species of organism. S is a photosynthesizing organism. Use this for questions 3 and 4.

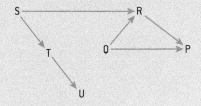

3. Which organism is **only** a primary consumer (herbivore)?

   A. P

   B. T

   C. U

   D. S

4. What is organism U?

   A. A producer

   B. A primary consumer

   C. A secondary consumer (carnivore)

   D. An autotroph

5. The diagram below shows part of an aquatic food web for a lake ecosystem.

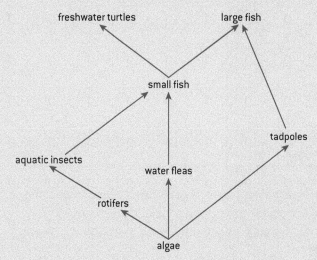

What is the maximum number of trophic levels represented in this food web?

   A. Three

   B. Four

   C. Five

   D. Six

3 BIODIVERSITY AND CONSERVATION

3.1 An introduction to biodiversity

Significant ideas:

→ Biodiversity can be identified in a variety of forms including species diversity, habitat diversity and genetic diversity.

→ The ability to both understand and quantify biodiversity is important to conservation efforts.

Applications and skills:

→ **Distinguish** between biodiversity, species diversity, genetic diversity and habitat diversity.

→ **Comment** on the relative values of biodiversity data.

→ **Discuss** the usefulness of providing numerical values of species diversity to understanding the nature of biological communities and the conservation of biodiversity.

Knowledge and understanding:

→ **Biodiversity** is a broad concept encompassing total diversity which includes diversity of **species**, **genetic diversity** and **habitat diversity**.

→ **Species diversity** in communities is a product of two variables, the number of species (richness) and their relative proportions (evenness).

→ Communities can be described and compared by the use of **diversity indices**. When comparing communities that are similar then low diversity could be evidence of pollution, eutrophication or recent colonization of a site. The number of species present in an area is often used to indicate general patterns of biodiversity.

→ **Habitat diversity** refers to the range of different habitats in an ecosystem or biome.

→ **Genetic diversity** refers to the range of genetic material present in a population of a species.

→ Quantification of biodiversity is important to conservation efforts so that areas of high biodiversity may be identified, explored, and appropriate conservation put in place where possible.

→ The ability to assess changes to biodiversity in a given community over time is important in assessing the impact of human activity in the community.

Types of biodiversity

You may think of biodiversity as the numbers of species of different animals and plants in different places. But it is a far more complex concept than that. Usually, high diversity indicates a healthy ecosystem but sometimes, particularly in the Arctic, there may be low diversity in an unpolluted, ancient site but this is because conditions are so harsh.

Biodiversity is the overall term for three different but inter-related types of diversity: **genetic diversity**, **species diversity** and **habitat diversity**.

If we think about biodiversity then the spread of individuals between species is more important than the total number of individuals in a habitat. **Species diversity** reflects this, as it looks at both the range (variety) and number of organisms (abundance). Species diversity is not just the total number of organisms in a place but the number of organisms within each of the different species.

Consider the examples in figure 3.1.1.

Forest A	Forest B
15 different species	15 different species
100 individuals of one species and one individual of each of the other 14 species.	Seven individuals of each of the 15 species.
Total individuals = 114	Total individuals = 105
Greater number of individuals	
Low species diversity because the individual species may not be able to breed = disappear	Greater species diversity – because the species are evenly spread and the populations more viable

▲ **Figure 3.1.1** Biodiversity in two forests

This variation in diversity alters from habitat to habitat. Some habitats such as coral reefs and rainforests have high species diversity. Urban habitats and polar regions have much lower species diversity by comparison.

Another way to measure diversity is to look at genetic diversity – that is the amount of variation that exists between different individuals within different populations of a species. A small population normally has a lower genetic diversity than a larger one because of the smaller gene pool.

Species are made up of both individuals and populations. Naturally, each individual in a species has a slightly different set of genes from any other individual in that species. And if a species is made up of two or more different populations in different places, then each population will have a different total genetic make-up. Therefore to conserve the maximum amount of genetic diversity, different populations of a species need to be conserved.

Not all species have the same amount of genetic diversity. Almost the entire world population of grey seals exist on the Farne islands off the north-east coast of England with a few small, scattered populations in

'It should not be believed that all beings exist for the sake of the existence of man. On the contrary, all the other beings too have been intended for their own sakes and not for the sake of something else.'

Maimonides, *The Guide for the Perplexed* 1:72, c. 1190

Key term

Species diversity in communities is a product of two variables, the number of species (richness) and their relative proportions (evenness).

TOK

The term 'biodiversity' has replaced the term 'Nature' in much writing on conservation issues. Does this represent a paradigm shift?

Key term

Genetic diversity is the range of genetic material present in a gene pool or population of a species.

▲ **Figure 3.1.2** Grey seal

Grey seals have one large population within which individuals interbreed. The whole population is adapted to the same conditions so they all have very similar genetic make-up and low genetic diversity. What may happen if disease strikes?

other places. This is an example of an organism with a small amount of genetic diversity.

Organisms where large genetic diversity exists include the European red fox, which is found right across Europe. Humans also have high genetic diversity as we exist planet-wide in many populations.

Each population has a slightly different genetic make-up so there is more variety within the species.

Humans can alter genetic diversity by artificially breeding or genetically engineering populations with reduced variation in their genotypes or even identical genotypes – clones. This can be an advantage if it produces a high-yielding crop or animal but a disadvantage if disease strikes and the whole population is susceptible. This domestication and plant breeding has led to a loss of genetic variety, hence the importance of 'gene banks'.

For many conservationists, more genetic variation is a good thing. So they would want to maximize genetic diversity. That means having many species and much variability within each species. Then species have a better chance of adapting to change in their habitats. But that could mean stopping succession as species diversity falls later in succession (2.4) so interfering with the natural process.

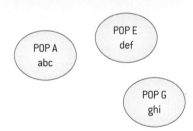

▲ **Figure 3.1.3** Red fox

Tropical rainforests are high in **habitat diversity** as there are many ecological niches due to the layering of the forests. Tundra has a lower level of habitat diversity (see 2.4, biomes).

Biodiversity – an indicator of ecosystem health

It is difficult to determine if one ecosystem is more healthy than another. Biodiversity is often used as a measure as high biodiversity usually equates with high ecosystem health.

Key term

Habitat diversity is the range of different habitats per unit area in a particular ecosystem or biome.

A habitat with high biodiversity has these advantages:

- resilience and stability due to the range of plants present of which some will survive drought, floods, insect attack, disease

- genetic diversity so resistance to diseases

- some plants there will have deep roots so can cycle nutrients and bring them to the surface making them available for other plants.

But high biodiversity does not always equate to having a healthy ecosystem:

- diversity could be the result of fragmentation (break up) of a habitat or degradation when species richness is due to pioneer species invading bare areas quickly

- managing grazing can be difficult as plant species have different requirements and tolerance to grazing

- some stable and healthy communities have few plant species so are an exception to the rule.

Diversity indices

We can only accurately compare two similar ecosystems or two communities within an ecosystem. Communities can be described and compared by the use of diversity indices (see 2.5).

When comparing communities that are similar then low diversity could be evidence of pollution, eutrophication or recent colonization of a site. The number of species present in an area is often used to indicate general patterns of biodiversity but only tells us part of the story.

It is important to repeat investigations of diversity in the same community over a period of time and to know if change is a natural process due to succession or due to impact from human activity. This could either increase or decrease biodiversity which would tell us if conservation efforts are succeeding or not.

Hotspots

Biodiversity is not equally distributed on Earth. Some regions have more biodiversity than others – more species and more of each species than in other areas. There are hotspots where there are also unusually high numbers of endemic species (those only found in that place). Where these hotspots are is debated but about 30 areas have been recognized.

- They include about ten in tropical rainforest but also regions in most other biomes.

- They tend to be nearer the tropics because there are fewer limiting factors in lower latitudes.

- They are all threatened areas where 70% of the habitat has already been lost.

- The habitat contains more than 1,500 species of plants which are endemic.

- They cover only 2.3% of the land surface.

- They tend to have large densities of human habitation nearby.

> **TOK**
>
> Biodiversity index is not a measure in the true sense of the word, but merely a number (index) as it involves a subjective judgement on the combination of two measures; proportion and richness. Are there examples in other areas of the subjective use of numbers?

> **Key term**
>
> A biodiversity **hotspot** is a region with a high level of biodiversity that is under threat from human activities.

These hotspots are shown in figure 3.1.4; between them they contain about 60% of the world's species so have very high species diversity.

Critics of naming hotspots say that they can be misleading because they:

- focus on vascular plants and ignore animals

- do not represent total species diversity or richness

- focus on regions where habitats, usually forest, have been lost and ignore whether that loss is still happening

- do not consider genetic diversity

- do not consider the value of services, for example water resources.

But they are a useful model to focus our attention on habitat destruction and threats to unique ecosystems and the species within them.

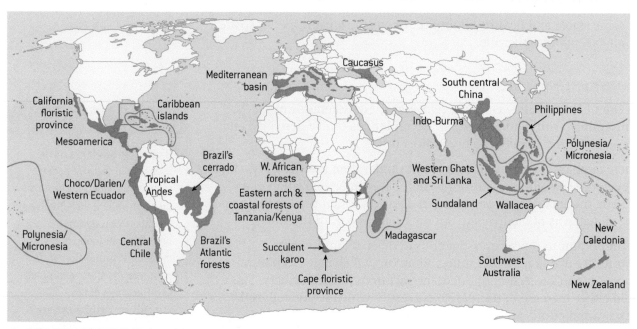

▲ **Figure 3.1.4** World map of regions with biodiversity hotspots

Questions

Hotspot	Plant species	Endemic plant species	Endemics as a percentage of world total
Atlantic Forest, Brazil	20,000	8,000	2.7
California Floristic Province	3,488	2,124	0.7
Cape Floristic Region, South Africa	9,000	6,210	2.1
Caribbean Islands	13,000	6,550	2.2
Caucasus	6,400	1,600	0.5
Cerrado	10,000	4,400	1.5
Chilean Winter Rainfall – Valdivian Forests	3,892	1,957	0.7
Coastal Forests of Eastern Africa	4,000	1,750	0.6
East Melanesian Islands	8,000	3,000	1.0

Hotspot	Plant species	Endemic plant species	Endemics as a percentage of world total
Eastern Afromontane	7,598	2,356	0.8
Guinean Forests of West Africa	9,000	1,800	0.6
Himalaya	10,000	3,160	1.1
Horn of Africa	5,000	2,750	0.9
Indo-Burma	13,500	7,000	2.3
Irano-Anatolian	6,000	2,500	0.8
Japan	5,600	1,950	0.7
Madagascar and the Indian Ocean Islands	13,000	11,600	3.9
Madrean Pine–Oak Woodlands	5,300	3,975	1.3
Maputaland–Pondoland–Albany	8,100	1,900	0.6
Mediterranean Basin	22,500	11,700	3.9
Mesoamerica	17,000	2,941	1.0
Mountains of Central Asia	5,500	1,500	0.5
Mountains of Southwest China	12,000	3,500	1.2
New Caledonia	3,270	2,432	0.8
New Zealand	2,300	1,865	0.6
Philippines	9,253	6,091	2.0
Polynesia–Micronesia	5,330	3,074	1.0
Southwest Australia	5,571	2,948	1.0
Succulent Karoo	6,356	2,439	0.8
Sundaland	25,000	15,000	5.0
Tropical Andes	30,000	15,000	5.0
Tumbes–Choco–Darien	11,000	2,750	0.9
Wallacea	10,000	1,500	0.5
Western Ghats and Sri Lanka	5,916	3,049	1.0

▲ **Figure 3.1.5** Biodiversity hotspots showing the total and endemic plant species

1. How many of these hotspots can you locate on a world map?
2. Why is the number of plant species important for biodiversity?
3. What are the criteria for defining hotspots?
4. What is an endemic species?
5. For what reasons do you think these hotspots are threatened areas?

Practical Work

* Investigate the differences in biodiversity in pools and riffles.
* Investigate global biodiversity.
* Investigate plant biodiversity in two local areas.
* Evolution/natural selection simulation.

3.2 Origins of biodiversity

Significant ideas:

→ Evolution is a gradual change in the genetic character of populations over many generations achieved largely through the mechanism of natural selection.

→ Environmental change gives new challenges to species, which drives evolution of diversity.

→ There have been major mass extinction events in the geological past.

Applications and skills:

→ **Explain** how plate activity has influenced evolution and biodiversity.

→ **Discuss** the causes of mass extinctions.

Knowledge and understanding:

→ **Biodiversity** arises from evolutionary processes.

→ Biological variation arises randomly and can either be beneficial to, damaging to, or have no impact on the survival of the individual.

→ **Natural selection** occurs through the following mechanism:

- within a population of one species there is genetic diversity, which is called variation

- due to natural variation some individuals will be fitter than others

- fitter individuals have an advantage and will reproduce more successfully

- the offspring of fitter individuals may inherit the genes that give the advantage.

→ This **natural selection** will contribute to evolution of biodiversity over time.

→ Environmental change gives new challenges to the species, those that are suited survive, and those that are not suited will not survive.

→ **Speciation** is the formation of new species when populations of a species become isolated and evolve differently.

→ **Isolation of populations** can be caused by environmental changes forming barriers such as mountain building, changes in rivers, sea level change, climatic change or plate movements. The surface of the earth is divided into **crustal, tectonic plates** which have moved throughout geological time. This has led to the creation of both land bridges and physical barriers with evolutionary consequences.

→ The distribution of continents has also caused climatic variations and variation in food supply, both contributing to evolution.

→ **Mass extinctions** of the past have been caused by a combination of factors such as tectonic movements, super-volcanic eruption, climatic changes (including drought and ice ages), and meteor impact all of which resulted in new directions in evolution and therefore increased biodiversity.

How new species form

Charles Darwin, proposed **the theory of evolution** which is outlined in **The Origin of Species**, published in 1859. Evidence, in the form of the fossil record, discovery of the structure of DNA and mechanisms of mutations, all supports the theory. It is summarized below.

- Each individual is different (except identical twins who have the same genotype) due to their particular set of inherited genes and to mutations.

- Each will be slightly differently adapted (or fitted) to its environment.

- Resources are limited for any population and there will be competition for these resources, eg for food, light, space, water.

  - For example, a giraffe with a slightly longer neck than the other giraffes may be able to reach tree leaves that are out of reach to the others and so get more food. This gives that giraffe an advantage.

  - These small differences mean that some individuals will be more successful. They will survive to breed more than others and so pass their genes on to the next generation.

- Over time these changes show and the whole population gradually changes.

This is **natural selection** where those more adapted to their environment have an advantage and flourish and reproduce but those less adapted do not survive long enough to reproduce. So the fittest survive – '**survival of the fittest**' is the term you may have heard.

Over many generations, if a population is separated from others and isolated, the differences may increase to such an extent that, should the populations be reunited, they will be unable to interbreed. Then a new species has formed and **speciation** has occured.

Key term

Speciation is the gradual change of a species over a long time. When populations of the same species become separated, they cannot interbreed and if the environments they inhabit change they may start to diverge and a new species forms. Humans can speed up speciation by artificial selection of animals and plants and by genetic engineering but the natural process of speciation is a slow one. Separation may have geographical or reproductive causes.

TOK

The theory of evolution by natural selection tells us that change in populations is achieved through the process of natural selection. Is there a difference between a convincing theory and a correct one?

▲ **Figure 3.2.1** Possible ancestors in evolution of *Homo sapiens* species

Isolation may be on an island but could be isolation on a mountain or in a body of water (lake or even pond). Some populations mix freely but they are isolated in other ways: their mating seasons are not synchronized or their flowers mature at different times.

Physical barriers

Species can develop into two or more new species if their population is split by some kind of physical barrier, for example a mountain range or ocean. The physical barrier will split the gene pool: the genes of the populations on both sides of the barrier will not be mixing anymore and the two populations can develop in different directions. Examples of speciation due to physical barriers:

1. Large flightless birds only occur on those continents that were once part of Gondwana (Africa, Australia including New Zealand, South America). However, because Gondwana split up a very long time ago, the large flightless birds are not closely related.

▲ **Figure 3.2.2** Large flightless birds (rhea, ostrich, emu, kiwi, tinamou, moa, elephant bird, cassowary)

2. Australia (and Antarctica) split off from Gondwana a long time ago. At that time, both marsupials and early placental mammals lived in the same area. In South America and Africa, the placental mammals prevailed and outcompeted the marsupials. In Australia the opposite happened and now marsupials are mostly only found in Australia and nearby Papua New Guinea (though a few still survive in South America). Placental mammals are rare in Australia. These mammals came by sea (seals), air (bats) or were introduced by humans (dogs, cats, rabbits and rats).

3. The cichlid fish in the lakes of East Africa are one of the largest families of vertebrates. In Lake Victoria there are 170 species of cichlids (99% endemic); in Lake Tanganyika are 126 species (100% endemic); in Lake Malawi are 200 species (99% endemic). These populations have probably been isolated from each other for millions of years and have different selection pressures within them – slightly different environments. So the fish have adapted to the different environments. As long as the population is large enough, they will continue to diverge from other populations but some isolated populations can be too small and die out.

▲ **Figure 3.2.3** A marsupial – great grey kangaroo and joey

Lake Tanganyika species   **Lake Malawi species**

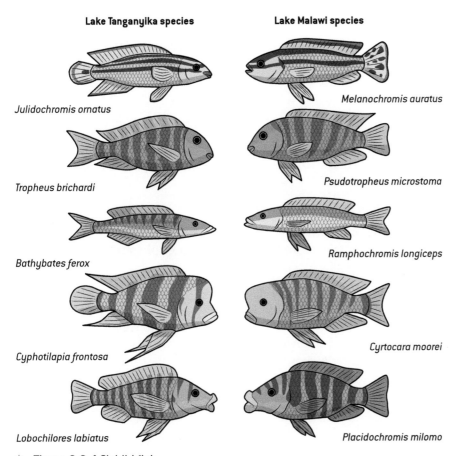

Julidochromis ornatus

Melanochromis auratus

Tropheus brichardi

Psudotropheus microstoma

Bathybates ferox

Ramphochromis longiceps

Cyphotilapia frontosa

Cyrtocara moorei

Lobochilores labiatus

Placidochromis milomo

▲ **Figure 3.2.4** Cichlid fish

Land bridges

These allow species to invade new areas. North and South America were separated for a long time and therefore have rather different species. Obviously that is no longer the case as they are now joined by the recently (geologically) formed land bridge of Central America, which allowed species to move to the North or to the South. Bears for example moved from North to South America.

Land bridges may result from the lowering of seawater levels instead of continental drift. Examples include the (now disappeared) land bridges between England and Europe (now under the English Channel) and North America and Asia (Bering Straits).

Continental drift

Continental drift has also resulted in new and diverse habitats. During their drifting over the globe, the continents have moved to different climate zones. The changing climatic conditions and therefore food supplies forced species to adapt and resulted in an increase in biodiversity. A dramatic example of this are the changes on Antarctica. Once this continent had a tropical climate and was covered by forest. When Antarctica moved southwards, the forest gradually disappeared, the snow and ice-covered landscape was formed and new, cold adapted species arrived or evolved. The only remains of the former forest are some tree fossils.

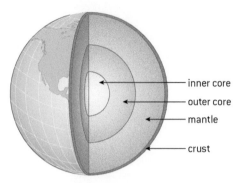

▲ **Figure 3.2.5** Simplified diagram of the Earth

How plate activity influences biodiversity

The lithosphere is divided into seven large tectonic plates and many smaller ones and these drift around, moving at about 50 to 100 mm per year. This is called **continental drift** and the study of the movement of the plates is **plate tectonics**. Where the plates meet, they may:

- Slide past each other (eg the San Andreas fault line, California).

- Diverge (eg the mid-Atlantic ridge where the plates are moving slowly apart). This could cause the physical separation of populations.

- Converge – these convergent plates may:

  - collide and both be forced upwards as mountains (eg formation of the Himalayas and Alps). This creates physical barriers.

  - collide and one (the heavier oceanic plate) sinks underneath the lighter continental plate. This is a subduction zone where deep ocean trenches and volcanic island chains are formed (eg where the Nazca plate under the Pacific meets the west coast of South America and the Andes form). This creates land bridges and myriad new niches.

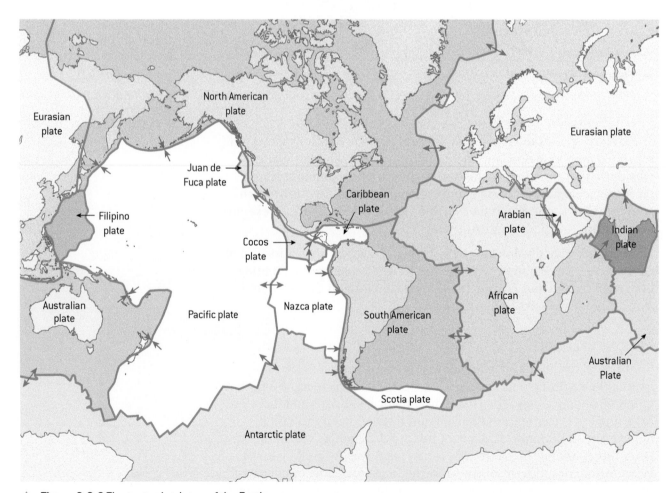

▲ **Figure 3.2.6** The tectonic plates of the Earth

Jigsaws and continents

Looking at a world map or a globe, we all notice that the west coast of the African continent and the east coast of the South American continent appear to fit together like pieces of a jigsaw. Many others have noticed this in the past, since the first world maps were drawn. Abraham Ortelius, from Antwerp, noticed (see figure 3.2.7).

He was a cartographer and drew one of the first atlases in 1570. Francis Bacon (1620), a British philosopher and scientist, did too, as did Benjamin Franklin (USA), and Antonio Snider-Pellegrini (France) who, in 1858 drew a map of how they may have fitted together.

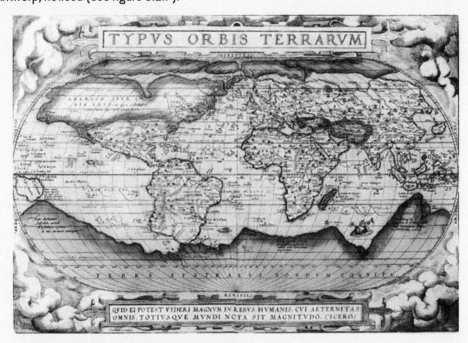

▲ **Figure 3.2.7** Ortelius' world map in 1570

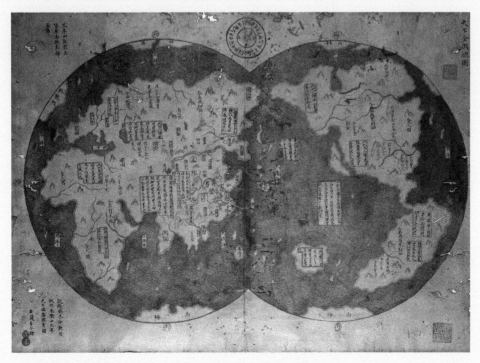

▲ **Figure 3.2.8** World map of 1320 possibly from Zhu Ziben

For at least a millennium, sailors have made long ocean voyages using the Sun and stars to guide them in rudimentary navigation. There is debate about the dates of the earliest maps of the world but Zhu Ziben in the Yuan Dynasty in China may have produced the map in figure 3.2.8 as long ago as 1320. This not only shows the shape of the Earth but also the continents including Australia and Antarctica and the Americas.

In 1912, Alfred Wegener also noticed the jigsaw fit of the continents but additionally he noted that fossils on the west coast of Africa were similar to those on the east coast of South America and coal fields in Europe matched those in North America. He proposed that this was because the continents had once been joined and had floated apart. This was thought outrageous and criticism was heaped upon it. In 1930, Wegener died during an expedition to the Arctic and still few believed his hypothesis. It took until the late 1960s for his idea to be accepted when research was done on the deep ocean floor vents at the edges of the plates.

If the shapes of the continents have been known by some for so long, why do you think Wegener was laughed at when he proposed that the continents move?

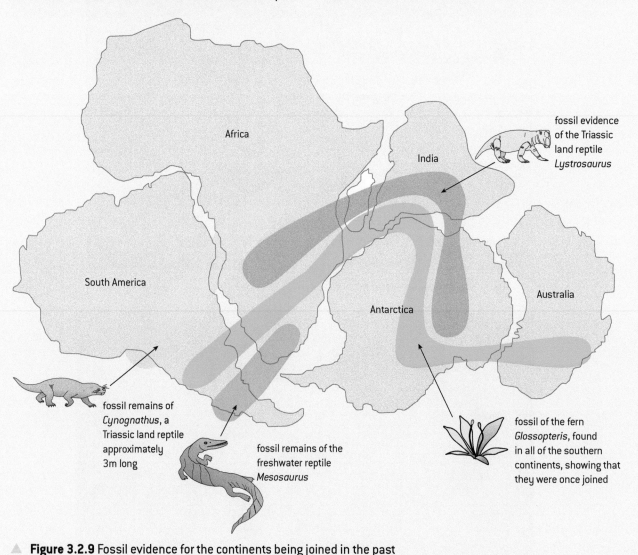

▲ **Figure 3.2.9** Fossil evidence for the continents being joined in the past

Similar groups of animals

There are striking similarities between groups of animals on various continents and this has not gone unnoticed. Here are a few. You can probably think of others.

- Llamas and camels are both domesticated animals fulfilling similar roles – llamas in South America, camels in Africa (one hump) and central Asia (two humps). They are distant cousins. Their common ancestor was a rabbit-sized mammal. Both were domesticated about 5,000 years ago and are used by humans as pack animals and for meat, milk and an indication of wealth. This suggests that these areas of the globe must have been connected in the past.

- Kangaroos fulfil a similar ecological role in Australia as cattle do in the rest of the world. They are both large herbivores, eating grass and converting it to meat. The kangaroos are marsupials (as are wombats and koalas) and are one of the three types of mammal. All mammals suckle their young with milk that the mother produces. The placental mammals (having a placenta to provide food for the embryo in the uterus) are the most successful in evolutionary terms. The marsupials (with a pouch in which the young embryo – joey – continues to develop) and monotremes (eg duck-billed platypus) which lay eggs instead of giving birth to live young are the others. They are mostly confined to Australasia and it seems that the placental mammals outcompeted the other two groups in most other continents. Australia must have broken off from the mainland areas early on to allow for this dominance.

- African and Indian elephants are another example and there are plant species that also show much similarity between continents.

A little bit about the geological timescale

The Earth formed about 4.6 billion (4.6×10^9) years ago. The first life forms are thought to have been simple cells like bacteria, living about 4 billion years ago. Some 65 million years ago, the dinosaurs became extinct. The human species has been recognizably human for the last 200,000 years. To put this timescale into a perspective that we humans can appreciate, if you consider geological time as a 24-hour day, humans appeared at a few seconds to midnight.

Background and mass extinctions

Background extinction rate is the natural extinction rate of all species and we think it is about **one species per million species per year** so between 10 and 100 species per year. This is estimated from the fossil record because we do not know how many species there are alive today. Some species will become extinct before we even know that they exist.

If species are becoming extinct at a rate far greater than the background rate, what else is happening? Mass extinctions have occurred over geological time and there have been five major ones. We know this from the fossil record when suddenly fossils disappear from the rock strata and there is an abrupt increase in rates of extinction. We think this is due to a rapid change of climate, perhaps a natural disaster (volcanic eruption, meteorite impact) which results in many species dying out as

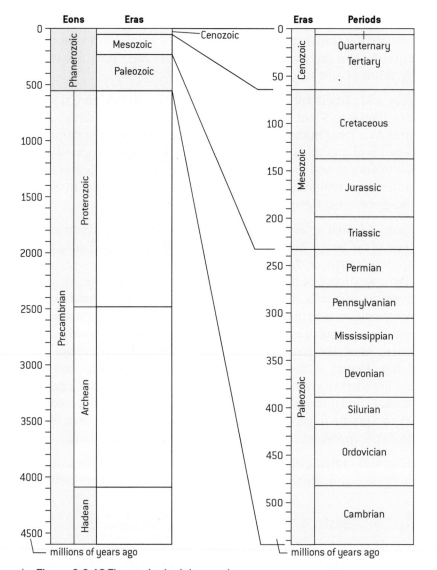

Eons	Eras		Eras	Periods

Figure 3.2.10 The geological timescale

Geological Timescale copyright 2005—geology.com

Lonesome George

▲ **Figure 3.2.11** Lonesome George: born before 1912 – died 24 June 2012

Lonesome George was the world's most famous reptile. He was over 100 years old and was a Pinta Island giant tortoise from the Galapagos Islands and the last one of his subspecies. He was found in 1971 and no other tortoise had been recorded on the island since 1906. George's fellow tortoises were taken for meat by sailors or goats destroyed their habitat and food source. He was a symbol of what went wrong for many species whose fate is in the hands of humans.

- Do you think it matters that Lonesome George was the last of his kind? There are other subspecies of giant tortoise.

- Is anyone to blame for this?

- What could have been done?

they cannot survive the change in conditions. Most biologists now think we are in the sixth mass extinction – called the **Holocene extinction event** as the Holocene is the part of the Quaternary geological period we are in now (see figure 3.2.10). This started at the end of the last ice age about 9,000 to 13,000 years ago when large mammals such as the woolly mammoth and sabre-toothed tiger became extinct probably through hunting. But the rate has accelerated in the last 100 years. While it may be due to climate change, the big difference is that it has been caused by one species – *Homo sapiens*.

We think there are about 5,000 mammal species alive today. Their background extinction rate would be one per 200 years but the past 400 years have seen 89 mammalian extinctions, much more than the background rate. Another 169 mammal species are listed as critically endangered or the 'living dead'. 'Living dead' species are ones which have such small populations that there is little hope they will survive. Or

To think about

Biodiversity headlines

Scientists find 24 new species in Suriname (2007) including a fluorescent purple toad and 12 kinds of dung beetles

From a devilish-looking bat to a frog that sings like a bird, scientists have identified 126 new species in the Greater Mekong area, the WWF said in a new report detailing discoveries in 2011. But threats to the region's biodiversity mean many of the new species are already struggling to survive, the conservation group warned. 'The good news is new discoveries. The bad news is that it is getting harder and harder in the world of conservation and environmental sustainability,' Nick Cox, manager of WWF-Greater Mekong's Species Programme, said.

13 of the coolest new species discovered In 2013: Chocolate Frogs, Walking Sharks, Glue-Spitting Velvet Worms and more

Figure 3.2.12 The olinguito was the first carnivorous mammal species discovered in the Americas since 1978.

Two new species of wobbegongs, otherwise known as carpet sharks, have been found in Western Australian waters(2008)

Does it surprise you that we know so little about life on Earth?

they have lost a species that they depend upon, for example a pollinator insect for a flowering plant, a prey species for a predator.

The last five mass extinctions were spread over 500 million years. You may have heard of some of them, particularly the one when the dinosaurs became extinct. This was the Cretaceous–Tertiary also called the K-T extinction or boundary. It happened about 65 million years ago. In this mass extinction, most of the large animals on land and sea and small oceanic plankton died out but most small animals and plants survived. The causes have been argued about for years but the general view now is that it was caused by:

- A volcanic eruption: The volcanoes of the Deccan plateau in what is now India erupted for a million years at the time of the K-T boundary.

- And a meteor impact putting huge amounts of dust into the atmosphere. The evidence for the impact is the Chicxulub crater in the Yucatan peninsula, Mexico. The crater is 180 km in diameter and the igneous rock underneath contains high levels of iridium, a mineral common in extraterrestrial object such as meteorites but not common on earth.

- The result of climate change over a long period – the eruption and the meteor impact would have caused dust clouds to block much of the incoming solar radiation. Plants would have been unable to photosynthesize and so died and food webs would have collapsed.

All the mass extinctions are in the table below.

Mass extinction	MYA (million years ago)	Geological period	Estimate of losses (a family contains up to 1,000 species)
6th	now	Holocene	unknown
5th	65	Cretaceous–Tertiary	17% families and all large animals including dinosaurs
4th	199–214	End Triassic	23% families, some vertebrates
3rd	251	Permian–Triassic	95% of all species, 54% of families
2nd	364	Devonian	19% families
1st	440	Ordovician–Silurian	25% families

▲ **Figure 3.2.13** The six mass extinctions

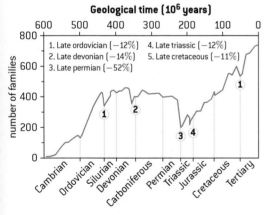

▲ **Figure 3.2.14** The five major extinction episodes of life on Earth shown by family diversity of marine vertebrates and invertebrates (after E. O. Wilson, 1988)

After each of the first five mass extinctions, there was a burst of adaptive radiation of the remaining species which adapted to fill the ecological niches left vacant. But this is all within a geological and evolutionary timescale so did not happen overnight but over tens of millions more years.

The sixth mass extinction

Many scientists believe we are currently in the sixth mass extinction. This is caused by the actions of humans and may be far greater in both extent and rate – humans are pushing more animals to extinction faster than happened in any of the previous mass extinctions.

We have wiped out many large mammal and flightless bird species – woolly mammoths and ground sloths, moas in New Zealand, dodos in Mauritius. One estimate is that 25% of all plant and animal species will have gone extinct between 1985 and 2015.

The UN estimated that 12 October 1999 was the day when there were six billion people alive on Earth and seven billion on 12 March 2012. Humans alter the landscape on an unprecedented scale. Some organisms do well in the environments that we create (urban rats, domesticated animals, some introduced species) but most do not. We call the successful ones **weedy species**, both animals and plants. Many weedy species will probably survive, and thrive, in the current mass extinction. But others, many never identified, are likely to die out. The question we should ask is whether we are a weedy species or not. It has taken 5 to 10 million years to recover biodiversity after past mass extinctions. That would be more than 200,000 generations of humankind and there have been about 7,500 generations since the emergence of *Homo sapiens*. And just to put that into perspective, human civilization started only 500 generations ago.

The previous mass extinctions were due to physical (abiotic) causes over extended periods of time, usually millions of years. The current mass extinction is caused by humans (anthropogenic) so has a biotic cause and is happening over a few decades, not millennia. Humans are the direct cause of ecosystem stress because we:

- Transform the environment – with cities, roads, industry, agriculture.
- Overexploit other species – in fishing, hunting and harvesting.
- Introduce alien species – which may not have natural predators.
- Pollute the environment – which may kill species directly or indirectly.

WWF, the Worldwide Fund for Nature (www.panda.org) produces a periodic report called the **Living Planet Report** on the state of the world's ecosystems. The 2012 report painted a grim picture of loss and degradation, and of decline in the overall living planet index (measure of ecosystem health) of 30% between 1970 and 2008. But also improvements in temperate oceans and terrestrial ecosystems. The major losses are in the tropics.

Practical Work
* Research number of species on Earth today and in the past.
* Investigate how plate activity has influenced evolution and biodiversity using named species as examples.

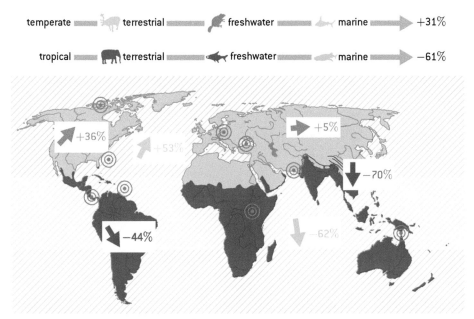

▲ **Figure 3.2.15** The living planet index from 1970 to 2008 reported in the WWF Living Planet Report 2012

The Living Planet Index measures trends in the Earth's biological diversity. It follows populations of 2,500 vertebrate species (fish, amphibians, reptiles, birds, mammals) from all around the world and so gives a numerical index of changes in biodiversity.

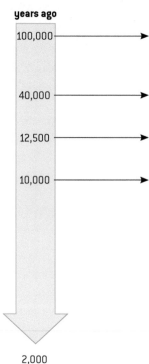

years ago

100,000

40,000

12,500

10,000

2,000

▲ **Figure 3.2.16** The two phases of the sixth mass extinction

Stage 1: modern humans spread over the Earth: *Homo sapiens* outcompeted Neanderthal man in Europe and probably led to his extinction.

- In Australia, the megafauna (largest animals) disappeared after the arrival of humans.

- North America: the arrival of humans resulted in the hunting to extinction of woolly mammoths and mastodons.

Stage 2: humans became farmers about 10,000 years ago when there were about one to ten million humans living on Earth.

- The invention of agriculture meant that humans did not compete with other species in the same way for food but could manipulate their environment and other species for their own use.

- By importing goods to where they were living, they could also exceed the carrying capacity (how many individuals of a population the environment can support) of the local environment and so live in communities.

- Humans concentrated on using a few species – crop plants and domestic animals and consider the others as weeds or pests (to be eliminated).

- Humans colonized Madagascar, larger animals disappeared soon after.

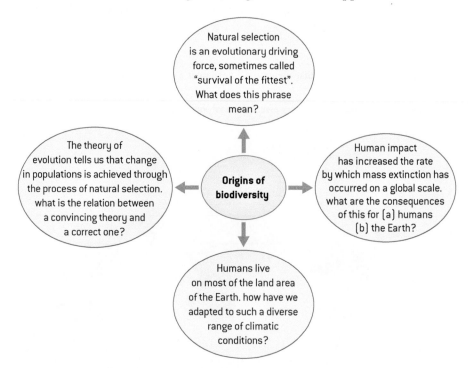

▲ **Figure 3.2.17** The origins of biodiversity

Only in Africa did the large mammals survive. What do you think the reasons for that may be?

3.3 Threats to biodiversity

Significant ideas:

→ While global biodiversity is difficult to quantify, it is decreasing rapidly due to human activity.

→ Classification of species conservation status can provide a useful tool in conservation of biodiversity.

🧪 Applications and skills:

→ **Discuss** the case histories of three different species: one that has become extinct due to human activity, another that is critically endangered, and a third species whose conservation status has been improved by intervention.

→ **Describe** the threats to biodiversity from human activity in a given natural area of biological significance or conservation area.

→ **Evaluate** the impact of human activity on the biodiversity of tropical biomes.

→ **Discuss** the conflict between exploitation, sustainable development and conservation in tropical biomes.

🌐 Knowledge and understanding:

→ **Estimates** of the **total number of species** on the planet vary considerably. They are based on mathematical models, which are influenced by classification issues and lack of finance for scientific research, resulting in many habitats and groups being significantly under-recorded.

→ The current rates of **species loss** are far greater now than in the recent past, due to increased human influence. The human activities that cause species extinctions include habitat destruction, introducing invasive species, pollution, overharvesting and hunting.

→ The **International Union for Conservation of Nature (IUCN)** publishes data in the **Red List of Threatened Species** in several categories. Factors used to determine the conservation status of a species include: population size, degree of specialization, distribution, reproductive potential and behaviour, geographic range and degree of fragmentation, quality of habitat, trophic level and the probability of extinction.

→ **Tropical biomes** contain some of the most globally biodiverse areas and their unsustainable exploitation results in massive losses in biodiversity and their ability to perform globally important ecological services.

→ Most tropical biomes occur in Less Economically Developed Countries (LEDCs) and therefore there is conflict between exploitation, sustainable development and conservation.

Total world biodiversity

How many species are there on Earth today? That should be a straightforward question as surely we have explored just about every region of the Earth and logged and catalogued what is there. But, in fact, we have very little idea about many groups of organisms there are and certainly no clear idea of how many are becoming extinct.

A conservative estimate of species alive today is about 7 million species excluding bacteria.

● Most are animals and most are terrestrial.

● 2/3rds are in the tropics, mostly tropical rainforests.

● 50% of tropical rainforests have been cleared by humans. We are clearing, burning or logging about 1 million square kilometres every 5–10 years of the original 18 million square kilometres. Many countries have no primary forest (not been degraded or destroyed by humans) left.

But only between about 1.4 and 1.8 million species of organisms have been described and named, so are 'known to science'. How we reach an estimate of species depends on the species. It is easier to see large animals but harder for smaller ones. Big, furry animals grab our attention. It is relatively easy to identify the big animals or ones that do not run away. Most mammals and birds are known. It is also easier to find plants as they cannot move. But many groups of smaller organisms such as insects, nematodes, fungi and bacteria have not been found, identified and named.

Some think that there are 50 times more species on Earth than have been named so there are up to 100 million species. Others think there are about eight to ten million species on Earth. That is quite a range of estimates.

Beetles (Coleoptera) are the group with the most identified and named species. Beetles are insects and have about 25% of all named species so up to 2 million. One way of finding and identifying insects is 'fogging' the canopies of rainforest trees with short-lived insecticides. Then the organisms fall out of the trees and are collected and counted and the numbers are extrapolated.

Groups	Species found	Total estimated species	Percentage identified and named
ANIMALS:			
Vertebrates	46,500	50,000	93
Molluscs	70,000	200,000	35
Arthropods: Insects	840,000	8,000,000	11
Arachnids	75,000	750,000	10
Crustaceans	30,000	250,000	12
PROTOZOA	40,000	200,000	20
ALGAE	15,000	500,000	3
FUNGI	70,000	1,000,000	70
PLANTS	256,000	300,000	85

▲ **Figure 3.3.1** Numbers of identified species and total estimated species for various groups

But the number of species alive on Earth has not been constant over geological time. Some become extinct while others evolve into new species.

Extinction, when a species ceases to exist after the last individual in that species dies, is a natural process. Eventually all species become extinct and the average lifespan for a species varies. Most mammals have a

TOK

Fogging the canopy with insecticide kills the insects there. Science can involve killing individuals to find out more for the greater good. To what extent can this be justified? Are there absolute limits beyond which we should not go?

species lifespan of one million years, some arthropods of ten million years or more. The rate at which extinctions occur is not constant and is made up of the background extinction rate and mass extinctions when a sudden loss of species occurs in a relatively short time.

Current extinction rates

How large is the loss of species due to human activities? How long is this taking to happen? These are tricky questions to answer because we can only estimate. Also, in geological time, species evolved and became extinct long before humans appeared. So there is a natural background extinction rate (3.2). But we do know that habitat loss caused by humans is the major cause of species extinction.

We think that current extinction rate is about **100 species per million species per year**.[1] Other estimates range from 30,000 to 60,000 (out of all alive today) species a year and are between 100–10,000 times greater than background rates.

E.O. Wilson, a well-known biologist from Harvard, thinks that the current rate of extinction is 1,000 times the background rate and that it is caused by human activities. He suggests that 30–50 per cent of species could be extinct within 100 years. The rate is estimated to be about three species per hour.

But the rate is not equally spread over the Earth, it is far greater in some areas which are called **hotspots** (3.1). Species are not distributed evenly over the Earth but highly concentrated in some areas. Up to 50% of animals and plants are in one of the 30 hotspots which together make up only 2% of the land area on Earth. These areas are very vulnerable to habitat loss and many species within them are endemic (only found there). Tropical rainforests and coral reefs are particularly vulnerable and some species are more likely to become extinct than others. Those we humans like to hunt or eat or wear or those that are dangerous to us or our crops may be more vulnerable to extinction than others.

But extinction rates are not linear. When half a habitat is lost, animals and plants remain in the other half. There are fewer of them but the species are still there. Only when nearly all that habitat goes will extinction rates increase rapidly. So the current rate can only increase greatly.

But the better news is that we think that protecting only 5% of the habitat could preserve 50% of the species within it. However, remember the hotspots (3.1).

Factors that help to maintain biodiversity

Complexity of the ecosystem

You may have studied food webs and food chains already. The more complex a food web, the more resilient it is to the loss of one species or reduction in its population size. If one type of prey or food source or predator is lost, the others will fill the gaps left. This resilience of more

---

[1] Pim, S.L. and Raven, P. Biodiversity: Extinction by numbers *Nature* 403, 843–845 (24 February 2000) | doi:10.1038/35002708

complex communities and ecosystems is a good thing for biodiversity overall. But it may not be good for species diversity as one species is lost completely but the community continues.

Stage of succession

When plants and animals colonize a bare piece of land, there are few species colonizing it at first. Species diversity increases with time until a climax community is reached when the species composition is stable. It may fall slightly once the climax community is reached but normally, species diversity increases as succession proceeds. So communities in young ecosystems that are undergoing succession may be more vulnerable than those in older ones which are more resilient and stable.

Limiting factors

If it is difficult for the organisms in an ecosystem to get enough raw materials for growth, eg water is limiting in a desert, then any change that makes it even harder may result in species disappearing. If the abiotic factors required for life are available in abundance (water, light, heat, nutrients), the system is more likely to manage if one is reduced.

Inertia

Inertia is the property of an ecosystem to resist change when subjected to a disruptive force. Along with resilience and stability, it is key to helping planners know which site will either resist change or recover most quickly.

Factors that lead to loss of biodiversity

Natural hazards are naturally occurring events that may have a negative impact on the environment (and humans). Above a certain level the impact is so bad they are considered as natural disasters. Natural hazards are often considered disasters when they impact humans badly. They can be considered 'Acts of God' as they are outside the control of humans. We can mitigate their effects but are unlikely to stop them happening.

Natural hazard	Natural disaster
Volcanic eruptions	Eruption of Mount St Helens in Washington State, USA in 1980
Earthquakes	Haiti 2010
Floods	Yangtze River floods in China that left 14 million people homeless, the southern Asia floods of 2007
Wild fires	Clear 4 million to 5 million acres (1.6 million to 2 million hectares) of land in the USA every year[2]
Hurricanes	Hurricane Katrina in New Orleans, USA in 2005

▲ **Figure 3.3.2** Examples of recent natural hazards

Environmental disasters are usually thought of as caused by human activity and would include loss of tropical rainforest on a massive scale and oil spills.

---

[2] http://www.ecoplum.com/greenliving/view/274?_dispatch=gcontents/view/274#sthash.JLzhn46v.dpuf

Figure 3.3.3 Tsunami wave –a natural hazard

Figure 3.3.4 Forest fire

The major cause of loss of biodiversity is **loss of habitat**. There is loss of diversity on a small and large scale due to human activities. In many parts of the world, humans have destroyed or changed most of the original natural habitat.

- In the Philippines, Vietnam, Sri Lanka and Bangladesh, where human population levels are high, we have lost most of the wildlife habitat and most of the primary rainforest.

- In the Mediterranean region, only 10% of original forest cover remains.

- In Madagascar, by 2020, it is reckoned that no moist forest will be left except for the small area under protection. Madagascar is the only place where lemurs occur and there are many endemic species. Loss of their habitat is highly likely to make them extinct.

Habitat destruction and degradation occur when we develop or build on a piece of land.

Fragmentation of habitat is the process whereby a large area is divided up into a patchwork of fragments, separated from each other by roads, towns, factories, fences, power lines, pipelines or fields. The fragments are isolated in a modified or degraded landscape and they act as islands within an inhospitable sea of modified ecosystems. There are edge effects to the islands with higher edge to area ratios as the fragments get smaller and greater fluctuations of light, temperature and humidity at the edge than the middle. Invasion of the habitat by pest species or humans increases and there is the possibility that domestic and wild species come into contact and spread diseases between populations.

Pollution caused by human activities can degrade or destroy habitats and make them unsuitable to support the range of species that a pristine ecosystem can support. These include:

- Local pollution, for example spraying of pesticides may drift into wild areas, oil spills may kill many seabirds and smaller marine species.

- Environmental pollution by emissions from factories and transport can lead to acid deposition or photochemical smog.

- Run-off of fertilizers into waterways can cause eutrophication, toxic chemicals can accumulate in food chains.

- Climate change alters weather patterns and shifts biomes away from the equator.

Overexploitation has escalated as human populations have expanded and technology has allowed us to get better and better at catching, hunting and harvesting. Chain saws have replaced hand saws in the forests, factory ships with efficient sonar and radar find and process fish stocks and bottom trawling scoops up all species of fish whether we then eat them or not. We are just too good at getting our food. The Grand Banks off Newfoundland were one of the richest fishing grounds of the world but are now fished out. If we exceed the maximum sustainable yield of any species (the maximum which can be harvested each year and replaced by natural population growth), then the population is not sustainable (see 1.4 and 8.4). The difficulty is knowing what this figure is and how much is left. But growing rural poverty mixed with improved methods of hunting and harvesting mean that more and more humans living at a subsistence level overexploit the environment. With lower population densities, traditional customs and practices prevented overexploitation but with the choice being between starvation and eating bush meat (wild animal meat) or felling the last tree, there is no real choice.

Nowadays we also 'harvest' the natural environment for animals to fill the ever-increasing demands of the pet trade. As personal wealth increases people want new and different, exotic pets. Victims of this trade include many primates, birds and reptiles.

Introducing non-native (exotic/alien) species can drastically upset a natural ecosystem. Humans have done this through colonization of different countries, bringing their own crops or livestock.

Sometimes it works:

- potatoes from the Americas to Europe
- rubber trees from the Amazon to South-East Asia.

Sometimes it is a disaster:

- Rhododendrons were introduced to Europe from Nepal by plant collectors as they have large flowers. But they have escaped into the wild and taken over many areas as they outcompete the native plants and are toxic.
- Dutch Elm disease came from imported American logs to Europe and decimated elm populations.
- Sudden oak death was also imported in this way.
- Australia has been particularly unfortunate with rabbits, cane toads, red foxes, camels, blackberry, prickly pear and the crown of thorns starfish to name just a few. The very different flora and fauna of Australia was well adapted to the environment but unable to compete with the aggressive invasive species.

Sometimes the species is introduced by accident or escapes from gardens or zoos.

Spread of disease may decrease biodiversity. The last population of black-footed ferrets in the wild was wiped out by canine distemper in 1987 and the Serengeti lion population is reduced due to distemper. Diseases of domesticated animals can spread to wild species and vice versa, particularly if population densities are high. In zoos, disease is a constant threat where species are kept close together. Diseases tend to be species specific (only hit one species) but if they mutate they can infect across the species barrier and this is an ever-present threat. Recent examples are:

- A swine flu outbreak in 2010 – swine flu is endemic in pigs and can sometimes pass from pigs to humans.

- Bird flu (avian flu) is a virus adapted to both birds and humans and a very pathogenic strain (H5N1) has spread through Asia to Europe and Africa since 2003.

- Foot and mouth disease (FMD) is a virus that affects all animals with cloven hooves (cattle, sheep, pigs, deer, goats) and spreads easily, but rarely to humans. A major outbreak in the UK in 2001 led to the government slaughtering up to 10 million animals to try to halt the spread of FMD.

Modern agricultural practices also reduce diversity with monocultures, genetic engineering and pesticides. Fewer and fewer species and varieties of species are grown commercially and more pest species are removed.

To do

Rabbits in Australia

Rabbits are not indigenous to Australia. In 1859, rabbits were introduced into Australia for sport and for meat by shipping a few rabbits from Europe. With no predator there, they multiplied exponentially and after ten years, there were estimated to be two million. The rabbits ate all the grass and forage so there was none for sheep and the farming economy collapsed. Rabbits caused erosion as the topsoil blew away but also probably caused the extinction of many Australian marsupials which could not compete with them.

It was so bad that in 1901–7, Western Australia built a rabbit-proof fence from north to south over 3200 km long to keep the rabbits out. Other control methods are shooting, trapping and poisoning.

In 1950, the **myxomatosis virus** was brought from Brazil to control the rabbits and released. The epidemic that followed killed off up to 500 million of the 600 million rabbits. But the rabbits were not eliminated as a very few were resistant to the virus. They bred and now rabbits are again a problem with the population recovering to 200–300 million within 40 years.

In 1996, the government released Rabbit Hemorrhagic Disease or RHD as a second disease to attempt to control the rabbits. But neither of these biological control measures has eliminated the rabbits.

1. What factors contributed to the success of the rabbits in Australia?

2. What is a biological control measure?

3. Research the introduction of cane toads and prickly pear (*Opuntia*) into Australia.

4. What do all these three species have in common in terms of being a pest?

▲ **Figure 3.3.5** Tropical rainforest in Malaysia showing canopy and emergent trees

Vulnerability of tropical rainforests

Tropical rainforests contain over 50% of all species of plants and animals on Earth, yet the rainforest only covers 6% of the land area on Earth. One hectare of rainforest may have 300 species living in it. They also produce about 40% of the oxygen that animals use, hence they are sometimes called the lungs of the Earth. Overall tropical rainforests have high species diversity and habitat diversity yet some areas have even more than others. These are called hotspots. Malaysia has one of the richer rain forests: 8,000 flowering plant species of 1,400 genera including 155 genera of the huge Dipterocarp trees (eg mahogany) which have enormous commercial importance. Denmark, of similar size to the Malay Peninsula, has 45 species of mammals compared with Malaysia's 203.

And they have been lost at a massive rate within the last 50 years. They covered up to 14% of the Earth's land surface in 1950 but have been cleared by humans at unprecedented rates. Some estimate that they will be gone completely in another 50 years due to human activity. Perhaps 1.5 hectares of rainforest is cleared per second. There is about 50% of the Earth's timber in tropical rainforests and timber is the next biggest resource after oil in the world today. But ranching and logging commercially are not the biggest threats to the remaining rainforest, it is people. At least two billion humans live in the wet tropics and many of these rely on the rainforests for subsistence agriculture. A low density of human population is sustainable as they clear a small area of forest, grow crops for two or three years then move on to the next as the soil is exhausted. This is called shifting cultivation. This works as long as there is enough time for the forest to regenerate before the same area is cleared again (up to 100 years). With increasing population pressures and loss of forests to clearance, the forest does not fully regrow before it is cleared and there is a gradual degradation of nutrients and biodiversity.

It is because of their biodiversity and their vulnerability that so much is written and spoken about rainforests. Of course, clearance of the forest does not mean that nothing then grows on that land. There may be animals grazing grasslands, oil palm plantations, subsistence agriculture as well as urban development. Forest cleared and then not managed does regrow in time. But it is estimated that it takes 1,000 years for the biodiversity of the primary forest (before logging or clearing) to be recovered and the secondary forest that does grow up is impoverished in many ways.

The rainforests are so diverse because of the many ecological niches they provide. High levels of heat, light and water are present all year and this means that photosynthesis is rapid and not limited by lack of raw materials. The four layers of the rainforest allow for many different habitats and niches and these are filled by diverse species. All year round growth means that food can be found at any time of the year. Some rainforests are old both in living and geological terms. The lack of disturbance may mean that the system has had time to become more complex.

The fast rate of respiration and decomposition means that the forests appear to be very fertile with high levels of biomass in standing crop (trees and other plants). But most nutrients are held in the plants, not the soil or leaf litter. Once the plants are cleared or burned, fertility reduces rapidly because the heavy rainfall washes the nutrients and soil away and the vegetation is not there to lock up the nutrients nor to protect the soil. So more forest is cleared to get the short-term fertility for crop growth.

What makes a species prone to extinction

Some species are more vulnerable to extinction than others. Even in the same ecosystem, some species survive habitat loss or degradation, others do not. These factors make a species more likely to be in danger of extinction:

Narrow geographical range

If a species only lives in one place and that place is damaged or destroyed, the habitat has gone. It may be possible to keep breeding populations of the species in zoos or reserves but that is not the real solution. There are so many examples of species lost or threatened by this – the Golden Lion Tamarin, birds on oceanic islands, fish in lakes that dry up, lemurs.

Small population size or declining numbers – low genetic diversity

A small population has a smaller genetic diversity and is less resilient to change – cannot adapt as well. As individual numbers fall, there is more inbreeding until populations are either so small that they are the 'living dead' or they become extinct. Large predators and extreme specialists are commonly in this category, eg snow leopard, tiger, Lonesome George (3.2).

Low population densities and large territories

If an individual of a species requires a large territory or range over which to hunt and only meets others of the species for breeding, then habitat fragmentation can restrict its territory. If there is not a large enough area left for each individual or if they are unable to find each other because there is a city/road/factory/farm splitting up the territory, they are less likely to survive as a species, eg the giant panda.

Few populations of the species

If there are only one or two populations left then that is their only chance of survival. It only takes that one population to be wiped out and the species has gone. e.g. lemurs

A large body

As we mentioned earlier in the book, the 10% rule (where only about 10% of the energy is passed on to the next trophic level and 90% is lost to the environment) means that large top predators are rare. Whether aquatic or terrestrial, they tend to have large ranges, low population densities and need a lot of food. They also compete with humans for food (eg wolves and tigers), may be a danger to humans and are hunted for sport.

Low reproductive potential

Reproducing slowly and infrequently means the population takes a long time to recover. Whales fall into this category. Many of the larger species of seabirds, for example gannets, albatrosses or some species of penguins, only produce one egg per pair per year, and do not breed until several years old.

▲ **Figure 3.3.6** Island species are prone to extinction eg on Sipadan Island, Sabah

Seasonal migrants

Species that migrate have it tough. Not only do they often have long migration routes (swallows between southern Africa and Europe, songbirds between Canada and central and south America) and a hazardous journey, they need the habitats at both end of the migration route. If one is destroyed, they get there to find no food or habitat. Barriers on their journey can prevent them completing it (salmon trying to swim upriver to spawn, African large herbivores following the rains).

Poor dispersers

Species that cannot move easily to new habitats are also in trouble. Plants can only rely on seed dispersal or vegetative growth to move. That takes a long time and climate change resulting in biome shift may mean the plant dies out before it can move. Non-flying animals, eg the flightless birds of New Zealand, are mostly extinct as they cannot escape introduced hunters nor fly to another island.

Specialized feeders or niche requirements

The giant panda (about 2,000 left) mostly eats bamboo shoots in forests of central China. The koala only eats eucalyptus leaves and lives in the coastal regions of southern and western Australia. Bog plants such as the sundew or pitcher plant can only survive in damp places.

Edible to humans and herding together

Overhunting or overharvesting can eradicate a species quickly, especially if that species lives in large groups – herds of bison in North America, shoals of fish, flocks of passenger pigeons. Then, with human technology, it is easier to exploit the species as, once located, many can be caught.

Also under threat from humans are tigers. Although they are not edible as such all their body parts are used in traditional Chinese medicine and so demand for them is very high.

Populations suffering from any of the above environmental stresses can be prone to extinction. Often though, more than one pressure operates on an organism, or that the nearer to extinction an organism becomes the greater the number of increasing pressures there are.

▲ **Figure 3.3.7** Giant panda

▲ **Figure 3.3.8** Tiger hunting   ▲ **Figure 3.3.9** Minke whale harvest

Island organisms

Island organisms can be particularly vulnerable to extinction. Dependent on the size of the island:

- Populations tend to be small.

- Islands have a high degree of endemic species.

- Genetic diversity tends to be low in small unique island populations.

- Islands tend to be vulnerable to the introduction of non-native predators to which they have no defense mechanism.

The **dodo** is a classic example of an island species that became extinct.

Sadly, many of these characteristics of extinction-prone species can be found in the same species. Large top carnivorous mammals (tigers, leopards, bears, wolves) have many of these characteristics. Species that live in large groups and are an attractive food source (bison, cod) have many too.

The **minimum viable population size** that is needed for a species to survive in the wild is a figure that scientists and conservationist consider. But there is no magic number. It depend on so many factors such as genetic diversity in the individuals left, rate of reproduction, mortality rate, growth rate, threats to habitat and so on.

For large carnivores, 500 individuals is sometimes thought of as the absolute minimum below which there is little hope for the species. Humans have hunted many big animal species to very endangered levels, eg:

- There may be 400 tigers in Sumatra and about 3,500 in the world today (and an estimate of 40,000 in India alone at the start of the 20th century).

- Some 700 blue whales are thought to live in the Antarctic oceans. There were, it is estimated, 250,000 about 60 years ago.

The IUCN Red List

The IUCN is the International Union for the Conservation of Nature and Natural Resources but is usually called the World Conservation Union. It is an international agency, was founded in 1948 (when the universal declaration of human rights was adopted by the UN general assembly) and brings together 83 states, 110 government agencies, more than 800 non-governmental organizations (NGOs), and some 10,000 scientists and experts from 181 countries in partnership.

The Union's mission is to influence, encourage and assist societies throughout the world to conserve the integrity and diversity of nature and to ensure that any use of natural resources is equitable and ecologically sustainable.

> **To think about**
>
> What are the consequences of extinction of tigers or blue whales? Do you think that humans should try to protect these and other iconic species?

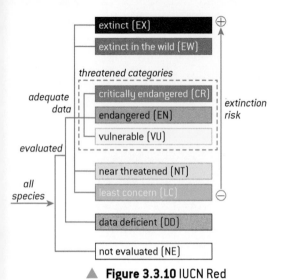

▲ **Figure 3.3.10** IUCN Red List categories

The IUCN Red List

The IUCN monitors the state of the world's species through the **Red List** of Threatened Species and supports the Millennium Ecosystem Assessment as well as educating the public, advising governments and assessing new World Heritage sites.

The Red List determines the conservation status of a species based on several criteria: population size, degree of specialization, distribution, reproductive potential and behaviour, geographic range and degree of fragmentation, quality of habitat, trophic level and the probability of extinction.

The **Red Lists** of threatened species are objective lists of species under varying levels of threat to their survival. The data is gathered scientifically on a global scale and species under threat are put into one of the Red List categories which assess the relative danger of extinction. There is also a Red List of extinct and extinct in the wild species. The lists are regularly updated and inform government policies on trade in endangered species and conservation measures. The IUCN Red List has 40,000 species on it and about 40% of these are listed as threatened.

The IUCN Red List criteria are:

EXTINCT (EX) – when there is no reasonable doubt that the last individual has died. A taxon* is presumed Extinct when exhaustive surveys in known and/or expected habitat have failed to record an individual.

EXTINCT IN THE WILD (EW) – when it is known only to survive in cultivation, in captivity or as a naturalized population (or populations) well outside the past range.

CRITICALLY ENDANGERED (CR) – when it is considered to be facing an extremely high risk of extinction in the wild.

ENDANGERED (EN) – when it is facing a very high risk of extinction in the wild.

VULNERABLE (VU) – when it is facing a high risk of extinction in the wild.

NEAR THREATENED (NT) – when it has been evaluated against the criteria but does not qualify for Critically Endangered, Endangered or Vulnerable now, but is close to qualifying for or is likely to qualify for a threatened category in the near future.

LEAST CONCERN (LC) – when it is widespread and abundant.

DATA DEFICIENT (DD) – when there is inadequate information to make a direct, or indirect, assessment of its risk of extinction based on its distribution and/or population status.

NOT EVALUATED (NE) – when it has not yet been evaluated against the criteria.

*A taxon is a taxonomic unit, eg family, genus, species.

Questions for all case studies:

1. What are the consequences of the extinction of this species?

2. In your judgement, does it matter if it becomes extinct?

3. How much effort should be made to preserve it and why?

Currently endangered species

1. Carnaby's cockatoo

▲ **Figure 3.3.11** Carnaby's cockatoo

Description

Carnaby's cockatoo is a large, black bird found only in the southwest of Western Australia and threatened by destruction of its habitat. We think about 50% of the population size in 1950 is now left.

It begins breeding when four years old and produces only two eggs each year. Fewer than one bird from each nest survives each year and the death rate during the first few years of life is extremely high. They live for 40–50 years.

Carnaby's cockatoos are difficult to breed in captivity and their population is declining. They are protected by law and are officially listed as **endangered** and likely to continue to decline in numbers.

Ecological role

These cockatoos have specialized habitat requirements. They breed in holes in mature (over 100 years old) salmon gum trees using the same breeding holes each year.

They feed in open heathland on seeds and insect larvae.

They migrate to populated coastal areas in late summer and autumn.

Pressures

*   The cockatoos' habitat has been lost to wheat farming during the last 100 years and more recently to gravel mines, firebreaks and agriculture. Regeneration of mature trees for nesting is slow. As the young trees begin to grow they are grazed by rabbits and sheep.

*   Nest hollows are cleared for firewood or in tidying up back yards.

*   Competition from invasive species for the nesting sites.

*   Poaching – prized specimens for bird collectors and illegal robbing of eggs and baby birds from nests is a lucrative activity. Enforcement of the law is difficult in remote areas.

2. Rafflesia

▲ **Figure 3.3.12** Rafflesia flower

Description

There are 15–19 species of this tropical parasitic plant. All of which are found in the jungles of Southeast Asia.

Ecological role

These plants are single sexed (either male or female) and pollination must be carried out when the plants are in bloom (flowering). Therefore a male and female in the same area must both be ready for pollination at same time. The seeds are dispersed by small squirrels and rodents and they must reach a 'host' vine.

Pressures

The rafflesia plants are very vulnerable because they need very specific conditions to survive and carry out their life cycle. They are vulnerable due to deforestation and logging which destroy their habitat. Humans damage them and fewer plants means less chance of breeding.

Method of restoring populations

In Sabah, Sumatra and Sarawak there are rafflesia sanctuaries.

There are many locating, protecting and monitoring programmes being set up. Education: there are 'SAVE RAFFLESIA' campaigns.

3. Tiger

▲ **Figure 3.3.13** A Siberian tiger in Philadelphia Zoo

Description

One of the big cats, found in eastern and southern Asia. Carnivore. Territorial, solitary and endemic to regions with the highest human populations on Earth.

Ecological role

Top carnivore, eats various deer and large herbivores.

Pressures

At the beginning of the 20th century, it was estimated that there were around 100,000 wild tigers. Today, the number is less than 8,000. The number of tiger sub-species has reduced from 8 to 5 over the last century. Malaysia has about 600, Indonesia 600, Bangladesh has 300, while Vietnam and Russian have about 200 each. Only India has a substantial number with more than 60 per cent of the world tiger population. The Javan, Balinese, and Caspian tigers have all become extinct during the 20th century; most researchers believe the Sumatran Tiger has a maximum of 50 years left before extinction. Tigers in India are thought to be

lost at a rate of 1 per day. Many researchers feel the wild tiger population is closer to 2,000 and they predict total extinction by the year 2025 due to the loss of its habitat.

Forests across all of Asia have been destroyed for timber or for the conversion to agriculture. Along with the direct death of tigers, this has also caused population fragmentation. This means that tigers in one area cannot mate with tigers in nearby areas, causing repeated breeding with the same group of animals. Over time, the gene pool is severely weakened and tigers are born with birth defects and mutations.

Habitat destruction has been caused primarily by the uncontrolled growth of the human population in LEDCs. A second major cause is poaching. A boned out tiger is estimated to be worth US$50,000–100,000. The sale of one tiger skeleton can yield profits equivalent to more than what some local people can earn in 10 years. Even though it is now illegal to kill a tiger, wild tigers are still being poached today because their bones, whiskers and other body parts can be sold on the black market for a lot of money. Tiger parts are commonly used in traditional Chinese medicine or sold as luxuries. A tiger skin sells for US$15,000 while only one kilogram of tiger bone is worth US$1,250. Worse and more shocking still, a bowl of tiger penis soup costs US$320 at some Taiwanese restaurants.

Method of restoring populations

Recently, conservation efforts have picked up. The World Wildlife Fund and many other organizations have begun innovative and aggressive approaches, some beginning as early as the 1970s. The WWF now work in nearly all 14 of the countries that tigers inhabit. Conservation sites such as Tiger Island have been set up. The tiger conservation plan includes strengthening international treaties, supporting surveys and monitoring tiger populations, pushing for the enforcement of laws controlling the illegal trade in tiger parts, working with the traditional Chinese medicine community to reduce the use of tiger bone and other parts by finding alternative products, supporting anti-poaching efforts where tigers are most in danger, and using state of the art ecology methods to target critical populations of tigers in certain areas. Tigers are also now covered under the strictest protection of the Convention on International

Trade in Endangered Species (CITES). However, the value and demand for their parts make such enforcement efforts very difficult. Without conservation efforts, the future of the tiger really does look bleak and most of the efforts going in today do not seem to be enough. Many feel the extinction of the tiger is inevitable.

Recovered species

1. Australian saltwater crocodile

▲ **Figure3.3.14** Australian saltwater crocodile

18 out of 23 species of crocodiles worldwide were once endangered but have since recovered sufficiently to be removed from the list. Many species are thriving.

Description

The Australian saltwater crocodile can grow up to 5m long. It is a bulky reptile with a broad snout. It lays up to 80 eggs each year which take up to 3 months to hatch. The crocodiles take 15 years to mature. The Australian saltwater crocodile was listed as a protected species in Australia in 1971, and is protected under CITES which banned trade in endangered animals.

Ecological role

The habitat of the saltwater crocodile is estuaries, swamps and rivers. Nests are built on river banks in a heap of leaves. The eggs are food for goannas, pythons, dingoes and other small animals. Older crocodiles eat young crocodiles, mud crabs, sea snakes, turtle eggs and catfish. Baby crocodiles eat tadpoles, crabs and fish. The Australian saltwater crocodile is a top predator.

Pressures

The saltwater crocodile was over exploited for skin (leather), meat and body parts through illegal hunting, poaching and smuggling. It was hunted for sport. Crocodiles were also deliberately killed because of attacks on humans.

Method of restoring populations

To restore the crocodile populations there was a sustainable use policy with limited culling of wild populations, ranching (collecting eggs and hatchlings from the wild and raising them in captivity) and closed-cycle farming (maintaining breeding adults in captivity and harvesting offspring at 4 years of age). The exploitation of farmed animals reduces the hunting of wild crocodiles. Visitors tour areas to see wild crocodiles, so they are now a valued species.

This policy was supported by the Species Survival Commission (SSC) of IUCN but was accused by others of treating crocodiles inhumanely.

2. Golden lion tamarin

▲ **Figure3.3.15** Golden lion tamarin

CASE STUDY

Description

Small monkey. Endemic to Atlantic coastal rainforests of Brazil. Amongst rarest animals in the world. About 1,000 in the wild and 500 in captivity. Some say there are only 400 in the wild. Life expectancy about 8 years.

Ecological role

Omnivores. Prey to large cats, birds of prey. Live territorially in family groups in the wild in tropical rainforest in the canopy.

Pressures

Only 2% of their native habitat is left. Poaching can get US$20,000 per skin. Predation is great in the wild and their food source is not dependable as well as habitat destruction.

Method of restoring populations

A captive breeding programme (breeding in zoos) for the last 40 years or more. Over 150 institutions are involved in this and exchange individuals to increase genetic diversity. Some are reintroduced to the wild but with only a 30% success rate as the habitat is threatened and predators (including humans) take many.

It seems unlikely that this species would be alive today if not for captive breeding. The long-term future is uncertain in the wild.

Extinct species

1. Thylacine (Tasmanian tiger)

▲ **Figure3.3.16** The last thylacine

Description

The thylacine was a marsupial, similar in appearance to a wolf, but with a rigid tail. During the last few hundred years it was found only on the island of Tasmania in Australia but prior to that it had existed on the mainland. The thylacine was a strong fighter with jaws that could open more widely than any other known mammal. Thylacines had a life expectancy of 12–14 years and gave birth to 2–4 young a year.

Ecological role

The thylacines' habitat was open forest and grassland but they became restricted to dense rainforest as the population declined. Thylacines lived in rocky outcrops and large, hollow logs. They were nocturnal. Their typical prey were small mammals and birds, but they also ate kangaroos, wallabies, wombats and echidnas, and sheep.

Pressures

Thylacines were outcompeted by dingoes on the mainland of Australia and became extinct there hundreds of years ago. In Tasmania they were hunted by farmers whose stock was the thylacine's prey. Private and government bounties were paid for scalps, leading to a peak kill in 1900. Bounties continued until 1910 (government) and 1914 (private). There was an intense killing spree by hunting, poisoning and trapping. Shooting parties were organized for tourists' entertainment. The severely depleted population was affected by disease and competition from settlers' dogs.

The last wild thylacine was killed in 1930 and the last captive animal died at the Hobart Zoo on September 7, 1936. The thylacine was legally protected in 1936. It was classified by IUCN as endangered in 1972 and is now listed as extinct.

Consequences of disappearance

The thylacine was a carnivorous marsupial from a unique marsupial family. In Tasmania, where there were no dingoes, the thylacine was a significant predator.

Introduced dogs have taken over the ecological role of the thylacine.

CASE STUDY

2. The dodo

▲ **Figure 3.3.17** Dodo

Description

Large flightless bird endemic to the island of Mauritius.

Ecological role

No major predators on Mauritius so the dodo had no need of flight. The dodo was a ground-nesting bird.

Pressures

In 1505 Portuguese sailors discovered Mauritius and used it as a restocking point on their voyages to get spices from Indonesia. Ate the dodo as a source of fresh meat. Later, island used as a penal colony (jail) and rats, pigs and monkeys were introduced. These ate dodo eggs and humans killed them for sport and food. Crab-eating macaque monkeys introduced by sailors also seem to have had a large impact, stealing dodo eggs. Conversion of forest to plantations also destroyed their habitat. Known to be extinct by 1681.

Consequences of disappearance

Island fauna impoverished by loss of the dodo. Became an icon due to its apparent stupidity (just standing there and being clubbed to death – as it had no fear of humans, never having had need to fear predators) and its untimely extinction. Very few skeletons remain 'Dead as a dodo' is now a common saying.

To do

1. What is **biodiversity**?

2. What is an endangered species? List several examples.

3. What are the **two** main reasons for extinction of species?

4. List four examples of **human activity** that threaten species.

5. Consider the case studies of organisms in the previous pages and the list of characteristics that make species more prone to extinction. Make a table to compare the characteristics shared by these species. What do they have in common?

3.4 Conservation of biodiversity

Significant ideas:

→ The impact of losing biodiversity drives conservation efforts.

→ The variety of arguments given for the conservation of biodiversity will depend on environmental value systems.

→ There are various approaches to the conservation of biodiversity, with associated strengths and limitations.

Applications and skills:

→ **Explain** the criteria used to design and manage protected areas.

→ **Evaluate** the success of a named protected area.

→ **Evaluate** different approaches to protecting biodiversity.

Knowledge and understanding:

→ Arguments about species and habitat preservation can be based on aesthetic, ecological, economic, ethical and social justifications.

→ **International, governmental and non-governmental organizations (NGOs)** are involved in conserving and restoring ecosystems and biodiversity, with varying levels of effectiveness due to their use of media, speed of response, diplomatic constraints, financial resources and political influence.

→ Recent **international conventions on biodiversity** work to create collaboration between nations for biodiversity conservation.

→ Conservation approaches include **habitat conservation**, **species-based conservation** and a mixed approach.

→ **Criteria** for consideration when **designing protected areas** include: size, shape, edge effects, corridors, and proximity to potential human influence.

→ Alternative approaches to the development of protected areas are species-based conservation strategies that include

- the Convention on International Trade in Endangered Species (CITES)
- captive breeding and reintroduction programmes and zoos
- selection of 'charismatic' species to help protect others in an area (flagship species)
- selection of **keystone species** to protect the integrity of the food web.

→ Community support, adequate funding and proper research influences the success of conservation efforts.

→ The location of a conservation area in a country is a significant factor in the success of the conservation effort. Surrounding land use for the conservation area and distance from urban centres are important factors for consideration in conservation area design.

The relatively new discipline of conservation biology has brought together experts from various disciplines in the last twenty years. All are concerned about the loss of biodiversity and want to act together to save species and communities from extinction. Their rationale is that:

- diversity of organisms and ecological complexity are good things;

- untimely extinction of species is a bad thing;

- evolutionary adaptation is good; and

- biological diversity has intrinsic value and we should try to conserve it.

Some acronyms to help you make sense of this sub-topic:

UN   United Nations

IUCN   International Union for the Conservation of Nature

UNEP   United Nations Environment Programme

CITES Convention on International Trade in Endangered Species

UNDP   United Nations Development Programme

WWF   Worldwide Fund for Nature

WRI   World Resources Institute

NGO   non-governmental organization

GO   governmental organization

MDG   millennium development goals

Why conserve biodiversity?

There are many reasons to conserve biodiversity – economic, ecological, social and aesthetic.

The value of biodiversity

Direct

Food sources

- We eat other species whether animal or plant. Even if we do only eat a small number of food crops, there are many varieties of these.

- We need to preserve old varieties in case we need them in the future.

- Pests and diseases can wipe out non-resistant strains.

- Breeders are only one step ahead of the diseases and require wild strains from which they may find resistant genes.

For example, wheat, rice and maize provide one half of the world's food. In the 1960s, wheat stripe rust disease wiped out a third of the yield in the US. It was the introduction of resistant genes from a wild strain of wheat in Turkey that saved the crops. Maize is particularly vulnerable to disease, as it is virtually the same genetically worldwide. A perennial maize was found in a few hectares of threatened farmland in Mexico. The plants contain genes that confer resistance to four of the seven major maize diseases and could give the potential of making maize a perennial crop that would not need sowing.

Natural products

- Many of the medicines, fertilizers and pesticides we use are derived from plants and animals.

- Guano is a fertilizer high in phosphate which is seabird droppings.

- Oil palms give us oil for anything from margarine to toiletries.

TOK

Before you read the section on why to conserve biodiversity, think about this yourself for a few moments and make a list of reasons why we should conserve and reasons why we should not. These questions may help you make your lists:

- Do humans need other species? In what ways?

- Do other species exist for human use? Do they only exist for human use?

- Do other species have a right to exist? Does a great ape have more rights than a mosquito?

- Are your reasons based on rational thought or emotion or both? Does this affect how valid they are?

To do

1. Read the reasons to preserve biodiversity below. List each under the headings: economic, ecological, social and aesthetic reasons. Do some fit in more than one column?

2. Under the same headings list reasons NOT to preserve biodiversity. Discuss your list with your fellow students.

- Rubber (latex) is from rubber trees, linen from flax, rope from hemp, cotton from cotton plants, silk from silkworms.

- Honey, beeswax, rattan, natural perfumes, timber are all from plants or animals.

Indirect
Environmental services
We are just beginning to give monetary value to environmental services (8.2). If a value can be placed on these processes, then we could quantify the natural capital of biodiversity.

- Soil aeration depends on worms.

- Fertilization and pollination of some food crops depend on insects.

- Ecosystem productivity gives our environment stability and recycles materials. Plants capture carbon and store it in their tissues. They release oxygen which all organisms need for respiration.

- Agriculture captures 40% of the productivity of the terrestrial ecosystem.

- Soil and water resources are protected by vegetation.

- Climate is regulated by the rainforests and vegetation cover.

- Waste is broken down and recycled by decomposers.

For example, the preservation of as many species as possible, and as natural or semi-natural habitat as possible, may render the environment more stable, and less likely to be affected by the spread of disease (plant, animal or human) or some other environmental catastrophe. Natural, multi-layered, varied forest was removed from tropical islands, and replaced with a single species – coconut plantations. This was followed by serious damage when the islands were affected by tropical storms.

Scientific and educational value
We investigate and research the diversity of plants and animals. The encyclopaedia of life www.eol.com intends to document all the 1.8 million species that are named on the Earth. We do this because we believe we should know.

Biological control agents
Some species of living things help us control invasive species without the use of chemicals, eg myxamatosis disease and rabbits in Australia.

Gene pools
Wild animals and plants are sources of genes for hybridization and genetic engineering.

Future potential for even more uses
With new discoveries to come, there will be many more practical reasons to appreciate biodiversity. They can be environmental monitors as an early warning system. Miners used to take canaries into the mines with them. If the birds died, they knew there was a toxic gas around and they should get out. Indicator species, eg lichens, can show air quality.

Human health

- The first antibiotics (such as penicillin) were obtained from fungi.

- A rare species of yew (*Taxus*) from the Pacific Northwest of the USA has recently been found to produce a chemical that may prove of value in the treatment of certain forms of cancer.

- The rosy periwinkle, from the Madagascan forest, is curing children with leukaemia.

Human rights

If biodiversity is protected, indigenous people can continue to live in their native lands. If we preserve the rainforests, indigenous tribes can continue to live in them and continue to make a livelihood.

Recreational

Many people take vacations in areas of outstanding natural beauty and national parks, go skiing, scuba diving and hiking. This brings extra finance to an area and provides employment.

Ecotourism

Biodiversity is often the subject of aesthetic interest. People rely on wild places and living things in them for spiritual fulfilment.

Ethical/intrinsic value

Each species has a right to exist – a bioright unrelated to human needs. Biodiversity should be preserved for its own sake and humans have a responsibility to act as stewards of the Earth.

Biorights self-perpetuation

Biologically diverse ecosystems help to preserve their component species, reducing the need for future conservation efforts on single species.

Conservation and preservation of biodiversity

Many people use these two terms interchangeably but they do have different meanings.

Conservation biologists do not necessarily want to exclude humans from reserves or from interacting with other organisms. They will even consider harvesting or hunting of species as long as it is sustainable. They recognize that it is very difficult to exclude humans from habitats and development is needed to help people out of poverty. But they want the development not to be at the expense of the environment. So a conservation biologist would look for ways to create income for local people from ecotourism or management of a reserve to allow education of the public by opening it up to access (an anthropocentric viewpoint).

Preservation biology has an ecocentric viewpoint which puts value on nature for its own intrinsic worth, not as a resource that humans can exploit. Deep green ecologists argue that, whatever the cost, species should be preserved regardless of their value or usefulness to humans. So the smallpox virus should not be destroyed according to preservation biologists even though it causes disease in humans. Preservation is often a more difficult option than conservation.

> **Key terms**
>
> **Conservation biology** is the sustainable use and management of natural resources.
>
> **Preservation biology** attempts to exclude human activity in areas where humans have not yet encroached.

To do

Look back in sub-topic 1.1 to find a pioneer conservationist and a pioneer preservationist. With which one do you agree?

Efforts to conserve and preserve species and habitats rely on citizens, conservation organizations and governments to act in various ways. Because conservation is mostly seen as for the 'public good', politicians act to change public policy to align with conservation aims. There will be tension between what is seen as good for the economy and human needs and good for the environment but the two can coincide and this is being recognized more and more. A report on climate change (The Stearns Review in 2006 for the UK government) proposed that 1% of GDP per annum should be invested in climate change mitigation activities in order to avoid a drop of 20% in GDP later on.

There is a danger in thinking that you, just one individual, can have little or no impact on what is a global issue. But do not think that. You can have an effect locally in a specific place as an individual and perhaps a larger effect as a member of a group. The group may act locally or globally or both and major changes have happened in thinking about and acting for the environment in the last few decades. 'Think globally, act locally' is a slogan you may have heard. It was first used in the early 1970s and is as true today as it was then. Your environmental conscience starts at home, whether in using less plastic packaging and fossil fuel or acting to help a local nature reserve. You may not be personally able to save a whale but you can act to save some form of biodiversity.

Sustainable development – meeting the needs of the present without negatively impacting the needs of future generations and biodiversity (WRI/IUCN/UNEP 1992) – can and has been applied to conservation projects. Providing revenue to local people within a national park is an example of this. There is a danger of 'greenwash' though. Greenwash is when organizations give the impression that they have changed their practices to have less impact on the environment but, in fact, they have changed nothing. When mining has extracted all the minerals it economically can, areas can be restored after mine closure. But just bulldozing spoil heaps and adding grass seed is not enough.

How conservation organizations work – IGOs, GOs and NGOs

Organizations that work to conserve or preserve biodiversity and the environment may be local or global or both. There are national branches and regional and local offices of the large international organizations. They also fall into three categories depending on their constitution and funding.

Intergovernmental organizations (IGOs) –

- composed of and answering to a group of member states (countries)
- also called international organizations
- eg the UN, IPCC.

Governmental organizations (GOs) –

- part of and funded by a national government
- highly bureaucratic
- research, regulation, monitoring and control activities
- eg Environmental Protection Agency of the USA (EPA), Environmental Protection Department of China.

Non-governmental organizations (NGOs) –

- not part of a government
- not for profit
- may be international or local and funded by altruists and subscriptions
- some run by volunteers
- very diverse
- eg Friends of the Earth, Greenpeace.

To do

Which is which?

Consider these logos of some international agencies and organizations that do environmental work. Work out which organization it is from the logo and then do research to find out:

1. Is it an IGO, GO or NGO?
2. What are its main aims?
3. How does it accomplish these aims?

United Nations Educational, Scientific and Cultural Organization

GREENPEACE

UNEP WWF

▲ **Figure 3.4.1** some conservation organization logos

The organizations you have investigated operate in different ways. In their dealings with people, some work at government level whilst others work with local people 'in the field'. To bring about change, some work conservatively by careful negotiation, whilst others are more radical and draw attention to issues using the media.

4. Place the organizations shown in figure 3.4.1 on the following axes and add others that you know of locally and nationally.

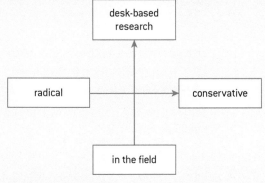

▲ **Figure 3.4.2**

Comparison of IGOs, GOs and NGOs

	IGO and GO	NGO
Use of media	• Media liaison officers prepare and read written statements • Control /works with media (at least one TV channel propagates the official policy in even the most democratic regimes) so communicates its decision/attitudes/policies more effectively to the public	• Use footage of activities to gain media attention • Mobilize public protest to put pressure on governments • Gain media coverage through variety of protests (eg protest on frontlines /sabotage); sometimes access to mass media is hindered (especially in non-democratic regimes)
	• They both provide environmental information to the public of global trends, publishing official scientific documents and technical reports gathering data from plethora of sources	
Speed of response	• Considered, slow – they are bureaucratic and can take time to act as they depend on consensus often between differing views • Directed by governments, so sometimes may be against public opinion	• Can be rapid – usually its members already have reached consensus (or else they wouldn't have joined in the first place)
Political diplomatic constraints	• Considerable – often hindered by political disagreement especially if international • Decisions can be politically driven rather than by best conservation strategy	• Unaffected by political constraints – can even include illegal activity • Idealistic/driven by best conservation strategy/often hold the high moral ground over other organizations/may be extreme in actions or views
Enforceability	• International agreements and national or regional laws can lead to prosecution	• No legal power – use of persuasion and public opinion to pressure governments
Public image	• Organized as businesses with concrete allocation of duties • Cultivate a sober/upright/measured image based on scientific/businesslike approach	• Can be confrontational/radical approach to an environmental issue like biodiversity
	• Both lead and encourage partnership between nations and organizations to conserve and restore ecosystems and biodiversity	
Legislation	• Enforce their decisions via legislation (may even be authoritarian sometimes)	• Serve as watchdogs (suing government agencies/businesses who violate environmental law)
	• Both seek to ensure that decisions are applied	
Agenda	• Provide guidelines and implement international treaties	• Use public pressure to influence national governments/lobby governments over policy/legislation • Buy and manage land to protect habitats, wildlife etc.
	• Both may collaborate in global, transnational scientific research projects, both may provide forum for discussion	

	IGO and GO	NGO
Funding	• Fund environmental projects by monies coming from national budget	• Manage publicly owned lands • Fund environmental projects by monies coming from private donations
Extent of influence geographically	• Global or national in extent	• Focus more on local and/or national information, aiming at education, producing learning materials and opportunities for schools and public
Monitoring activities	• IGOs monitor regional and global trends • NGOs also monitor/research species and conservation areas at a variety of levels	

The good news – international conventions on biodiversity

No nation today can afford to ignore the rest of the world. Authoritarian nation states are on the decline and the rise of the information age and freedom of information via the worldwide web, human rights and democracy mean that the individual has more power than ever before.

• International cooperation was formalized with the United Nations and its huge number of specialist agencies which have a major impact.

• UNEP set up the intergovernmental panel on climate change (IPCC).

• UNEP drove the Montreal Protocol for phasing out the production of CFCs. Countries are getting together with economic groupings – ASEAN, the EU, the African Union, OPEC to name but a few and some of these organizations are working for sustainable development and environmental protection.

• They are complemented by the NGOs, often started as grassroots movements (locally by individuals) but are now having global impacts, eg Greenpeace, WWF.

Institutional inertia (an inability to get going) has been a block to change but the power of people wanting change, and voting for politicians that say they will deliver it, may be able to stop the swing of the pendulum and slow down or even reverse degradation of our planet by human activities.

Timeline of key dates for biodiversity conservation

1961   World Wildlife Fund set up by IUCN and Julian Huxley

1966   **Species Survival Commission** published **Red Data Lists**

1973   Convention on the International Trade of Endangered Species of Fauna and Flora (**CITES**).

1980   **World Conservation Strategy**

1980   Brandt Commission published – beginnings of sustainable development

1982   UN World Charter for Nature

1987   Brundtland Commission on Our Common Future – first defined sustainable development

1991   **Caring for the Earth: A Strategy for Sustainable Living**

1992   **Earth Summit Rio de Janeiro** produced **Agenda 21**, **Convention on Biological Diversity** (**CBD**) and the Rio Declaration leading to **BAPs** (biodiversity action plans)

        Earth Council **Global Biodiversity Strategy**

2000   **UN Millennium Summit and the MDGs (millennium development goals)**

2002   **World Summit on Sustainable Development** held in Johannesburg

2005   **World Summit**, New York

2010   International Year of Biodiversity

2012   **Rio+20** – UN conference on sustainable development (UNSD)

The World Conservation Strategy (WCS) and subsequent milestones

The WCS was published by IUCN, UNEP and WWF in 1980. It was a ground-breaking achievement which presented a united, integrated approach to conservation for the first time. It called for international, national and regional efforts to balance development with conservation of the world's living resources. Its aims were to:

- maintain essential ecological processes and life support systems;
- preserve genetic diversity;
- ensure the sustainable utilization of species and ecosystems.

Many countries adopted the WCS and developed their own strategies for addressing national issues.

In 1982, the UN World Charter for Nature was adopted which had as its principles:

- Nature shall be respected and its essential processes shall not be impaired.
- The genetic viability on the Earth shall not be compromised; the population levels of all life forms, wild and domesticated, must be at least sufficient for their survival, and to this end necessary habitats shall be safeguarded.
- All areas of the Earth, both land and sea, shall be subject to these principles of conservation; special protection shall be given to unique areas, to representative samples of all the different types of ecosystems and to the habitats of rare or endangered species.
- Ecosystems and organisms, as well as the land, marine and atmospheric resources that are utilized by man, shall be managed to achieve and maintain optimum sustainable productivity, but not in such a way as to endanger the integrity of those other ecosystems or species with which they coexist.

- Nature shall be secured against degradation caused by warfare or other hostile activities.

In 1991, Caring for the Earth: A Strategy for Sustainable Living was updated and launched in 65 countries.

It stated the benefits of **sustainable use of natural resources**, and the benefits of **sharing resources** more equally amongst the world population.

1992: Rio Earth Summit

World leaders agreed on a sustainable development agenda – Agenda 21– and the Earth Council **Global Biodiversity Strategy**. The aim of the strategy was to help countries integrate biodiversity into their national planning. The three main objectives were:

- the conservation of biological variation;
- the sustainable use of its components; and
- the equitable sharing of the benefits arising out of the utilization of genetic resources.

The implementation of **Agenda 21** was intended to involve action at international, national, regional and local levels. Some national and state governments have legislated or advised that local authorities take steps to implement the plan locally, as recommended in Chapter 28 of the document. Such programmes are known as **Local Agenda 21** or LA21.

2000 UN Millennium Summit: Millennium Development Goals (MDGs)

Largest-ever gathering of world leaders who agreed to a set of time bound and measurable goals for combating poverty, hunger, disease, illiteracy, environmental degradation and discrimination against women. The aim was to achieve the goal by 2015.

2002: The Johannesburg Summit on sustainable development

Supposed to consolidate the Rio Earth Summit but little action came out of its deliberations.

2005 World Summit, New York

Outlined a series of global priorities for action, and recommended that each country prepare its own national strategy for the **conservation of natural resources** for long-term human welfare. National conservation strategies have been written as a result of the international meetings. Integrating conservation with development is now a high priority – this was not the case few decades ago.

2010 Biodiversity target

Adopted by EU heads of state in 2001 to halt biodiversity decline by 2010; was shown to have largely failed.

2013, Rio+20

How to build a green economy and how to improve cooperation for sustainable development resulted in a non-binding paper 'The future we want'.

To do

Focus on the WWF and Greenpeace

Go to www.panda.org and www.greenpeace.org/international/

Find out

1. What WWF stands for.

2. Why do you think WWF has a panda as their logo?

3. For both WWF and Greenpeace, find out:

   a. Who they are.

   b. What they do.

   c. Where they work.

   d. How they work.

4. Find the nearest WWF project to where you live.

5. What are Greenpeace's main **biodiversity** campaigns?

6. Choose one and identify one aspect (eg whales) from question 5. Describe an action Greenpeace has taken to draw international attention to the issue.

TOK

There are various approaches to the conservation of biodiversity. How can we know when we should be disposed to act on what we know?

Approaches to conservation

There are three basic approaches to conservation: species-based, habitat based and a mixture of both. Conservation is more successful when it involves research, adequate funding and the support of the local community.

Species-based conservation

This type of conservation focuses on conserving the species but does not look at conserving the habitat in which it lives. Here are five examples of species-based conservation approaches.

1. CITES (the Convention on International Trade in Endangered Species of Wild Fauna and Flora)

Many species are becoming endangered because of international trade. CITES is an international agreement between governments to address this problem. Governments sign up to CITES voluntarily and have to write their own national laws to support its aims. Its aim is to ensure that international trade in specimens of wild animals and plants does not threaten their survival. Covered by CITES are threatened species from elephants to turtles, orchids to mahogany.

CITES has dramatically reduced the trade in endangered species of both live animal imports (eg tortoises) or animal parts (eg elephant tusk, rhino horn).

The species are grouped in the CITES Appendices according to how threatened they are by international trade.

Appendix I – species cannot be traded internationally as they are threatened with extinction.

Appendix II – species can be traded internationally but within strict regulations ensuring its sustainability.

Appendix III – a species included at the request of a country which then needs the cooperation of other countries to help prevent illegal exploitation.

The appendices include some whole groups, such as primates, cetaceans (whales, dolphins and porpoises), sea turtles, parrots, corals, cacti and orchids and many separate species or populations of species.

It is a great example of what can be done voluntarily to conserve species. About 5,000 animal species and 28,000 plant species are on its lists. Since 1975, it has been one of the most effective international wildlife conservation agreements in the world.

2. Captive breeding and zoos

Zoos, aquaria and other captive breeding facilities keep many examples of species, but of course they cannot keep every species. Even if they can breed the species they have in captivity or keep their DNA, it is not a straightforward process to reintroduce a species into the wild if they are wiped out in their native habitat.

Programmes to reintroduce populations or establish new ones are expensive and difficult. A few are successful, eg for the Californian condor, Przewalski's horse in Mongolia and black-footed ferret in Wyoming, USA. The best programmes are highly worthwhile and the best hope of saving a species that is extinct in the wild or in severe decline. Success is more likely if there are incentives to local people to support a reintroduction programme. Released animals may need extra feeding or care or even be recaptured if their food supply runs out or they are in danger of dying. The reintroduction of the golden lion tamarin to the remaining Atlantic Coast Forest of Brazil has been a rallying point for local people and reintroduction of the Arabian oryx in Oman has given employment to the local Bedouin tribes.

Many programmes are not successful and are seen as an unnecessary waste of money, poorly run or unethical. Often the less successful programmes are those where the animal has become used to humans. Orphaned orang utans have to be taught how to climb and socialize with each other. It takes patience and perseverance to get them to live in the wild after years in captivity and many do not make the jump successfully.

Rare plants can be reintroduced after raising seedlings in controlled conditions. But these may be dug up by collectors once planted out, outcompeted by other plants or eaten by herbivores.

Sometimes it is impossible to reintroduce a species to its native habitat because the habitat has gone. Then the only way to keep that species may be in captivity – animals in zoos, game farms and aquaria, plants in arboretums, botanic gardens and seed banks.

Zoos have had a bad press and sometimes this is justified when animals are kept in close confinement in small cages or treated with cruelty. Most zoos are open to the public and so need to keep the megafauna (giraffe, elephant, hippo, rhino, polar bear) as these are what we pay to see, not the beetles or worms. The best zoos look after the animals very well, many have educational centres and breeding programmes and exchange animals with each other to widen the gene pools for reproductive success. Aquaria may keep the large cetaceans – killer whale and porpoises, and many fish and invertebrate species.

Frozen zoos are stores of animal tissue (eg sperm, eggs, embryos, skin) that is cryopreserved (frozen at very low temperature for long-term storage, often in liquid nitrogen at –196 °C). In theory, animals which have become extinct in the wild could be raised from this material and then breed in zoos or be reintroduced to nature reserves. In practice, there are many issues involved in this process. For one example, see the Frozen Ark Project http://www.frozenark.org/.

3. Botanical gardens and seed banks

The largest botanical garden in the world is the Royal Botanical Gardens in Kew, London. There grow 25,000 plant species (10% of the world's total) and about 10% of these are threatened in the wild. Around the world are some 1,500 botanical gardens and they not only grow plants but identify and classify them and carry out research, education and conservation.

Seed banks are where seeds are stored, frozen and dry, for many years. They are gene banks for the world's plant species and an insurance policy for the future. If the plant species is lost in the wild, and up to 100,000 plant species are in danger of extinction, the seeds may be preserved for future use. The seed bank is a way of preserving the genetic variation of a species. Some crop plants which are widely grown are of just a few varieties and the seed bank should contain many more varieties of the species.

Figure 3.4.3 Seeds for storage in a seed bank, Kew Gardens

BEHIND THE EAGLE
STAND THE FORESTS

Figure 3.4.4

National Tiger Conservation Authority

राष्ट्रीय व्याघ्र संरक्षण प्राधिकरण

Figure 3.4.5

There are seed banks around the world, holding national and international collections. A major concern for some is that of who owns the seeds in the seed bank. As they need high levels of technology to maintain them, they tend to be in MEDCs yet may contain seeds from many countries. A seed bank in Svalbard, Norway is funded by the Global Crop Diversity Trust as well as the Norwegian government and has been built within the permafrost so needs no power to keep the seeds frozen. This is the ultimate safety net against loss of other seed banks through civil strife or war.

4. Flagship species

These species are charismatic, instantly recognized, popular and can capture our imagination. Look back at the WWF logo in figure 3.4.1. Most of the flagship species are large and furry but they may not have a significant role in the ecosystem. If they disappeared, the rest of the community would be little affected.

What they do have is instant appeal and are used to ask for funds from the public; these funds are then used to protect the habitat which will include the other species that may be under more threat.

Disadvantages of naming flagship species:

* they take priority over others
* if they were to become extinct, the message is that we have failed
* they may be in conflict with local peoples, eg man-eating tigers.

Sometimes these are called **umbrella species** – one that conservationists use to gain support to conserve that species, and in return it greatly helps the other species in the same habitat – those under the umbrella. A species can be both keystone (see below) and umbrella, eg the lemurs of Madagascar.

5. Keystone species

A keystone species is one that plays a critical role in maintaining the structure of the ecosystem in which they live.

All species are not equal in that some have a bigger effect on their environment than others, regardless of their abundance or biomass. They act like the keystone in an arch, holding the arch together. This means that their disappearance from an ecosystem can have an impact far greater than and not proportional to their numbers or biomass.

Loss of the small population of the keystone species could destroy the ecosystem or imbalance it greatly and far more than the loss of other species. The difficulty for researchers is in identifying the keystone species.

The characteristics they tend to have are as a predator or engineer in the ecosystem. A small predator can keep a herbivore population in check, without which the herbivores would increase and eat all the producers so causing the loss of all food in the ecosystem.

Examples:

* Sea otter eating sea urchins in kelp forests. If there are no sea otters, the urchins need only eat the holdfast (anchor) of the kelp and it floats away.

To do

While we can watch stunning wildlife documentaries on DVD, they are not the same as being there and watching a real elephant feed. If you believe that animals and plants provide you with a service, then zoos, aquaria and botanical gardens are there for humans to enjoy.

Is keeping animals just so humans can go and look at them a pointless exercise?

Or does it have value in itself?

Do we have the right to capture and cage other species even if we treat them well?

If it is a choice between allowing a species to become extinct in the wild or keeping the last few individuals in a zoo, which is right?

- Beavers are engineers, making dams which turn a stream into a swampy area. Swamp-loving species then move in. Without the beaver and its dams, the habitat changes.

- Elephants in the African savanna are also engineers, removing trees and then grasses can grow.

Habitat conservation

Designing protected areas

Where a conservation area is within a country is a significant factor in the success of the conservation effort. Surrounding land use for the conservation area and distance from urban centres are important factors for consideration in conservation area design.

Many protected areas or nature reserves were set up in the past on land that no-one else wanted. It may have been on poor agricultural land, land far away from high human population density or land that was degraded in some way. The haphazard nature of this meant that early reserves may not have been large enough or inappropriate to the needs of the species they were aiming to protect. UNESCO's **Man and the Biosphere programme** (MAB), started in 1970, created a world network of international reserves, now with 480 reserves in over 100 countries. The MDGs and sustainability and conservation are their aims.

Questions that conservationists now ask themselves when planning a protected area are:

- How large should it be to protect the species? Are there species that need protection in the middle of a large reserve?

- Is it better to have one larger or many smaller reserves? What about the edge effects?

- How many individuals of an endangered species must be protected?

- What is the best shape?

- If there are several reserves, how close should they be to each other?

- Should they be joined by corridors or separate?

To do

Choose an animal species that is **threatened**. Browse the Red Lists or WWF if you cannot make up your mind. Find the following information:

- Download image(s) of your chosen species.

- What is the Red List category of your species? (It must be **CR, EN** or **VU**.)

- What is the global distribution of your species? Find a map if possible.

- Estimated current population size?

- In which CITES Appendix is your species?

- Threats facing your species (ecological, social and economic pressures on the species).

- Suggested conservation strategies to remove threat of extinction.

- Pay particular attention to how local people and government actions are involved in helping your species to recover.

Present your assignment to your class and hand out an information sheet.

The large or small debate is known as the **SLOSS debate** (single large or several small).

Single large (SL) OR	Several small (SS)
Contains sufficient numbers of a large wide-ranging species – top carnivores	Provide a greater range of habitats
Minimizes edge effects	More populations of a rare species
Provides more habitats for more species	Danger of a natural or human-made disaster (fire, flood, disease) wiping out the reserve and its inhabitants is reduced as some reserves may escape the damage
Both used for education and so further long-term goals of conservation through education	

The consensus is that large or small will depend on the size and requirements of the species that the reserve is there to protect. Sometimes only small is possible and better than nothing. But, whatever size, edge effects should be minimized so that the circumference to area ratio is low.

Edge effects occur at **ecotones** (where two habitats meet and there is a change near the boundary). There are more species present in ecotones as there are the ones from each habitat and some opportunists so there is increased predation and competition. There is a change in abiotic factors as well, eg more wind or precipitation. Long thin reserves have a large edge effect, a circular one has the least. But, in practice, the shape is determined by what is available and parks are irregular in area. Dividing up the area with fences, pylons, roads, railways, farming should be avoided if possible as this fragments the habitat. But sometimes, it is easier for governments to put road and rail links across national parks as there is less opposition than from privately owned land.

Corridors which are strips of protected land may link reserves. These allow individuals to move from reserve to reserve and so increase the size of the gene pool or allow seasonal migration. This worked well in Costa Rica where two parks were linked by a 7,700 ha corridor several kilometres wide and at least 35 species of bird use it to fly from park to park. But corridors have disadvantages as a disease in one reserve may be spread to the other and it may be easier for poachers and hunters to kill the animals in the corridor which is harder to protect than the reserves. Exotic or invasive species may also get into a reserve via the corridors.

The MAB reserves have, if possible, buffer zones. This is a zone around the core reserve which is transitional. Some farming, extraction of natural resources, eg selective logging, and experimental research can go on in the buffer zone.

The core reserve is undisturbed and species that cannot tolerate disturbance should be safer there.

Practical Work

* Investigate a specific example of a protected area and evaluate it.

* Research the international conventions on conservation and biodiversity that have been adopted over the past decades. Evaluate their successfulness.

* Evaluate the success of a named protected area.

* Investigate and evaluate the criteria used to design and manage two named protected areas.

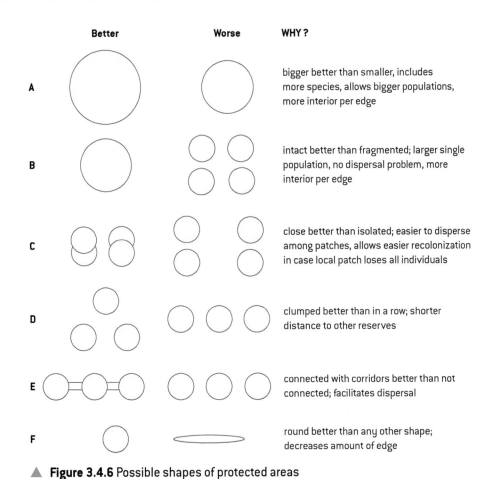

	Better	Worse	WHY ?
A			bigger better than smaller, includes more species, allows bigger populations, more interior per edge
B			intact better than fragmented; larger single population, no dispersal problem, more interior per edge
C			close better than isolated; easier to disperse among patches, allows easier recolonization in case local patch loses all individuals
D			clumped better than in a row; shorter distance to other reserves
E			connected with corridors better than not connected; facilitates dispersal
F			round better than any other shape; decreases amount of edge

▲ **Figure 3.4.6** Possible shapes of protected areas

To do

1. What is meant by the word **conservation**?

2. Here are ways to conserve species on the edge of extinction. Explain how this can be done with named examples. *You may need to look up some on the web.*

   a) Legislation and conserving habitats.

   b) Zoos and captive breeding.

   c) Flagship species.

   d) Nature reserves.

REVIEW

Discuss in what ways might our approaches to conservation alter biodiversity in the future.

To what extent can conservation restore habitats and species?

Discuss in what ways your EVS influences your attitude towards conserving other species.

BIG QUESTIONS

Biodiversity and conservation

To what extent is loss of biodiversity sustainable?

Comment on the factors that may make your own EVS differ from that of others with respect to conservation.

Reflective questions

→ The term "biodiversity" has replaced the term "Nature" in much writing on conservation issues. Does this represent a paradigm shift?

→ Biodiversity index is not a measure in the true sense of a word, but merely a number (index) as it involves a subjective judgement on the combination of two measures; proportion and richness. Are there examples in other areas of the subjective use of numbers?

→ The theory of evolution by natural selection tells us that change in populations is achieved through the process of natural selection. Is there a difference between a convincing theory and a correct one?

→ There may be long term consequences when biodiversity is lost. Should people be held morally responsible for the long term consequences of their actions?

→ There are various approaches to the conservation of biodiversity. How can we know when we should be disposed to act on what we know?

→ Within the human population distinct characteristics have evolved within different populations through natural selection and exposure to the environmental conditions that were unique to the regions of those populations. How has globalization altered some of the environmental factors that were formerly unique to different human populations?

→ Human impact has increased the rate at which some mass extinctions have occurred on a global scale. In which regions are the highest extinction rates and why?

→ Why is the science of taxonomy important in understanding species extinction?

→ Why is international scientific collaboration important in the conservation of biodiversity?

→ Why does conservation need to involve the local community to create meaningful and sustainable change?

→ Why are NGOs important in global conservation agreements, assessing global status of species' numbers and influencing governments?

Quick review

1. The table below gives the approximate number of bird species found at different altitudes in tropical South America.

Altitude (m)	Number of species
0–500	2000
500–1000	1950
1000–1500	1550
1500–2000	1100
2000–3000	950
3000–4000	500
4000–5000	200

Data from a diagram in Gaston K and Spicer J, *Biodiversity: an Introduction*, Blackwell Science, 1998

   a) Describe and explain the relationship between altitude and number of species shown in the table. [3]

   b) Define *habitat diversity* and *species diversity*. [2]

   c) Outline three characteristics that an area should have if it is to be designated a nature reserve or similar protected area. [3]

   d) An area of forest has been made a nature reserve. It is surrounded by farmland with several towns. Describe some of the changes that might occur in the area following its protection in this way. [2]

   e) Briefly describe a named protected area or nature reserve that you have studied and explain how it has been managed to protect its biodiversity. [3]

2. a) Explain the term *biodiversity*. [2]

   b) List four arguments for the preservation of biodiversity. [2]

   c) Describe the processes that may lead to the formation of new species. [3]

   d) Discuss the relative advantages and disadvantages of a species-based conservation strategy compared to the use of reserves. [4]

   e) The diagram below shows the layout of various conservation reserves. The reserves represent "islands" containing protected ecosystems surrounded by unprotected areas affected by human activities.

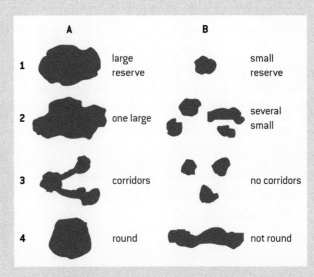

For each pair (1) to (4), explain why the areas represented in column A might be considered better for conservation than the areas in column B. [4]

3. Outline three reasons for the relative value of tropical rainforests in contributing to global biodiversity. [3]

4. The table below gives the number of flowering plant species for several tropical regions in the Americas, together with the area of each of the regions in km².

Region	Surface area in km²	Estimated total number of species
Amazon Basin	7,050,000	30,000
Northern Andes	383,000	40,000
Atlantic coastal forests of Brazil	1,000,000	10,000
Central America including Mexico	2,500,000	19,000

Data from: Andrew Henderson and Steven Churchill, 'Neotropical plant diversity', *Nature* (1991), Vol. 351, pp. 21–22. © Nature

   a) Which region has the greatest number of species per unit area? [1]

   b) Which region has the lowest number of species per unit area? [1]

   c) Explain the range of biodiversity shown in the data above. [2]

4 WATER, FOOD PRODUCTION SYSTEMS AND SOCIETY

4.1 Introduction to water systems

Significant ideas:
→ The **hydrological cycle** is a system of water flows and storages that may be disrupted by human activity.

 ### Applications and skills:
→ **Discuss** human impact on the hydrological cycle.
→ **Construct** and **analyse** a hydrological cycle diagram.

 ### Knowledge and understanding:
→ **Solar radiation** drives the hydrological cycle.
→ Fresh water makes up only a small fraction (approximately 2.6% by volume) of the Earth's water storages.
→ **Storages** in the hydrological cycle include organisms and various water bodies, including oceans, groundwater (aquifers), lakes, soil, rivers, atmosphere and glaciers and ice caps.
→ **Flows** in the hydrological cycle include evapotranspiration, sublimation, evaporation, condensation, advection (wind-blown movement), precipitation, melting, freezing, flooding, surface run-off, infiltration, percolation and stream-flow or currents.
→ **Human activities** such as agriculture, deforestation and urbanization have a significant impact on surface run-off and infiltration.
→ **Ocean circulation systems** are driven by differences in temperature and salinity. The resulting difference in water density drives the ocean conveyor belt, which distributes heat around the world, and thus affects climate.

The Earth's water budget

The Earth is not referred to as the 'Blue Planet' without due reason. From space the presence of water on Earth is very obvious. About 70% of the Earth's surface is covered by water.

- About 2.6% of all water is fresh water and

- 97% of the water on the planet is ocean water (salt water).

Fresh water is in quite short supply.

- About 68.7 % of this water is in the polar ice caps and glaciers,

- 30.1% in ground water.

- Water on the surface of the Earth in lakes, rivers and swamps is only about 0.3% of the total.

You may think the atmosphere holds a lot of water but:

- Only 0.001% of the total Earth's water volume is as water vapour in the atmosphere.

- According to the US Geological Survey, if all of the water in the atmosphere rained down at once, it would only cover the ground to a depth of 2.5 cm.

<aside>
Key term

The **water budget** is a quantitative estimate of the amounts of water in storages and flows of the water cycle.
</aside>

<aside>
To do

Using the data given, construct a table to list fresh water storages on Earth in decreasing order of size.
</aside>

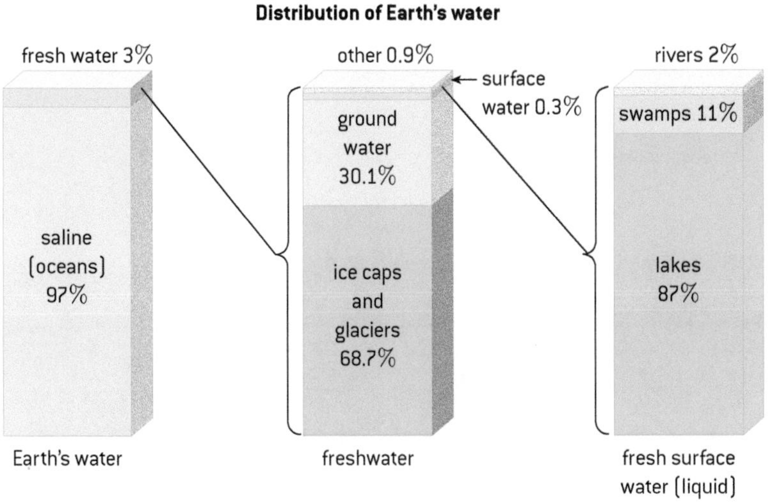

Distribution of Earth's water

▲ **Figure 4.1.1** Distribution of Earth's water

The turnover times (time it takes for a molecule of water to enter and leave that part of the system) are very variable.

- In the oceans, it takes 37,000 years

- Icecaps 16,000 years

- Groundwater 300 years

- Rivers 12–20 days

- Atmosphere only 9 days.

So water can be considered either a renewable resource or a non-renewable resource depending on where it is stored. See Figure 4.1.2.

Renewable	Middle ground	Non-renewable
Atmosphere	Groundwater aquifers	Oceans
Rivers	An aquifer is a layer of porous rock, sand or gravel underground that holds water. In aquifers, it takes longer than a human lifetime to replenish the water extracted	Icecaps

Water can easily slide from being renewable to being non-renewable if poorly managed

▲ **Figure 4.1.2** Status of water storages

The water (hydrological) cycle

Energy from **solar radiation** and the force of gravity drive the water cycle. The water cycle drives the world's weather systems.

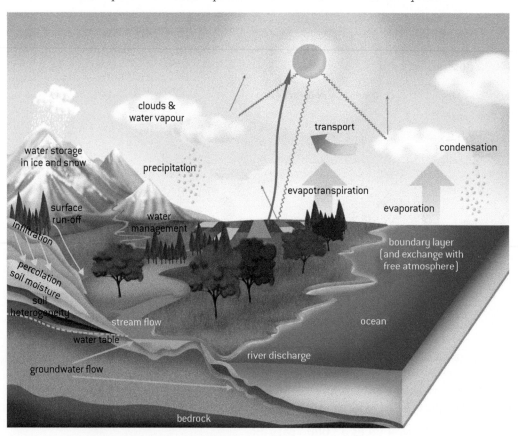

▲ **Figure 4.1.3** The water cycle

The water cycle consists of storages of water and the flows of water between the various storages. These flows may be transfers or transformations.

Transfers, when it stays in the same state, are:

- Advection (wind-blown movement)
- Flooding
- Surface run-off

- Infiltration and percolation (when water runs into and through soil or rocks)
- Stream flow and current.

Transformations, when it changes state to or from liquid water, are:

- Evapotranspiration – liquid to water vapour
- Condensation – water vapour to liquid
- Freezing – into solid snow and ice.

The storages include the:

- oceans,
- soil,
- groundwater (aquifers),
- lakes,
- rivers and streams,
- atmosphere, and
- glaciers and ice caps.

To do

1. Draw a systems diagram of the water budget and cycle showing the storages and flows given in the table. Make your storage boxes and width of flow arrows correspond to the proportions of these volumes. Label all storages and flows.

Storages	Water volume ($km^3 \times 10^3$)
Snow and ice	27,000
Ground water and aquifers	9,000
Lakes and rivers	250
Oceans	1,350,000
Atmosphere	13
Soil	35,000
Flows	
Precipitation over oceans	385
Precipitation over land	110
Ice melt	2
Surface run-off	40
Evapotranspiration from land	70
Evaporation from sea	425

2. There are six storages of water shown in the diagram. List them in order of decreasing size (largest size first) and calculate the percentage of the total hydrosphere stored in each.

3. Which of these storages can humans use and what percentage of all water is this?

TOK

The hydrological cycle is represented as a systems model. To what extent can systems diagrams effectively model reality, given that they are only based on limited observable features?

Human impact on the water cycle

Humans change the landscape and interrupt the movement of water by:

1. Withdrawals – for domestic use, irrigation in agriculture and industry.

2. Discharges – by adding pollutants to water, eg chemicals from agriculture, fertilizers, sewage.

3. Changing the speed at which water can flow and where it flows:

   a. In cities by building roads and channelling rivers underground or in concreted areas.

   b. Canalizing: straightening large sections of rivers in concrete channels to facilitate more rapid flow through sensitive areas.

   c. With dams, barrages and dykes, making reservoirs.

4. Diverting rivers or sections of rivers:

   a. Many are diverted away from important areas to avoid flood damage.

   b. Some are diverted towards dams to improve storage.

Examples of major changes caused by humans:

- Aral Sea – intense irrigation has almost stopped river flow into the sea and lowered the sea's level (it has shrunk in area by 90% in the last 50 years).

- Ganges basin – deforestation increases flooding as precipitation is not absorbed by vegetation.

- Run-off from urbanized areas causing local flooding.

Urbanization and flash floods

More and more humans live in cities and there are more of us than ever before. Flash floods occur when rainfall or snowmelt cannot infiltrate the soil and runs off on the surface. This could be due to land being hard-baked in hot, dry areas but more and more it is due to impermeable surfaces in cities.

In Manila, capital of the Philippines, 50% of the city was flooded in 2012 after record rainfall.

To do

Observe and record on your journey to school what the surfaces of the land you cover are made of. Could they absorb water? What has been done to channel the water and to where?

To research

Investigate two other examples (one local, one global) of where human activity has significantly impacted surface run-off and infiltration of water.

To do

Go to http://www.fao.org/nr/water/aquastat/countries_regions/jordan/index.stm

Read the information on Jordan and answer these questions.

a. Describe where most of the population live and explain why.

b. Summarize the agricultural industry of Jordan.

c. List the water supply sources available to Jordan. Explain which are sustainable.

d. Describe the political and the environmental issues around Jordan's water supplies.

e. Explain the water management strategies that Jordan is adopting.

Ocean currents and energy distribution

Ocean currents are movements of water both vertically and horizontally. They move in specific directions and some have names, and they are found on the surface and in deep water.

Ocean currents have an important role in the global distribution of energy. Without an understanding of ocean currents we cannot understand the global atmospheric energy exchanges.

Surface currents (upper 400 m of ocean) are moved by the wind.

The Earth's rotation deflects them and increases their circular movement.

Deep water currents, also called thermohaline currents, make up 90% of ocean currents and cause the oceanic conveyor belt (figure 4.1.4).

- They are due to differences in water density caused by (a) salt and (b) temperature.

- Warm water can hold less salt than cold water so is less dense and rises.

- Cold water holds more salt, is denser so sinks.

- When warm water rises, cold has to come up from depth to replace it. These are upwellings.

- When cold water rises, it too has to be replaced by warm water in downwellings.

- In this way, water circulates.

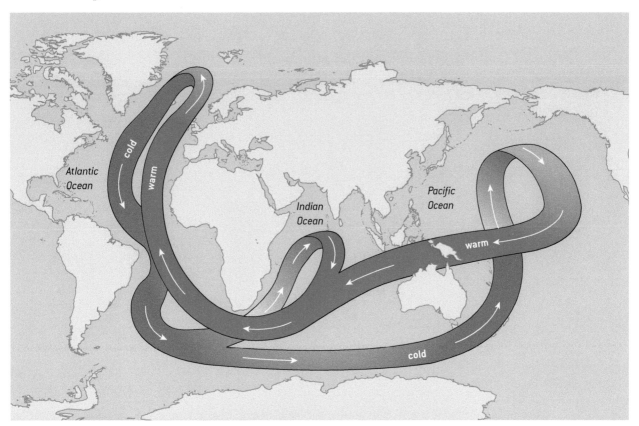

▲ **Figure 4.1.4** The great oceanic conveyor belt

To do

1. Explain how the issues of freshwater availability are of relevance to sustainability or sustainable development.

2. With reference to a named ecosystem, to what extent have the solutions emerging from water resources been directed at preventing causes of environmental impacts or problems, limiting the extent of impact from these causes, or restoring systems in which impacts have already occurred?

3. In what ways might the solutions to freshwater resource use alter your predictions for the state of human societies and the biosphere some decades from now?

Cold ocean currents run from the poles to the equator, for example:

- the Humboldt Current (off the coast of Peru)
- the Benguela Current (off the coast of Namibia in southern Africa).

Warm currents flow from the equator to the poles, for example:

- the Gulf Stream (in the North Atlantic Ocean)
- the Angola Current (off the coast of Angola).

Ocean currents and climate

Water has a higher specific heat capacity (the amount of heat needed to raise the temperature of a unit of matter by 1 °C) than land. This means that water masses heat up and cool down more slowly than landmasses. As a result, land close to seas and oceans has a mild climate with moderate winters and cool summers.

Ocean currents also affect local climate:

- The warm Gulf Stream/North Atlantic Drift moderates the climate of Northwestern Europe, which otherwise would have a sub-arctic climate.

- The cold Benguela Current, under the influence of prevailing southwest winds, moderates the climate of the Namibian desert.[2]

- The Humboldt Current impacts the climate in Peru. (see ENSO below).

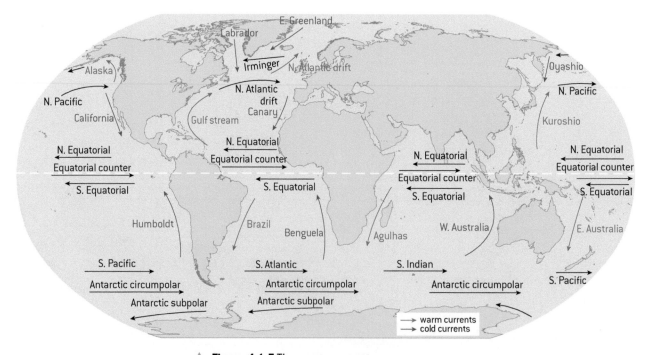

▲ **Figure 4.1.5** The ocean currents

---

[2] http://www.pbs.org/edens/namib/earth2.htm.

To do

El Niño Southern Oscillation (ENSO)

The El Niño Southern Oscillation (ENSO) is a phenomenon in the Pacific Ocean that has global consequences. Normally, air pressure in the eastern Pacific Ocean (South America) is higher than that in the western Pacific Ocean (Australia, Indonesia). This results in the trade winds that blow westward for most of the year. The trade winds blow the warm surface water westward.

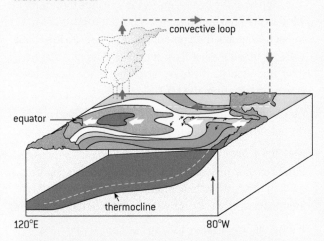

▲ **Figure 4.1.6** Trade winds blowing warm surface waters of the Pacific westward in a non-ENSO year

In some years however, the pressure difference across the Pacific Ocean is reversed.

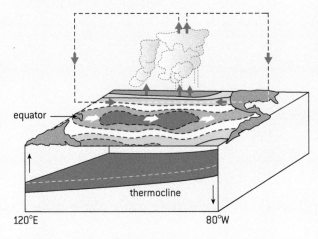

▲ **Figure 4.1.7** An ENSO year when the winds blow eastward across the Pacific

These pressure reversions are the Southern Oscillation. The changed pressure difference alters both the directions of the wind and of the warm surface current. This phenomenon was first discovered by Peruvian fishermen, who called it 'El Niño', the Boy Child, as it

usually occurred just before Christmas. Because the reversal of sea and air currents is caused by the changed pressure difference, the whole phenomenon is often called ENSO: El Niño Southern Oscillation. El Niño events occur about every two to eight years and last for about two years. They may be either weaker or stronger. Strong El Niño years are usually followed by several years of 'La Niña', when the ocean temperature of the eastern Pacific is unusually cold after being unusually hot in the El Niño years, as the system swings back and forth.

El Niño has local and global effects. Local effects include collapse of anchovy fish stocks, massive death of sea birds and storms and flooding in the coastal plain of Peru.

The anchovy fishery off the coast of Peru is extremely rich because of the occurrence of an **upwelling** when cold nutrient-rich waters come up from the ocean depths. Normally, productivity in oceans is quite low because of one of two limiting factors: light level and nutrient concentration. In the upper water levels light intensity is high, but nutrient levels are low and limit productivity. The available nutrients are taken in by phytoplankton and travel down the food chain. Dead organisms that are not eaten by some other organism sink and the nutrients stored within them end up on the ocean bottom. The lower water layers are therefore nutrient-rich. But here, the absence of light makes photosynthesis impossible. West of the Peruvian coast, the prevailing eastern trade winds push the surface water westward. This water is replaced by cold nutrient-rich water from the deep **Humboldt Current**, which originates in the Antarctic region and follows the South American coast to the north. The appearance of nutrient-rich water at the surface allows for high productivity, hence the high numbers of fish and their predators, the seabirds. During El Niño events, the upwelling disappears and the fish and seabirds starve.

During El Niño events the Peruvian coastal plain is subject to severe storms, accompanied by excessive rainfall. This is caused by the warm, extremely moist air being forced upward by the Andes Mountains.

The warm ocean water that moves east during El Niño events contains tremendous amounts of energy. The amount of energy is large enough to alter major air currents like the jet streams and, as a consequence, El Niño affects global weather. Examples of these changes are:

- Droughts in Australia, Indonesia, the Pacific Northwest of the United States and British Columbia (Canada). Forest fires are common in these areas.

- Heavy storms often resulting in flooding in California and the Midwest of the United States, Central Europe and eastern Asia.

- Absence of the monsoon in India. The Indian population depends on the monsoon rains for its food production.

While ENSO is a natural phenomenon, its effects are exacerbated by human pressures on the environment.

The Gulf Stream/North Atlantic Oscillation

The Gulf Stream is a current in the Atlantic Ocean that comes from the Gulf of Mexico across the Atlantic, where it is known as the North Atlantic Drift to western Europe. It carries warm waters in a current about 100 km wide and 1000 m deep on average. It makes Northern Europe warmer than it otherwise would be. As this water flows, some evaporates so its saltiness increases. By the time it reaches north of Britain and Scandinavian coasts, it is so much saltier, and so much more dense, than the surrounding sea water that it sinks and returns in the conveyor belt as the North Atlantic Deep Water back to where it started.

There is some evidence that the Gulf Stream current is slowing down, possibly due to global warming. Some think it may stop completely in a few decades. Melting of the Greenland ice sheet may be adding so much fresh water to the North Atlantic that the saltiness of the North Atlantic Drift is reduced and the sinking and return of the water to the Gulf of Mexico is slower. If this were the case, climate in Western Europe would be getting cooler but it is warming instead.

The North Atlantic Oscillation (NAO) is a weather phenomenon in the Atlantic as ENSO is in the Pacific. In high index NAO years, there is low pressure over Iceland and high over the Azores so westerly winds blow and winters are mild and summers cool and wetter. In low index NAO years, the pressure differences are lower, the westerlies reduced and winters are colder and summers have heat-waves. There is also an influence on eastern North America from the NAO.

Research the latest data on ENSO and NAO. Find out if there is an ENSO event now. What is the reduction in the Gulf Stream current now?

If human activity is altering climate, as most scientists believe, we are also exacerbating the effects of ENSO events as we put more pressure on the natural systems.

If the Gulf Stream were to stop suddenly (which we think has happened in the past), what would be the implications?

4.2 Access to freshwater

Significant ideas:

→ The supplies of freshwater resources are inequitably available and unevenly distributed, which can lead to conflict and concerns over water security.

→ Freshwater resources can be sustainably managed using a variety of different approaches.

Applications and skills:

→ **Evaluate** the strategies which can be used to meet increasing demand for freshwater.

→ **Discuss**, with reference to a case study, how shared freshwater resources have given rise to international conflict.

Knowledge and understanding:

→ **Access** to an adequate supply of freshwater varies widely.

→ **Climate change** may disrupt rainfall patterns and further affect this access.

→ As population, irrigation and industrialization increase, the **demand for freshwater increases**.

→ **Freshwater supplies may become limited** through contamination and unsustainable abstraction.

→ **Water supplies can be enhanced** through reservoirs, redistribution, desalination, artificial recharge of aquifers and rainwater harvesting. **Water conservation** (including grey-water recycling) can help to reduce demand but often requires a change in attitude by the water consumers.

→ The **scarcity of water resources** can lead to conflict between human populations particularly where sources are shared.

Water as a critical resource

Key points

These facts are from the World Water Council:

- 1.1 billion people live without clean drinking water.
- 2.6 billion people lack adequate sanitation (2002, UNICEF/WHO JMP 2004).
- 1.8 million people die every year from diarrheal diseases.
- 3,900 children die every day from waterborne diseases (WHO 2004).
- Daily per capita use of water in residential areas:
  - 350 litres in North America and Japan
  - 200 litres in Europe
  - 20 litres (or less) in sub-Saharan Africa.
- Over 260 river basins are shared by two or more countries mostly without adequate legal or institutional arrangements.
- Quantity of water needed to produce 1 kg of:
  - wheat: 1,000 l
  - rice: 1,400 l
  - beef: 13,000 l.

To do

1. Assuming there are 6.6 billion people alive on Earth, what is the percentage of the world's population that have no clean drinking water?

2. How many children die per year from waterborne diseases?

3. What is the percentage of daily per capita use of water in sub-Saharan Africa compared with that in Japan and North America?

4. How many days of water use by one person in Europe would it take to produce one kilogram of beef?

5. How many kilograms of wheat would be produced by the equivalent of one year of water consumption by one person in sub-Saharan Africa?

Although there is a lot of water on Earth, most of it is saline. We can remove the salt from water in desalination plants but the costs in terms of energy (especially burning fossil fuels) are large and it is only currently possible in wealthy countries which are water-stressed and near the sea, for example Israel, Australia, Saudi Arabia. A major issue with desalination is that salt is a by-product and is often returned to the ocean, increasing the density of the water which then sinks and damages ocean-bottom ecosystems. Unless we can find the technology to desalinate water cheaply, it is not a viable proposition at the moment.

The freshwater available to us is limited and the UN has applied the term 'Water Crisis' to our management of water resources today. There is not enough usable water and it can be very polluted. Up to 40% of humans alive today live with some level of water scarcity. This figure will only increase.

Humans use fresh water for:

- domestic purposes – water used at home for drinking, washing, cleaning;

- agriculture – irrigation, for animals to drink;

- industry – including manufacturing, mining;

- hydroelectric power generation;

- transportation (ships on lakes and rivers);

- marking the boundaries between nation states (rivers and lakes).

The WHO (World Health Organization) states that each human should have access to a minimum of 20 litres of fresh water per day but Agenda 21 says this should be 40 litres. Much of the world has access to far less than this recommended minimum whilst other areas have far more.

Water scarcity is not just a measure of how much water there is but of how we use it. There may be enough water in a region but it is diverted for non-domestic use. Agriculture uses water for irrigation and to provide water for livestock – usage rates are tens of times higher than domestic use. As human population expands, we need water to grow more food but, like food, it is not that there is not enough worldwide, it is that the distribution of it is uneven.

Egypt, for example, imports more than half its food as it does not have enough water to grow it, and in the Murray–Darling basin in Australia there is water scarcity for humans as so much is used for agriculture. Adding droughts and climate change, soil erosion and salinization to the story, you can see that water is and will become a major issue for nations and international organizations.

Many major rivers run through several countries.

- The Danube River basin is shared by 19 countries and 81 million people.

- The Tigris and Euphrates rivers carry water that is extracted by Iran, Iraq and Syria.

One country pollutes – the next country suffers, the question is who owns the water? Another Tragedy of the Commons. And so wars have been and will continue to be fought over water as it becomes increasingly needed and increasingly scarce.

Sustainability of freshwater resource usage

Sustainable use of resources allows full natural replacement of the resources exploited and full recovery of the ecosystems affected by their extraction and use.

▲ **Figure 4.2.1** World map of water scarcity

Sources of freshwater are **surface freshwater** (rivers, streams, reservoirs and lakes) and underground **aquifers**. The water can be extracted directly from surface or via wells from aquifers.

An aquifer is a layer of porous rock (holds water) sandwiched between two layers of impermeable rock (does not let water through). They are filled continuously by infiltration of precipitation where the porous rock reaches the surface, but this is only in limited areas.

Water flow in aquifers is extremely slow (horizontal flows can be as slow as 1–10 metres per century).[1] As a result, aquifers are often used unsustainably. Many aquifers are also 'fossil aquifers' – meaning the recharge source is no longer exposed at the surface and so they are never refilled. These can never be used sustainably.

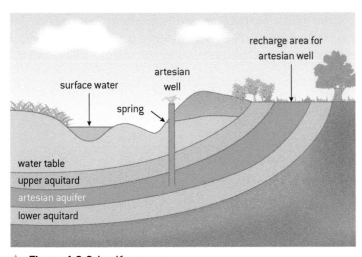

▲ **Figure 4.2.2** Aquifer structure

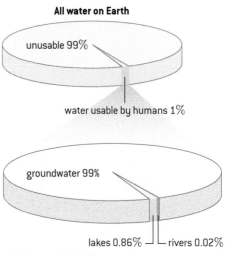

▲ **Figure 4.2.3** Water to humans

---

[1] M.B. Bush, *Ecology of a Changing Planet*, 3rd ed., Prentice Hall, Upper Saddle River, 2003, pages 187–188.

TOK

Aid agencies often use emotive advertisements around the water security issue. To what extent can emotion be used to manipulate knowledge and actions?

Global freshwater consumption is increasing strongly because the human population is increasing and because the average quality of life is improving. This increased freshwater use leads to two types of problems: water scarcity and water degradation. Degradation means that water quality deteriorates, making it less suitable for use. Let's have a more detailed look at the problems related to freshwater use and possible solutions:

Issues:

- Climate change may be disrupting rainfall patterns, even changing monsoon rains, causing further inequality of supplies.

- Low water levels in rivers and streams. The Colorado River in the USA, once a major river, now is not much more than a tiny stream when it enters the Gulf of Mexico, making navigation impossible.

- Slow water flow in the lower courses of rivers results in sedimentation, which makes the already shallow river even shallower and may extend deltas further into the sea.

- Underground aquifers are being exhausted. This simply means that the aquifer cannot be used anymore, which strongly affects agriculture. Buildings can be damaged when the soil is shrinking because the water has been taken away.

- Pumping rates from the aquifers are too fast; this causes a cone of exhaustion making the well unusable.

- Freshwater becomes contaminated and so unusable.

- **Irrigation** often results in soil degradation, especially in dry areas. Much of the water used in irrigation evaporates before it is absorbed by the crops. Dissolved minerals remain in the top layer of the soil, making it too saline (salty) for further agriculture. This process is called **salinization**.

- Fertilizers and pesticides used in agriculture often pollute streams and rivers.

- Industries release pollutants into surface water bodies.

- Industries and electricity plants release **warm water** into rivers. Warm water can hold less dissolved oxygen than cold water, so aquatic organisms that take their oxygen from the water (fish, crayfish) are negatively affected. Warm water outflow from power stations changes the species composition in the water.

Solutions:

- Increase freshwater supplies by:
  - reservoirs
  - redistribution
  - desalination plants removing salt from seawater
  - rainwater harvesting systems – large and small scale
  - artificially recharging aquifers.
- Reduce domestic use of freshwater by using more water-efficient showers, dishwashers and toilets.

- Wash cars in car washes with a closed water system. Not washing the car in the street also reduces pollution by oil.

- Grey-water recycling – grey water is water from showers, baths, household laundry, kitchen sinks etc. which can be reused on site for flushing WCs, garden irrigation.

- Irrigation: selecting drought resistant crops can reduce the need for irrigation. (Some areas may simply be unsuitable for growing crops; cattle grazing may be better.) Closed pipes instead of open canals and trickle systems instead of spraying water can both reduce evaporation. Alternatively using subsurface drip irrigation.

- Reduce the amount of pesticide and fertilizer used (using the smallest possible amount at the most appropriate time).

- Replace chemical fertilizers with organic ones – the release of nutrients is slower and more likely to be absorbed by the crop plants.

- Prevent over-spray, eg spraying pesticide or fertilizer directly in a stream.

- Use highly selective pesticides instead of generic pesticide, or use biological control measures.

- Industries can remove pollutants from their wastewater with water treatment plants. Often, they are forced by law to do so.

- Regulate maximum temperatures of released cooling water. Instead of releasing the warm cooling water, cooling towers that evaporate the water can be used.

Practical Work

* Unequal access to freshwater can cause conflict between countries that have an abundance of freshwater and those who have not. Research two examples where conflict has arisen over water supplies.

* Evaluate the strategies which can be used to meet our increasing demand for freshwater.

* Investigate the volume of water that you use (a) directly and (b) indirectly per day. Compare this with other users locally and globally.

To think about

Grey water

If you live in a house with running water, only purified and treated freshwater enters your home in one pipeline. This water is used for all the many processes in the home that need water. But it does not all have to be as clean as our drinking water.

Black water or sewage contains human waste and may carry disease-causing bacteria and other organisms (eg worms).

Grey water can be very lightly used water. Do you keep the tap running when you clean your teeth or run a glass of water? If so, most of this water is perfectly clean yet goes down the pipes and mixes with black water in a shared sewerage system. All this water is then cleaned to the highest standards before being recycled back into our piped water system.

This is a crazy waste of clean water and the two types, black and grey could be separated before they leave our homes. Grey water could be used to irrigate gardens, clean cars, flush WCs and anywhere that requires water but not drinking-quality water. It must be used at once though to avoid build-up of bacteria.

Water wars

The scarcity of water resources can lead to conflict between human populations particularly where sources are shared.

CASE STUDY

1. The geopolitics of Israel's water shortage

▲ **Figure 4.2.4** Aquifers of the region

Water shortage looms in Israel after prolonged drought, but supply to Jordan continuing

International Herald Tribune. Posted on March 21st, 2008.

"Israel is suffering its greatest drought in the past decade and will have to stop pumping from one of its main drinking water sources by the end of the summer, an official said Wednesday. Water Authority spokesman Uri Schor said when Israel has to stop pumping from the Sea of Galilee — the source of about 40 percent of its drinking water — it will have to step up extraction from already-depleted aquifers, underground water-bearing seams of rock. 'The situation is very, very bad,' Schor said. 'As we pump more from the aquifers, the quality of the water will go down.'

Israel's water problem stems from population growth and an improvement in quality of life that brings a greater desire to water lawns and gardens, Schor said. This winter was the fourth that Israel got less than average rain, with only about 50–60 percent of the average in most areas, he said. Critics of government policy note that agriculture uses a large proportion of Israel's water, receiving heavily subsidized water rates. Since Israel in any event does not grow much of the food it needs, they say irrigation for farming should be drastically curtailed. Israel's rainy winter season ends this month,

though there can be occasional rainy days through June. The rainy season begins around October.

Despite the shortage, Israel will probably not reduce the amount of water supplied to Jordan according to the peace treaty between the countries, Schor said. Jordan's drought is much worse than Israel's, he said. Water is a contentious issue in the dry region, and one of the disputes Israeli and Palestinian negotiators hope to overcome in talks to work out a peace agreement.

In an effort to stem a serious shortage of water, Israel will launch a conservation campaign, targeting mostly household use. As part of the efforts, Israel has in recent weeks reduced by more than 50 percent the drinking water supplied to farmers, increasing their need for recycled water, Schor said. Water officials will this weekend debate raising the cost of drinking water in another attempt to cut household use, he said.

Israel has two desalination plants that supply about one-third of water needed by municipalities and households, Schor said. Three other plants scheduled to be completed by 2013 will double that amount. The next one is due to be operational in 2009."

1. List Israel's water sources (stores) and demands (flows) on this water.

2. What are the long-term environmental problems of overuse of water? Can you explain this in terms of your water cycle diagram?

3. What are the reasons for strained relations between neighbouring countries over water supplies?

4. How could Israel solve these problems in both the short term and the long term?

5. If some countries in the region can afford to desalinate sea water but others cannot, should they share the water?

6. In your opinion what would be a **sustainable** solution to water management in this part of the world?

To do

Research the water politics of the Nile basin in North-east Africa. The Nile flows through 10 countries which all extract water from it.

1. Evaluate the claims of Egypt and Sudan for their use of Nile water.

2. Suggest solutions to the tensions.

2. China's Three Gorges dam update

▲ **Figure 4.2.5** Yangtze River and Three Gorges dam, China—where do you think the dam is located?

The Three Gorges hydroelectric dam on the Yangtze River, China was one of the largest engineering projects and the largest power station in the world. Its estimated cost was US$30 billion and it opened in 2012. The hydroelectric power produced from the dam could provide 9% of all China's output so the saving in carbon emissions would be huge. Up to a quarter of a million workers built the dam and 1.3 million people were resettled as the flood waters covered their homelands. Apart from electricity generation, the dam is designed to stop the river flooding. These floods have drowned up to 1 million people in the past 100 years and flooded the major city of Wuhan for months.

Criticisms of the dam are that it will cause environmental degradation, refugees, destroy archaeological sites, threaten some wild species and silt up.

The Siberian Crane is critically endangered and their winter habitat is the wetlands that the dam floods. The Yangtze River dolphin is probably extinct in the wild due to pollution and loss of water flow downstream. The Yangtze sturgeon is affected as well.

Upstream of the dam, siltation will increase as the silt will get caught behind the dam. This not only increases the danger of clogging the turbines but means it does not flow downstream. Lack of silt may make the downstream river banks more liable to erosion. Shanghai, at the mouth of the Yangtze, is built on a silt bed and this may erode increasingly without replacement from upstream.

Behind the dam, the water level rose by up to 100 metres and historical and cultural sites were flooded. The Ba people settled here 3,000 years ago and buried their dead in coffins in caves on cliffs.

Pollution by toxic wastes and materials used in building the dam may make the waters unsuitable for drinking and minimal sewage treatment may increase algal blooms and eutrophication in the dammed waters.

There is a risk of energy security as well. Having so much power from one dam means that, if it is destroyed, by terrorism or earthquake, all this power generation is lost. It also has to be transmitted long distances to where it is used.

To think about

To have information is useful. To turn it into knowledge is difficult. This information age feeds us a massive glut of information, but for all this information there is also a vast amount of misinformation. As a critical thinker, always be aware of what is not written as much as what is written. Omitting essential facts that do not support the case you are arguing for is an easy thing to do. Consider either the Three Gorges dam or any other project to manage water (eg the Aswan dam, Aral Sea, Ogallala Aquifer).

Either

Put together a case FOR the project and AGAINST the project using the information you research. You can omit some facts and exaggerate others. Pseudoscience does just this.

OR

Set up a class debate for and against the project.

4.3 Aquatic food production systems

Significant ideas:
→ Aquatic systems provide a source of food production.
→ Unsustainable use of aquatic ecosystems can lead to environmental degradation and collapse of wild fisheries.
→ Aquaculture provides potential for increased food production.

Applications and skills:
→ **Discuss** with reference to a case study the controversial harvesting of a named species.
→ **Evaluate** strategies that can be used to avoid unsustainable fishing.
→ **Explain** the potential value of aquaculture for providing food for future generations.
→ **Discuss** a case study that demonstrates the impact of aquaculture.

Knowledge and understanding:
→ **Demand** for aquatic food resources continues to increase as human population grows and diet changes.
→ Photosynthesis by phytoplankton supports a highly diverse range of food webs.
→ Aquatic (freshwater and marine) flora and fauna are harvested by humans.
→ The highest rates of **productivity** are found near the coast or in shallow seas where **upwellings** and nutrient enrichment of surface waters occurs.
→ Harvesting some species can be controversial eg seals and whales. Ethical issues arise over **biorights**, rights of indigenous cultures and international conservation legislation.

→ Developments in **fishing equipment** and changes to fishing methods have led to dwindling fish stocks and damage to habitats.
→ **Unsustainable exploitation** of aquatic systems can be mitigated at a variety of levels (international, national, local and individual) through policy, legislation and changes in consumer behaviour.
→ **Aquaculture** has grown to provide additional food resources and support economic development and is expected to continue to rise.
→ Issues around aquaculture include loss of habitats, pollution (with feed, antifouling agents, antibiotics and other medicines added to the fish pens), spread of diseases and escaped species (some involving genetically modified organisms).

'The world has quietly entered a new era, one where there is no national security without global security. We need to recognize this and to restructure and refocus our efforts to respond to this new reality.'

Lester R. Brown

Marine ecosystems and food webs

Marine ecosystems (oceans, mangroves, estuaries, lagoons, coral reefs, deep ocean floor) are usually very biodiverse and so have high stability and resilience.

Some are more productive than others. One half of marine productivity is in coastal regions above the **continental shelf.** Deep oceans where light does not reach have low productivity as the only food sources are chemotrophs (2.2) and dead organic matter that descends from above.

The **continental shelf** is the extension of continents under the seas and oceans. Where continental shelf exists it creates shallow water.

It is important because:

- It has 50% of oceanic productivity but 15% of its area.

- Upwellings (see 4.1 ENSO) bring nutrient-rich water up to the continental shelf.

- Light reaches the shallow seas so producers can photosynthesize.

- Countries can claim it as theirs to exploit and harvest.

The width of the shelf averages 80 km but varies from almost zero (eg coast of Chile or west coast of Sumatra where one tectonic plate is sliding under another) to nearly 1600 km for the Siberian Shelf in the Arctic. The North Sea between the British Isles and mainland Europe is all continental shelf. The depth at which the shelf stops and the sea bed slopes more steeply is remarkably constant at about 150 metres.

The UN Convention on the Laws of the Sea (UNCLOS) in 1982 designated continental shelves as belonging to the country from which they extend. It also designated a 200 nautical mile (370 km) limit from the low water mark of a shore as an exclusive economic zone belonging to that country. Outside that are international waters which no one country controls. This has huge impact on fishing these areas, who is allowed to fish there, who controls this and who cleans it up when there is a pollution problem. (See 'tragedy of the commons', later in this section.)

To do

Research the maritime claims of your own country and compare them with Ecuador and Peru in figure 4.3.1. Why is this important economically?

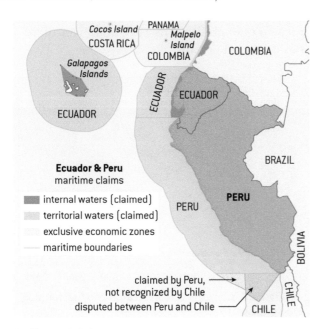

Ecuador & Peru maritime claims

- internal waters (claimed)
- territorial waters (claimed)
- exclusive economic zones
- maritime boundaries

claimed by Peru, not recognized by Chile
disputed between Peru and Chile

▲ **Figure 4.3.1**

Phytoplankton are single-celled organisms that can photosynthesize and are the most important producer in the oceans, producing 99% of primary productivity (2.3 and 2.4). They float in the sea. Zooplankton also float in the sea and are single-celled animals that eat the phytoplankton and their waste (dead organic matter – DOM) and, between them, these organisms support the complex food webs of the oceans.

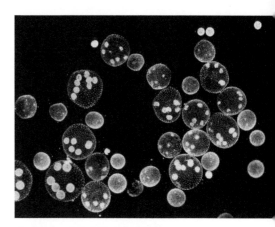

▲ **Figure 4.3.2** Phytoplankton and zooplankton range in size from 1 µm to 10 mm

Marine organisms can be classified as:

* Benthic – living on or in the sea bed

* Pelagic – living surrounded by water from above the sea bed to the surface.

Fisheries – industrial farming and hunting

About 90% of fishery activity is in the oceans and 10% in freshwater.

Fisheries include

* shellfish: oysters, mussels and molluscs (including squid)

* some vertebrates: eels, tuna etc.

Up to half a billion people make a livelihood in fisheries.

Three billion people gain 20% of their protein intake from fish and the rest of us about 15% of our protein.

Fish are a very important food for humans.

Fish are high in protein, low in saturated fats and contain various vitamins (A,B,D) that humans need for a healthy diet.

According to the FAO (Food and Agriculture Organization), more than 70% of the world's fisheries are fully exploited, in decline, seriously depleted or too low to allow recovery (under drastic limits to allow a recovery).[1] The global fish catch is no longer increasing even though technology has improved. Demand is high and rising but fishermen cannot find or catch enough fish. They are no longer there. But aquaculture has increased greatly.

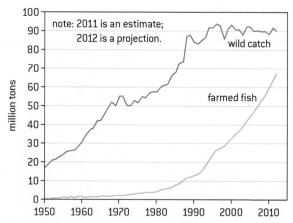

▲ **Figure 4.3.3** World wild fish catch and fish farming 1950–2012

Aquaculture

Human diets are changing and many people in MEDCs are becoming more health conscious, eating less meat from terrestrial animals and more fish and vegetables, as this change lowers saturated fat intake. Saturated fat intake is linked to cholesterol build-up in arteries

1   http://www.american.edu/TED/ice/canfish.htm

Key term

A **fishery** exists when fish are harvested in some way. It includes capture of wild fish (also called capture fishing) and aquaculture or fish farming.

Key term

Aquaculture is the farming of aquatic organisms in both coastal and inland areas involving interventions in the rearing process to enhance production. (FAO http://www.fao.org/aquaculture/en/)

and increased risk of heart attacks and strokes. Now, on average worldwide, each person eats about 20 kg of fish and 8 kg of meat per year. These figures used to be higher for meat and lower for fish and most of the extra fish comes from fish farms. This is because the wild caught fishing has reached its limit even though humans have better and better technology to find the fish. Either they are just not there because we have over exploited them or we cannot take more than a sustainable yield and that is not enough as human population grows.

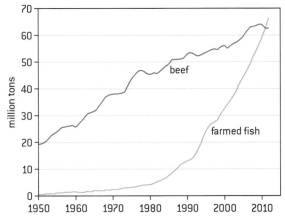

Figure 4.3.4 World farmed fish and beef production, 1950–2012

From figure 4.3.4, you can see that farmed fish production was more than beef production in 2011: 66 million tons (60 million tonnes) of farmed fish were produced in 2012. Now, we may eat more farmed fish than wild caught fish for the first time since we started fishing. If these are not sustainable, this cannot continue so the challenge is to make fish farming sustainable.

Ways in which fish farming is becoming more sustainable are:

- Fishmeal uses more trimmings and scraps which would have been wasted in the past.

- Livestock and poultry processing waste is substituted for fishmeal.

- United States Department of Agriculture has proven that eight species of carnivorous fish – white sea bass, walleye, rainbow trout, cobia, arctic char, yellowtail, Atlantic salmon and coho salmon – can get enough nutrients from alternative sources without eating other fish.[2]

Figure 4.3.5 shows the relative importance of China as a producer of farmed fish. China produces 62% of all farmed fish worldwide and most of that is carp or catfish. These are grown in rice paddies and their waste provides fertilizer for the rice. This system is used in many SE Asian countries and is of mutual benefit: the system produces rice and a healthy source of protein for the farmer.

---

[2] http://science.kqed.org/quest/2014/02/13/vegetarian-farmed-fish-may-be-key-to-sustainable-aquaculture/

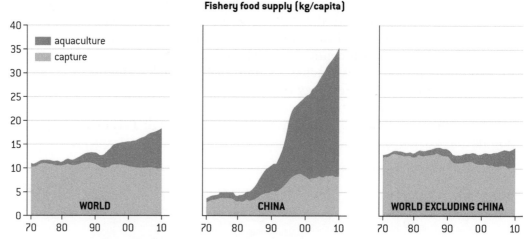

Figure 4.3.5 Relative contribution of aquaculture and capture fisheries to food fish consumption

But other aquaculture systems are less efficient. Shrimp and salmon are carnivores and are fed on fishmeal or fish oil produced from wild fish. Mangrove swamps have been replaced by fish farms – in the Philippines two-thirds of the mangroves have been lost in 40 years.

Impacts of fish farms include:

- loss of habitats
- pollution (with feed, antifouling agents, antibiotics and other medicines added to the fish pens)
- spread of diseases
- escaped species including genetically modified organisms which may survive to interbreed with wild fish
- escaped species may also outcompete native species and cause the population to crash.

To think about

Atlantic salmon and fish farming

www.atlanticsalmontrust.org

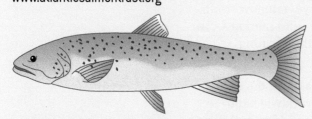

Figure 4.3.6 Atlantic salmon

Wild Atlantic salmon live in the North Atlantic and Baltic seas. Due to overfishing at sea and in rivers where they go to spawn, the commercial market for wild salmon crashed. Now they are farmed in fish farms (a form of aquaculture). Although commercially successful, these farms create pollution as the fish are kept in high densities and their uneaten food, feces and chemicals used to treat them enter the oceans. Sea lice are a problem in the farms and escape of some farmed salmon which then interbreed with wild salmon reduces the genetic fitness of the wild stock as farmed salmon are bred for fish farm life, not the wild. Because of this, the ability to survive of the salmon population in the wild is reduced.

As wild stocks of many fish species decline, fish farming has increased. Although there are environmental problems created by keeping fish of one species in high densities, the benefits of not chasing fish stocks across the oceans are large. Some have likened fishing in the seas to hunting on land. We now get most of our terrestrial meat from farmed livestock and little from hunting. It makes economic sense to do the same with our aquatic meat supply.

Unsustainable wild fishing industry

World fisheries were once thought to be inexhaustible but now the FAO estimates that 75% of fisheries are under threat of over-exploitation.

We are so good at finding and catching fish that, once found, fish populations can be depleted very quickly. Once we catch the larger specimens, we then catch smaller and smaller ones and do not leave them to mature and reproduce.

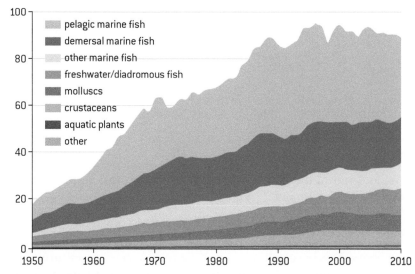

▲ **Figure 4.3.7** Global wild fish capture in million tonnes, 1950–2010, as reported by the FAO

Fish stocks are a resource under pressure, being exploited by overfishing – fishing at an unsustainable level. We are just too good at finding and catching fish on an industrial scale.

- Commercial fishing is informed by the latest satellite technology, GPS navigation and fish finding scanning technology of military quality.

- Fishing fleets have become larger and, with modern refrigeration techniques, including blast freezing, they can stay at sea for weeks or an entire season.

- Within a fleet there will be a suite of vessels including fishing vessels, supply vessels and factory ships that process the catch at sea.

- Indiscriminate fishing gear will take all organisms in an area, whether they are the target species or not.

- Trawlers drag huge nets over the seabed virtually clearcutting it.

All of this is a far cry from the hunter-gatherer ethos of fishermen going to sea in primitive craft with simple fishing equipment to land a fish catch for their own family or community.

When we talk of wars over resources, we are usually thinking about wars over fuel and energy supplies. However there are a significant number of serious international crises and near war events that have taken place over fishing and fishing rights in the last 60 years.

- 1970s: Iceland banned all foreign vessels from fishing in Icelandic waters. This led to three 'Cod Wars' between Britain and Iceland.

- 1994: British and French fishermen competed with Spanish fishermen for tuna in the Bay of Biscay.

- 1995: there was the turbot (also called halibut) war between Canada and Spain when Canada fired upon and captured the crew of a Spanish fishing boat, the Estai, having chased it from national into international waters. The Estai cut its trawl net when it was being pursued but the Canadians recovered it from the seabed and found it had a net with a mesh size smaller than that permitted and this caught the smaller turbot. Eventually, Canada and Spain agreed on a solution to the turbot war and the result was an increased regulation of fisheries.

What happens to the fish?

The world fish catch is just over 90 million tonnes per year of which about 20 million tonnes are not wild fish but from aquaculture (fish farming).

- 15% of animal protein eaten by humans comes from fish. In Japan nearly half of the animal protein in the diet is fish.

- About a quarter of the global catch goes into fish meal and fish oil products to feed animals … to feed us.

- About 20 million tonnes are 'by-catch' – unwanted fish and other marine animals that are thrown back into the oceans from the fishing boats. This by-catch may be the wrong species (non-target fish, whales, dolphins and porpoises, and even seabirds such as albatrosses), fish that are legally too small to catch, or fish over allotted quotas. Most of these are dead or dying when they are thrown back into the seas.

Draw a pie chart to illustrate these values.

Issues and solutions

> **TOK**
>
> Inuit people have an historical tradition of whaling. To what extent does our culture determine or shape our ethical judgements?

To think about

Collapse of the Newfoundland cod fishery

For centuries the Newfoundland cod fishery was one of the world's most productive fisheries, yielding 800,000 tons of fish and employing 40,000 people at its peak in 1968. Then its stocks plummeted as a result of overharvesting and habitat damage. In 1992, the fishery was closed in an effort to save it. But it may have been too late: two decades have passed, but stocks have not recovered.

This collapse was local in scale, but the issue is much larger. North Atlantic Ocean fisheries operating now catch half what they did 50 years ago, despite tripling their efforts. So many popular species – such as cod, tuna, flounder and hake – are now in serious decline. Cod stocks in the North Sea and to Scotland's west are on the verge of collapse.

The deterioration of oceanic fisheries can be reversed. Granting fishers an ownership stake in fish stocks is one way to help them understand that the more productive their fishery is, the more valuable their share. For example, fishers in Iceland and New Zealand have used marketable quotas, allowing them to sell catch rights, since the late 1980s. The upshot is smaller but more profitable catches and rebounding fish populations. The classic 'tragedy of the commons' problem is averted.

Because of the complexity of marine ecosystems, some scientists are pushing for management of whole ecosystems rather than single species. In addition, studies have shown that well-positioned and fully protected marine reserves, known as fish parks, can help replenish an overfished area. By giving fish a refuge to breed and mature in, reserves can increase the size and total number of fish both in the reserve and in surrounding waters. For example, a network of reserves established off St. Lucia in 1995 has raised the catch of adjacent small-scale fishers by up to 90 per cent. Preservation of nursery habitats like coral reefs, kelp forests and coastal wetlands is integral to keeping fish in the sea for generations to come.

Consumers can promote healthy fishery production by eating less fish and buying seafood from well-managed, abundantly stocked fisheries. The Seafood Lover's Guide from Audubon's Living Oceans programme is one valuable reference. Chilean seabass stocks are on the verge of collapse and illegal fishing abounds, so it makes the list of fish to avoid. The list also distinguishes between wild Alaska salmon, which comes from a healthy fishery, and farmed salmon, which is fed meal made from wild fish and thus does not relieve pressure on marine stocks. Proper labels are needed to allow consumers to make wise purchasing decisions. The Marine Stewardship Council, a new independent international accreditation organization, has thus far certified seven fisheries as being sustainably managed with minimal environmental impact.

The capacity of the world's fishing fleet is now double the sustainable yield of fisheries. Myers and Worm from Dalhousie University believe that the global fish catch may need to be cut in half to prevent additional collapses. Reducing by catch, creating no-take fish reserves, and managing marine ecosystems for long-term sustainability instead of short-term economic gain are all policy tools that can help preserve the world's fish stocks. If these are coupled with a redirection of annual fishing industry subsidies of at least $15 billion to alternatives such as the retraining of fishers, there could be a big payoff. It is difficult to overestimate the urgency of saving the world's fish stocks. Once fisheries collapse, there is no guarantee they will recover.

Fish stocks are shrinking because:

- industrialized nations subsidize their modern fleets by an estimated US$50 billion a year, and

- demand outstrips supply.

The tragedy of the commons

This metaphor illustrates the tension between the common good and the needs of the individual and how they can be in conflict. If a resource is seen as belonging to all, we all tend to exploit it and over-exploit it if we can. This is because the advantage to the individual of taking the resource (be it fish, timber, minerals, apples) is greater than the cost to the individual as the cost is spread amongst the whole population. In the short term it is worth taking all the fish you can because, if you do not, someone else will. This assumes that humans are selfish and not altruistic and it has caused much debate amongst economists and philosophers. The solution is often regulation and legislation by authorities which limits the amount of common good available to any individual. This may be by permit, set limits or by cooperation to conserve the resource.

Exploitation of the oceans is a good example of the tragedy of the commons. The Grand Banks off the coast of Newfoundland were once amongst the richest fishing grounds on Earth. Since the 1400s, they were fished by fleets from Spain, Portugal, England, France, and later

To do

Ethical issues arise over biorights, rights of indigenous cultures and international conservation legislation. **Discuss** with reference to a case study the controversial harvesting of a named species. For example, whales, seals, tuna.

For a really chilling look at the causes and problems of overfishing check out the documentary *End of the Line*.

Newfoundland, Canada, Russia, and the United States, and others. In the early 1990s, fish stocks crashed and by 1995, cod and flounder fishing there had been closed in an attempt to conserve the remaining stocks. So far, there has been little, if any, recovery of fish numbers.

To do

The United Nations Convention on the Laws of the Sea (UNCLOS) is an international agreement written over decades that attempts to define the rights and responsibilities of nations with respect to the seas and marine resources. Most, but not all, countries have signed and ratified the convention. It defines these categories of waters:

- Internal waters – next to a country's coastline where foreign ships may not travel and the country is free to set its own laws and regulate use.

- Territorial waters – up to 12 nautical miles (just over 22 km) from the coastline where foreign ships can transit on 'innocent passage' but not spy, fish or pollute. In island states, eg the Maldives, a boundary line is drawn around the whole archipelago.

- Beyond the territorial waters is another 12 nautical miles of contiguous zone where a state can patrol smuggling or illegal immigration activities.

- In the exclusive economic zones (EEZs), the state has sole rights to exploit all natural resources. Foreign nations may overfly or navigate through this zone. If a country's continental shelf is greater than 200miles from its coast, it also has exclusive rights to exploit this.

The international seabed authority, created by the UN, controls and monitors seabed exploitation in international waters.

Landlocked states are given a right of free access to and from the sea under UNCLOS rules.

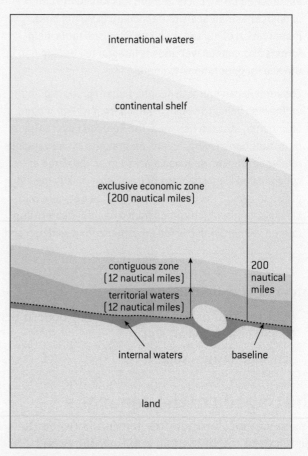

▲ **Figure 4.3.8** Ocean zones according to UNCLOS

Questions

Do you think that UNCLOS was a good idea?

Why would some nations not sign up to it?

What would have happened without UNCLOS?

To think about

What not to eat from the Marine Conservation Society www.mcsuk.org

Species	Reason	Alternatives
Atlantic cod (from overfished stocks)	Species listed by World Conservation Union IUCN. Some stocks close to collapse eg North Sea	Line caught fish from Icelandic waters
Atlantic salmon	Wild stocks reduced by 50% in last 20 years	Wild Pacific salmon. Responsibly and/or organically farmed salmon
Chilean seabass (Patagonian toothfish)	Species threatened with extinction by illegal fishing, also high levels of seabird by-catch	None
Dogfish/spurdog	Species listed by IUCN	None
European hake	Species heavily overfished and now scarce	South African hake (M.capensis)
European seabass	Trawl fisheries target pre-spawning and spawning fish also high levels of cetacean by-catch	Line caught or farmed seabass
Grouper	Many species are listed by IUCN	None
Haddock (from overfished stocks)	Species listed by IUCN	Line caught fish from Icelandic and Faroese waters
Ling (Molva spp)	Deep-water species and habitat vulnerable to impacts of exploitation and trawling	None
Marlin	Many species are listed by IUCN	None
Monkfish	Long-lived species vulnerable to exploitation. Mature females extremely rare	None
North Atlantic halibut	Species listed by IUCN	Line caught Pacific species. Also farmed N Atlantic halibut
Orange roughy	Very long-lived species vulnerable to exploitation	None
Shark	Long-lived species vulnerable to exploitation	None
Skates and rays	Long-lived species vulnerable to exploitation	None
Snapper	Some species listed by IUCN, others over-exploited locally	None
Sturgeon	Long-lived species vulnerable to exploitation. 5 out of 6 Caspian Sea species listed by IUCN	None although this species is now farmed
Swordfish	Species listed by IUCN	None
Tuna	All commercially fished species listed by IUCN except skipjack and yellowfin are over-fished	'Dolphin friendly' (EII monitored) skipjack or yellowfin. Preferably pole and line caught
Warm-water or tropical prawns	High by-catch levels and habitat destruction	Responsibly farmed prawns only

▲ **Figure 4.3.9** What not to eat

Source: Marine Conservation Society

Maximum sustainable yield (MSY)

A sustainable yield (SY) is the increase in natural capital, ie natural income that can be exploited each year without depleting the original stock or its potential for replenishment. Sustainable yield of an aquifer is the amount that can be taken each year without permanently decreasing the amount of water stored. For commercial ventures, it is **maximum sustainable yield** that is of interest – the highest amount that can be taken without permanently depleting the stock. In fisheries, this is a crucial amount – how many fish and of what size can be taken in any year so that the harvest is not impaired in subsequent years.

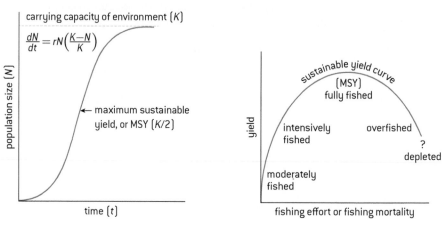

The carrying capacity for each species depends on

- its reproductive strategy
- its longevity
- the indigenous resources of the habitat / ecosystem.

Each breeding season / year, new individuals enter the population (either new offspring or immigrants). If the number recruited to the population is larger than the number leaving (dying or emigrating), then there is a net increase in population. If the difference in population from initial size to new population size is harvested the population will remain the same. This number is the maximum sustainable yield (MSY) for this population.

$$SY = \frac{\text{total biomass at time } t + 1}{\text{energy}} - \frac{\text{total biomass at time } t}{\text{energy}}$$

Or

$$SY = \text{annual growth and recruitment} - \text{annual death and emigration}$$

In practice, harvesting the maximum sustainable yield normally leads to population decline and thus loss of resource base and an unsustainable industry or fishery. There are several reasons for this.

- The population dynamics of the target species are normally predicted (modelled) rather than the species numbers being quantitatively measured (counted).
- It is often impossible to be precise about the size of a population.
- Estimates are made on previous experience.

- The model does not allow for monitoring of the dynamic nature of the harvest in terms of age and sex ratio. If the harvest primarily targets reproductive females this will have a much greater impact on future recruitment than the targeting of mature or old males. Targeting young immature fish will also equally impact on future recruitment rates and thus overall population size.

- Disease may strike the population.

So if you extract / harvest at the MSY, you will deplete a population in poor recruitment years.

A much safer approach is to adopt the harvesting of an **Optimal Sustainable Yield** (OSY). This usually requires less effort than MSY and maximizes the difference between total revenue and total cost. It has a much greater safety margin than MSY but still may have an impact on population size if there are other environmental pressures within a system.

Fishing quotas are often set as a percentage or proportion of the OSY per fleet per year. The quota is set as a weight of catch not number of fish.

To do

Research the history and current state of one of these fisheries:

- Pacific wild salmon
- North Sea herring
- Grand Banks fisheries
- Peruvian anchovy
- Or any other that is near to where you live.

Exam-style questions

The Inuit are indigenous aboriginal people of Northern Canada. The data below come from a study of a Inuit fish farming community. The Inuit fish in the open sea but have also sectioned off a large fjord (a long narrow inlet of the sea) which they use for farming salmon and shrimps. The shrimps eat microscopic plants in the sea called phytoplankton. Salmon and kawai (a wild fish) both eat shrimps.

Figure 1

	All units in kJ m^{-2} yr^{-1}
Insolation on fjord	185,000.0
Insolation on open sea	1,972,000.0
Farmed shrimp consumed by Inuit	26.0
Gross primary production by phytoplankton	3,470.0
Shrimp consumed by kawai	847.0
Respiratory loss by kawai (open sea)	572.0
Shrimp consumed by salmon (farmed)	461.0
Respiratory loss by salmon	410.0
Kawai consumed by Inuit	6.2
Salmon consumed by Inuit	4.3
Energy used in managing salmon farm	4.1
Energy used in fishing for kawai	6.7
Energy used in managing shrimp farm	14.0
Energy used in other human activities including trading furs	12.5

Practical Work

∗ Discuss with reference to a case study the controversial harvesting of a named species.

∗ Discuss a case study that demonstrates the impact of aquaculture.

∗ Investigate fishing rates in selected countries.

∗ Evaluate strategies that can be used to avoid unsustainable fishing.

(a) Use the data in figure 1 to complete the diagram below.   [6]

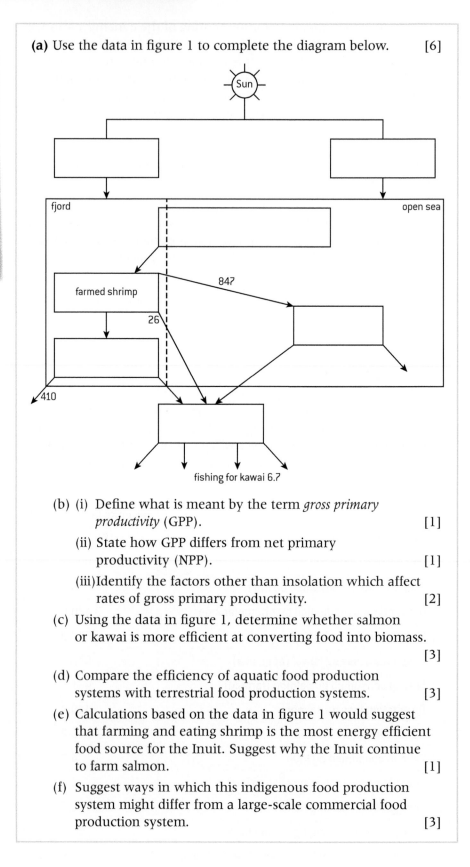

(b) (i) Define what is meant by the term *gross primary productivity* (GPP). [1]

     (ii) State how GPP differs from net primary productivity (NPP). [1]

     (iii) Identify the factors other than insolation which affect rates of gross primary productivity. [2]

(c) Using the data in figure 1, determine whether salmon or kawai is more efficient at converting food into biomass. [3]

(d) Compare the efficiency of aquatic food production systems with terrestrial food production systems. [3]

(e) Calculations based on the data in figure 1 would suggest that farming and eating shrimp is the most energy efficient food source for the Inuit. Suggest why the Inuit continue to farm salmon. [1]

(f) Suggest ways in which this indigenous food production system might differ from a large-scale commercial food production system. [3]

4.4 Water pollution

Significant idea:

→ Water pollution, both groundwater and surface water, is a major global problem the effects of which influence human and other biological systems.

Applications and skills:

→ **Analyse** water pollution data.

→ **Explain** the process and impacts of eutrophication.

→ **Evaluate** the uses of indicator species and biotic indices in measuring aquatic pollution.

→ **Evaluate** pollution management strategies with respect to water pollution.

Knowledge and understanding:

→ There are a variety of freshwater and marine **pollution sources**.

→ Types of **aquatic pollutants** include organic material, inorganic plant nutrients (nitrates and phosphates), toxic metals, synthetic compounds, suspended solids, hot water, oil, radioactive pollution, pathogens, light, noise and biological (invasive species).

→ A wide range of parameters can be used to directly test the quality of aquatic systems, including pH, temperature, suspended solids (turbidity), metals, nitrates and phosphates.

→ **Biodegradation of organic material** utilizes oxygen which can lead to anoxic conditions and subsequent anaerobic decomposition which leads to formation of methane, hydrogen sulphide and ammonia (toxic gases).

→ **Biochemical oxygen demand** (BOD) is a measure of the amount of dissolved oxygen required to break down the organic material in a given volume of water through aerobic biological activity. BOD is used to indirectly measure the amount of organic matter within a sample.

→ Some species can be indicative of polluted waters and be used as **indicator species**.

→ A **biotic index** indirectly measures pollution by assaying the impact on species within the community according to their tolerance, diversity and relative abundance.

→ **Eutrophication** can occur when lakes, estuaries and coastal waters receive inputs of nutrients (nitrates and phosphates) which result in an excess growth of plants and phytoplankton.

→ **Dead zones** in both oceans and freshwater can occur when there is not enough oxygen to support marine life.

→ Application of figure 1.5.6 to water pollution management strategies includes:

- reducing human activities producing pollutants (eg alternatives to current fertilizers and detergents);

- reducing release of pollution into the environment (eg treatment of wastewater to remove nitrates and phosphates);

- removing pollutants from the environment and restoring ecosystems (eg removal of mud from eutrophic lakes and reintroduction of plant and fish species).

Water pollution kills over 14,000 people per day (nearly 10 people every minute) through waterborne disease and poisoning.

Nearly half a billion people do not have access to clean, safe drinking water.

Water pollution affects MEDCS as well as LEDCs.

Types of water pollution

Both sea and freshwater can become polluted.

Pollutants can be:

1. anthropogenic (created by human activities) or natural (eg volcanic eruptions, algal blooms)

2. point source or non-point source (1.5)

3. organic or inorganic

4. direct or indirect.

Type	Pollutant	Example	Effects
Organic	Sewage	Human waste	Eutrophication (p233)
	Animal waste	Manure	Smell
	Biological detergents	Washing powders	
	Food processing waste	Fats and grease	
	Pesticides from agriculture	Insecticides, herbicides	Loss of biodiversity
	Chemicals from industry	PCBs, drugs, hormones	May be carcinogenic, growth-promoting hormones
	Pathogens	Waterborne and fecal pathogens	Disease
	Invasive species	Cane toads	Decimates indigenous species
Inorganic	Nitrates and phosphates	Fertilizers	Eutrophication, changes biodiversity
	Phosphates	Washing detergents	
	Heavy toxic metals	Industry and motor vehicles	Bioaccumulation and biomagnification in food chains, poisonous
	Hot water (thermal pollution)	Power stations	Changes physical property of water, kills fish, changes biodiversity
	Oil	Industry	Floats on surface, contaminates seabirds, reduces oxygen levels
	Radioactive materials	Nuclear power stations	Radiation sickness
	Light	Cities, hotels on beaches	Disrupt turtle nesting sites
	Noise	Aircraft	Upset whale navigation, change plant growth, upset bird cycles
Both	Suspended solids	Silt from construction sites	Damage corals and filter feeders
	Solid domestic waste	Household garbage	Plastics are especially bad – suffocate, cause starvation (Great Pacific Garbage Patch)
	Debris	Trash, shipwrecks	

Sources of freshwater pollution include:

1. Agricultural run-off, sewage, industrial discharge and solid domestic waste.

Sources of marine pollution include:

2. Rivers, pipelines, the atmosphere and human activities at sea, both operational and accidental discharges.

Measuring water pollution

Water pollution can be measured directly or indirectly by sampling.

See sub-topic 2.5 for this.

1. Biochemical oxygen demand

> **Key term**
>
> **Biochemical oxygen demand (BOD)** is a measure of the amount of dissolved oxygen required to break down the organic material in a given volume of water through aerobic biological activity by microorganisms.

2. Biotic indices and indicator species

> **Key terms**
>
> **Indicator species** are plants and animals that show something about the environment by their presence, absence, abundance or scarcity.
>
> A **biotic index** indirectly measures pollution by assaying the impact on species within the community according to their tolerance, diversity and relative abundance.

Indicator species are the most sensitive to change so they are the early warning signs that something may have changed in an ecosystem. For example, canaries were once taken down coal mines in Britain because they are more sensitive than humans to poisonous gases (carbon monoxide, methane). If gases were present in the mine the bird would die and so warn the miners in time for them to escape.

> **TOK**
>
> A wide range of parameters are used to test the quality of water and judgements are made about cause and effect. How can we know cause and effect relationships, given that we can only ever observe correlation?

> **To do**
>
> Four factories discharge effluent containing organic matter into rivers. The table shows the volume of discharge into the river and the resulting biological oxygen demand.
>
> | Factory | Volume of effluent/ 1000 l day^{-1} | BOD/g l^{-1} |
> |---------|--|----------------|
> | A | 14.0 | 27 |
> | B | 1.0 | 53 |
> | C | 3.0 | 124 |
> | D | 0.8 | 33 |
>
> 1. Explain whether these pollution data are for point sources or non-point sources.
>
> 2. Which pollution source, point source or non-point source, is easier to regulate?
>
>   Explain your choice.
>
> 3. Which factory is adding most to the BOD of the river into which it discharges?
>
> 4. If factory C discharges water at a temperature of 50 °C, give three possible effects on the organisms in the river.

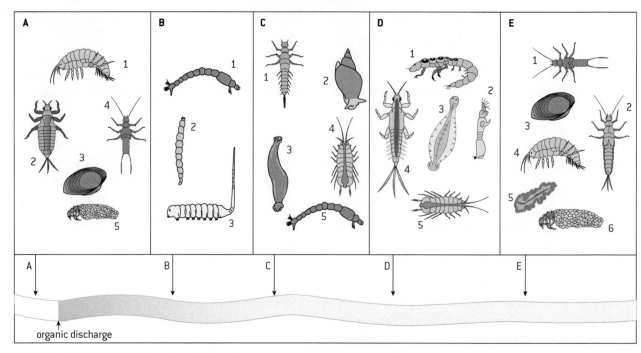

▲ **Figure 4.4.1** Invertebrates found in fresh water

A biotic index is a scale (1–10) that gives a measure of the quality of an ecosystem by the presence and abundance of the species living in it.

Figure 4.4.1 shows some typical invertebrates that might be found above and at various points below a **point source** sewage outfall pipe.

The pollutants are not directly measured but their effect on biodiversity is measured.

Figure 4.4.2 (left) shows some typical invertebrates that might be found above and at various points below a point source sewage outfall pipe.

Invertebrates are used to estimate levels of pollution, as they are sensitive to decreases in oxygen concentration in water, caused by the action of aerobic bacteria as they decompose organic matter. The presence of certain indicator species that can tolerate various levels of oxygen is used to calculate a biotic index, a semi-quantitative estimate of pollution levels.

Biotic indices based on indicator species are usually used at the same time as BOD. BOD gives a measure of pollution at the instant a water sample is collected whereas indicator species give a summary of recent history.

A diversity index can also compare two bodies of water (see 2.5 for Simpson's diversity index). What values might you expect for clean and polluted water?

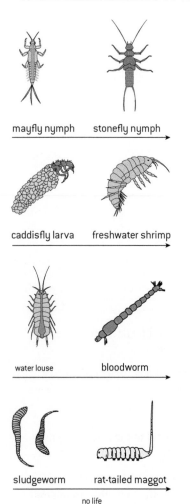

▲ **Figure 4.4.2**

mayfly nymph   stonefly nymph

caddisfly larva   freshwater shrimp

water louse   bloodworm

sludgeworm   rat-tailed maggot

no life

Eutrophication

Eutrophication is when excess nutrients are added to an aquatic ecosystem. It can be a natural process but anthropogenic eutrophication is more common.

When it is severe, it results in **dead zones** in oceans or freshwater where there is not enough oxygen to support life. In less severe cases, **biodegradation of organic material** uses up oxygen which can lead to **anoxic** (low oxygen) conditions and then anaerobic decomposition. This can release methane, hydrogen sulphide and ammonia which are all toxic gases.

Impacts of eutrophication

Anthropogenic eutrophication leads to unsightly rivers, ponds and lakes covered by green algal scum and duckweed. They also give off foul-smelling gases like hydrogen sulphide (rotten egg smell). Other changes include:

- oxygen-deficient (anaerobic) water
- loss of biodiversity and shortened food chains
- death of higher plants (flowering plants, reeds, etc.)
- death of aerobic organisms – invertebrates, fish and amphibians
- increased turbidity (cloudiness) of water.

The excess nutrients are nitrates and phosphates and they come from:

- detergents
- fertilizers
- drainage from intensive livestock rearing units
- sewage
- increased erosion of topsoil into the water.

These sources may be point sources, for example wastewater from cities and industry or animal feedlots, overflows from storm drains or sewers, and run-off from construction sites. Harder to identify are the sources of pollution from non-point sources. These may be run-off from crops and grassland, urban run-off, leaching from septic tanks or leaching from landfill sites or old mines.

The process of eutrophication

1. Fertilizers wash into the river or lake.
2. High levels of phosphate in particular allow algae to grow faster (as phosphate is often limiting).
3. Algal blooms form (mats of algae) that block out light to plants beneath them, which die.
4. More algae mean more food for the zooplankton and small animals that feed on them. They are food to fish which multiply as there is more food so there are then fewer zooplankton to eat the algae.
5. Algae die and are decomposed by aerobic bacteria.
6. But there is not enough oxygen in the water so, soon, everything dies as food chains collapse.

Key term

Eutrophication can occur when lakes, estuaries and coastal waters receive inputs of nutrients (nitrates and phosphates) which result in an excess growth of plants and phytoplankton.

▲ **Figure 4.4.3** Algal bloom in a village river in the mountains near Chengdu, Sichuan, China

7. Oxygen levels fall lower. Dead organic material forms sediments on the lake or river bed and turbidity increases.

8. Eventually, all life is gone and the sediment settles to leave a clear blue lake.

▲ **Figure 4.4.4** Dead fish resulting from toxins or oxygen depletion in Lake Binder, Iowa. Photo credit: Dr. Jennifer L. Graham, U.S. Geological Survey

▲ **Figure 4.4.5** A red tide in the Gulf of Mexico

The Gulf of Mexico has the largest dead zone in the USA caused by excess nitrates and phosphates from the Mississippi River basin agriculture. One solution proposed to this is a strategy of nutrient trading as a voluntary market-based reduction of nutrient use. It finds the most cost-effective ways to reduce nutrient use in both industry and on farms. It is a type of water quality trading scheme which allows those that can reduce nutrients at low cost to sell credits to those facing higher-cost nutrient reduction options. Nutrient trading, therefore, could allow sources of pollution to meet their pollution targets in a cost-effective manner and could create new revenue opportunities for farmers, entrepreneurs, and others who implement low-cost pollution reduction practices.

Red tides

In coastal waters, algal blooms (large numbers of phytoplankton) are sometimes caused by excess nutrients. If these phytoplankton are a species of dinoflagellate, the bloom looks red. This can be dangerous as the algae produce toxins which kill fish and accumulate in shellfish, which can make humans seriously ill.

In lakes or slow moving water bodies, eutrophication leads to a series of damaging changes, which severely reduces biodiversity. In fast moving water, eutrophication leads to a temporary reduction in biodiversity downstream which can be followed by a recovery and restoration of clean water.

▲ **Figure 4.4.6** Eutrophication in the Caspian Sea (lighter areas of algal bloom)

Eutrophication management strategies

Strategy for reducing pollution	Example of action
Altering the human activity producing pollution	1. Ban or limit detergents with phosphates (it is there to improve the performance of the detergent in hard water areas). 2. Use ecodetergents with no phosphates or new technology in washing machines. 3. Plant buffer zones between the fields and water courses to absorb the excess nutrients. 4. Stop leaching of slurry (animal waste) or sewage from their sources. 5. Educate farmers about more effective timing for fertilizer application.
Regulating and reducing the pollutants at the point of emission	Treat wastewater before release to remove phosphates and nitrates. Divert or treat sewage waste effectively. Minimize fertilizer dosage on agricultural lands or use organic matter instead.
Clean up and restoration	Pumping air through the lakes. Dredging sediments with high nutrient levels from the river and lake beds. Remove excess weeds physically or by herbicides and algicides. Restock ponds or water bodies with appropriate organisms.

▲ **Figure 4.4.7** Table of replace, regulate and restore models of pollution management for eutrophication

To do

1. Copy the table of strategies and actions on eutrophication and add a third column, entitled Evaluation. Complete this column.

2. Eutrophication can occur by point source river pollution due to nutrient input from an outfall pipe discharging sewage.

3. Copy the axes in figure 4.4.8 and draw curves to show changes in detritus (sewage), turbidity (suspended solids), bacterial growth, BOD, oxygen concentration, invertebrate and fish biodiversity, clean water species. Make a key to your curves.

4. Complete another copy of figure 4.4.8 to show the effects of high nutrient levels, eg from over-fertilized fields, being washed into a stream. Draw curves to show changes in nutrients (nitrate and phosphate), algal growth (bloom), detritus increase (due to higher plant death), bacterial growth, BOD, oxygen concentration, invertebrate biodiversity, clean water species.

5. Make a key to your curves.

Practical Work

* Countries with limited access to clean water often have higher incidences of water-borne illnesses. Research one example of this and the solutions being adopted.

* With respect to measuring aquatic pollution, compare a polluted and an unpolluted site (eg upstream and downstream of a point source).

* Evaluate the uses of indicator species and biotic indices in measuring aquatic pollution.

* Analyse water pollution data from a secondary source.

* Investigate the impact of intensive agriculture on a local aquatic ecosystem.

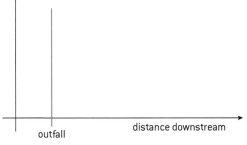

▲ **Figure 4.4.8**

Discuss the possible sustainable solutions to our freshwater crisis.

Comment on how global fisheries can be sustainably managed.

BIG QUESTIONS

Water, food production systems and society

Evaluate the hydrological cycle as a model.

To what extent are our solutions to ocean pollution mostly aimed at prevention, limitation or restoration?

Reflective questions

→ The hydrological cycle is represented as a systems model. To what extent can systems diagrams effectively model reality, given that they are only based on limited observable features?

→ Aid agencies often use emotive advertisements around the water security issue. To what extent can emotion be used to manipulate knowledge and actions?

→ Inuit people have an historical tradition of whaling. To what extent does our culture determine or shape our ethical judgements?

→ A wide range of parameters are used to test the quality of water and judgements are made about cause and effect. How can we know cause and effect relationships, given that we can only ever observe correlation?

→ Some rivers flow through more than one country. How might this lead to international disputes?

→ Different countries have different access to freshwater how can this cause conflict between countries?

→ Countries with limited access to clean water often have higher incidences of waterborne illnesses. Explain why this is the case and discuss whether responsibility for reducing these lies with the country or the international community.

→ Fish and many other species do not respect natural boundaries and for successful management of marine and some freshwater fisheries partnership between different nations is required. How can this work? Who decides?

Quick review

All questions are worth 1 mark

1. The amount of water present in the hydrosphere as freshwater is approximately

   A. 3%.

   B. 6%.

   C. 10%.

   D. 20%.

2. The percentage of the Earth's surface covered by oceans is about

   A. 90%.

   B. 70%.

   C. 50%.

   D. 45%.

3. Which of the following contains the greatest proportion of the world's freshwater?

   A. Organisms

   B. The atmosphere

   C. Ice caps and glaciers

   D. Streams, rivers and lakes

4. Which of the following is most likely to lead to an overall increase in the Earth's freshwater storages?

   A. Removal of forests

   B. Melting of polar ice caps

   C. Increase in evaporation rates from the oceans causing increase in precipitation over the continents

   D. Discovery of new underground aquifers

5. The diagram below represents the inputs and outputs of water from a lake over a period of time.

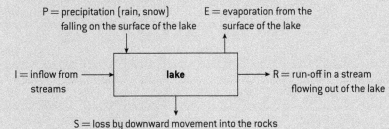

P = precipitation (rain, snow) falling on the surface of the lake

E = evaporation from the surface of the lake

I = inflow from streams

lake

R = run-off in a stream flowing out of the lake

S = loss by downward movement into the rocks

Assuming there is neither a rise nor a fall in the level of the lake, and there are no other inputs or outputs, what is the value of R?

   A. $P + I - E - S$

   B. $I + P + E + S$

   C. $(S + E) \times (I + P)$

   D. $I + P + E - S$

6. Which row correctly shows the storages of the water in order of decreasing size?

	LARGEST ⟶			SMALLEST
A.	Oceans	Ground water	Glaciers and ice caps	Lakes, rivers and atmosphere
B.	Oceans	Lakes, rivers and atmosphere	Glaciers and ice caps	Groundwater
C.	Lakes, rivers and atmosphere	Oceans	Ground water	Glaciers and ice caps
D.	Oceans	Glaciers and ice caps	Ground water	Lakes, rivers and atmosphere

7. Which of the following statements is correct?

   A. Less than 0.1% (by volume) of the Earth's water is freshwater.

   B. The hydrological cycle is independent of solar energy.

   C. The main reservoir of the Earth's freshwater is in ice caps and glaciers.

   D. Only abiotic storages are involved in the hydrological cycle.

5 SOIL SYSTEMS AND SOCIETY

5.1 Introduction to soil systems

Significant ideas:
→ The soil system is a dynamic ecosystem that has inputs, outputs, storages and flows.
→ The quality of soil influences the primary productivity of an area.

Applications and skills:
→ **Outline** the transfers, transformations, inputs, outputs, flows and storages within soil systems.
→ **Explain** how soil can be viewed as an ecosystem.
→ **Compare** and **contrast** the structure and properties of sand, clay and loam soils, with reference to a soil texture diagram, including their effect on primary productivity.

Knowledge and understanding:
→ The soil system may be illustrated by a **soil profile** that has a layered structure (horizons).
→ Soil system **storages** include organic matter, organisms, nutrients, minerals, air and water.
→ **Transfers** of material within the soil including biological mixing, leaching (minerals dissolved in water moved through soil) contribute to the organization of the soil.
→ There are **inputs** of organic material including leaf litter and inorganic matter from parent material, precipitation and energy. **Outputs** include uptake by plants and soil erosion.

→ **Transformations** include decomposition, weathering and nutrient cycling.
→ The structure and properties of **sand**, **clay** and **loam soils** differ in many ways, including: mineral and nutrient content, drainage, water-holding capacity, air spaces, biota and potential to hold organic matter. Each of these variables is linked to the ability of the soil to promote primary productivity.
→ A **soil texture triangle** illustrates the differences in composition of soils.

What is soil?

Soil is a complex ecosystem. It is made up of minerals, organic material, gases and liquids which forms the habitat for many animals and plants.

We tend to take the soil around us for granted, but it is more than just mud or dirt.

- All the food that we consume depends ultimately upon soil.

- Plants grow in soil and we either eat plants that grow directly in the soil or animals that have eaten plants.

- Soil is a habitat for many organisms. (In fact in some ecosystems the below ground biomass is greater than the above ground biomass.)

- As well as holding water and mineral nutrients that plants depend upon, soils act as an enormous filter for any water that passes through it, often altering the chemistry of that water.

- Soils store and transfer heat so affecting atmospheric temperature which in turn can affect the interactions between soil and atmospheric moisture.

- Soils are the part of the lithosphere where life processes and soil forming processes both take place.

We mentioned the four spheres of the Earth in sub-topic 1.2: the atmosphere, hydrosphere, lithosphere and biosphere. However there is also the pedosphere (soil sphere). This is a thin bridge between biosphere and lithosphere and is acted upon and influenced by the atmosphere, the hydrosphere and the lithosphere.

Storages	• Organic matter, organisms, nutrients, minerals, air and water.
Transfers within the soil	• Biological mixing, translocation (movement of soil particles in suspension) and leaching (minerals dissolved in water moved through soil).
Inputs	• Organic material including leaf litter and inorganic matter from parent material, precipitation and energy.
Outputs	• Uptake by plants and soil erosion.
Transformations	• Decomposition, weathering and nutrient cycling.

▲ **Figure 5.1.1** The components of the soil system

What is soil made from?

Soils are made up of four main components:

- Mineral particles mainly from the underlying rock.

- Organic remains that have come from the plants and animals.

- Water within spaces between soil grains.

- Air also within the soil grains.

It is also a habitat for plants and animals. Soil is a highly porous medium typically with a 50:50 mix of solids and pore spaces. The pore spaces contain variable amounts of water and air.

'Too many people have lost sight of the fact that productive soil is essential to the production of food.'

H H Bennett, 1943

To do

Make a systems diagram showing the relation between soil, the biosphere, hydrosphere and atmosphere. Add flows and storages to your diagram.

237

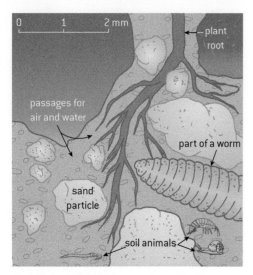

0   1   2 mm

plant root

passages for air and water

part of a worm

sand particle

soil animals

▲ **Figure 5.1.2** Cross-section of soil

It is the exact mix of these four portions that give a soil its character but it is not just these that make a soil what it is. The soils within any environment are the result of a mix of complex soil forming processes. Climate, parent rock material, the shape of the land, the organisms living on and within it and time all contribute to and affect the finished soil.

Fraction	Constituents	Function
Rock particles	Insoluble – eg gravel, sand, silt, clay and chalk. Soluble – eg mineral salts, compounds of nitrogen, phosphorus, potassium, sulphur, magnesium etc.	• provides the skeleton of the soil • derived from the underlying rock or from rock particles transported to the environment, eg glacial till.
Humus	Plant and animal matter in the process of decomposition.	• gives the soil a dark colour. • as it breaks down, it returns mineral nutrients back to the soil. • absorbs and holds on to a large amount of water.
Water	Water either seeping down from precipitation or moving up from underground sources by capillary action.	• dissolved mineral salts move through the soil and so become available to plants. • rapid downward movement of water causes leaching of minerals. • rapid upward movement can cause salinisation • large volumes of water in the soil can cause waterlogging leading to anoxic conditions and acidification.
Air	Mainly oxygen and nitrogen.	• Well-aerated soils provide oxygen for the respiration of soil organisms and plant roots.
Soil organisms	Soil invertebrates, microorganisms and large animals.	• large particles of dead organic matter are broken down by soil invertebrates like worms = smaller particles. • smaller particles are decomposed by soil microorganisms thus recycling mineral nutrients. • larger burrowing soil animals (eg moles) help to mix and aerate the soil.

▲ **Figure 5.1.3** Table of constituents of soil

What does a soil look like?

If you dig a trench in the ground, the side of the trench creates a soil profile, a cross-section. This profile changes as it goes down from the surface towards the underlying base rock. It is a record of the processes that have created the soil, its mineral composition, organic content, and chemical and physical characteristics such as pH and moisture.

Horizons

In cross-section soils have a profile which is modified over time as organic material leaches (washes) downwards and mineral materials move upwards. These processes sort the soil into distinctive horizons (zones/levels) that are often visible in the soil.

The top layer of the soil is often rich in organic material whilst the lower layers consist of inorganic material. The inorganic material is derived from the weathering of rocks. Within this, materials are sorted and layers are formed by water carrying particles either up or down – known as **translocation.** In hot, dry climates, where Precipitation<Evaporation (P<E), water is evaporating at the soil surface and water from lower soil layers moves upwards. When doing so, it dissolves minerals and takes them to the surface, where the minerals are left behind when the water evaporates. This also happens in irrigation and is called **salinization**. In colder and wetter climates, water flows down in the soil, dissolving minerals and transporting them downwards. This is **leaching** when P>E.

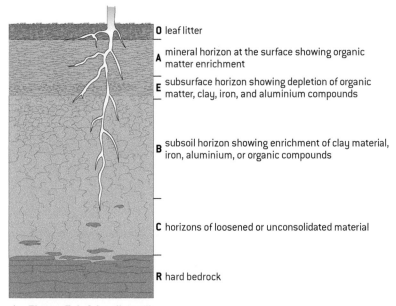

O leaf litter

A mineral horizon at the surface showing organic matter enrichment

E subsurface horizon showing depletion of organic matter, clay, iron, and aluminium compounds

B subsoil horizon showing enrichment of clay material, iron, aluminium, or organic compounds

C horizons of loosened or unconsolidated material

R hard bedrock

▲ **Figure 5.1.4** A soil profile

O Horizon: Many soils contain an uppermost layer of newly added organic material – this comes from organisms that die and end up on top of the soil. There, fungi, bacteria and many different kinds of animals will start to decompose the dead material.

A Horizon: Upper layer. In many soils, this is where humus builds up. Humus forms from partially decomposed organic matter and is often mixed with fine mineral particles. Often decomposition is incomplete and a layer of dark brown or black organic material is formed – the **humus layer**. In normal conditions, organic matter decomposes rapidly through the decomposer food web releasing soluble minerals that are then taken up by plant roots. Waterlogging reduces the number of soil organisms, which results in a build-up of organic matter and can eventually lead to the formation of peat soils.

B Horizon: This is the layer where soluble minerals and organic matter tends to be deposited from the layer above. In particular clay and iron salts can be deposited in this horizon.

C Horizon: This layer is mainly weathered rock from which the soil forms.

R horizon: Parent material (bedrock or other medium).

Not all soils contain all three A, B and C horizons; sometimes only two horizons can be distinguished while in other soils there may not be distinct layering. In some cases we cannot dig deep enough to find the C layer.

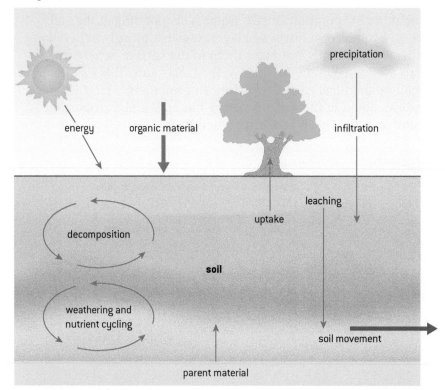

▲ **Figure 5.1.5** The soil system

Soil structure

The mineral portion of soil can be divided up into three particles based on size: sand, silt and clay. Most soils consist of a mixture of these soil particles and the soil texture therefore depends on the relative proportions of sand, silt and clay particles.

Particle diameter	Particle
< 0.002 mm	Clay
0.002–0.05 mm	Silt
0.05–2 mm	Sand

▲ **Figure 5.1.6** Sandy, clay and loam soil types

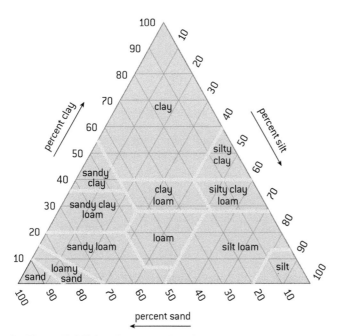

▲ **Figure 5.1.7** A soil texture triangle

The proportion of each of these particles gives soil its texture. Most soils also contain particles that are larger than 2 mm in diameter (pebbles and stones), but these are not considered in a description of soil texture.

It is possible to feel the texture of moist soil if you rub it between your fingers.

- Sandy soils are gritty and fall apart easily.

- Silty soils feel slippery like wet talcum powder and hold together better than sandy soils.

- Clay soils feel sticky and can be rolled up into a ball easily.

Most soils contain a mixture of different soil particles and can be described as sandy clay or silty clays. If a fairly equal portion of each size is present then the soil is said to be a loam.

It is possible to get a clearer picture of the soil proportions of soil particles by drying out a soil sample and passing it through a series of sieves of decreasing mesh size – first 2 mm, then 5 mm and finally 0.005 mm – separating the soil into its portion of clay, silt and sand particles. Alternatively you can place a sample of soil in a jam jar, fill it with water, shake it vigorously then leave it to settle out. The heaviest particles will settle first (sand) and the finer particles will settle last (clay).

Soil texture is an important property of a soil, as it determines the soil's fertility and the primary productivity. This is exemplified in figure 5.1.8.

Loam soils are ideal for agriculture.

- The sand particles ensure good drainage and a good air supply to the roots. The clay retains water and supplies nutrients – so they are fertile.

- The silt particles help to hold the sand and clay particles together and they can be worked easily.

Practical Work

* What value systems can you identify at play in the causes and approaches to resolving the issues addressed in soil conservation?

* In what ways might the solutions explored in soil systems and societies alter your predictions for the state of human societies and the biosphere some decades from now?

	Sandy soil	**Clay soil**	**Loam soil**
Composition (%)			
Sand	100	15	40
Silt		15	20
Clay		70	40
Mineral content	High	High	Intermediate
Potential to hold organic matter	Low	Low	Intermediate
Drainage	Very good	Poor	Good
Water-holding capacity	Low	Very high	Intermediate
Air spaces	Large	Small	Intermediate
Biota	Low	Low	High
Primary productivity	Low	Quite low	High

▲ **Figure 5.1.8** Comparison of three soil types

Porosity, permeability and pH

Related to a soil's texture is the amount of space between particles (**porosity**) and the ease at which gases and liquids can pass through the soil (**permeability**). Soils with a very fine clay particle texture have lots of micropores. This results in a combined large pore space made up of many microscopic pores but a low permeability as water molecules easily fill these spaces and adhere to the clay surface trapping the water as a film around the clay particles. In contrast, sandy soils have fewer macropores with a smaller total space. These spaces are too large for water's adhesive properties to work so sandy soils drain well.

The low permeability of clay means that it can also lock dissolved minerals between the pores making it hard for plant roots to access. Strangely the result can be a soil that is rich in minerals but has low fertility.

These features of clay soils also encourage high acidity, which has a great effect on the chemical characteristics. As the soil absorbs more water clay particles begin to fill up with positive hydrogen ions (H^+). This binds soil water tightly to the clay particles and makes the soil more acidic. It also reduces the amount of other positive ions that can bind, allowing potassium, magnesium and ammonium to be lost through leaching. Along with the loss of these important nutrients, as soil pH decreases, ions of aluminium and iron start to become more available to plants. Both of these are plant toxins.

Acidification of soils has had a major impact on forestry in Northern Europe where acid rain caused by industrial pollution has made the soil more acid. This in turn has meant more available aluminium and iron ions in the soil causing damage to evergreen forestry through needle death.

Soil sustainability

Fertile soil is a non-renewable resource. Once it is lost, it cannot be replaced quickly.

Soil formation takes a very long time. The natural soil renewal rate is about 1 tonne ha^{-1} yr^{-1} in natural ecosystems under the best conditions (wet, temperate climate), equivalent to 0.05–0.1 mm per year. And this is only after the initial chemical and physical weathering has occurred and fine material and soil organisms are present. So what this actually represents is the natural rate at which soils regain their fertility. As a consequence, soil use often exceeds soil formation and therefore soil should be considered a non-renewable resource/natural capital.

Fertile soil has enough nutrients for healthy plant growth. The main nutrients are **nitrates, phosphates and potassium** (**NPK**) and there are many micronutrients that plants also need. These nutrients can be leached out of soil or removed when a crop is harvested. They have to be replaced in agricultural soils via chemical fertilizer, growing legumes (2.1), crop rotation or through the application of organic matter (eg manure, compost).

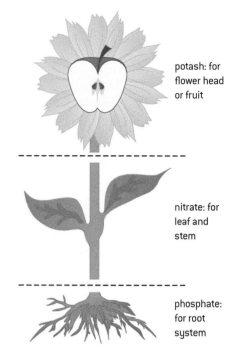

potash: for flower head or fruit

nitrate: for leaf and stem

phosphate: for root system

▲ **Figure 5.1.9**

To do

1.

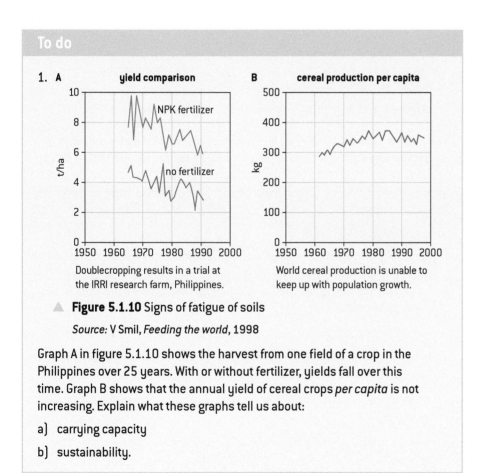

A **yield comparison**

NPK fertilizer

no fertilizer

t/ha

1950 1960 1970 1980 1990 2000

Doublecropping results in a trial at the IRRI research farm, Philippines.

B **cereal production per capita**

kg

1950 1960 1970 1980 1990 2000

World cereal production is unable to keep up with population growth.

▲ **Figure 5.1.10** Signs of fatigue of soils

Source: V Smil, *Feeding the world*, 1998

Graph A in figure 5.1.10 shows the harvest from one field of a crop in the Philippines over 25 years. With or without fertilizer, yields fall over this time. Graph B shows that the annual yield of cereal crops *per capita* is not increasing. Explain what these graphs tell us about:

a) carrying capacity

b) sustainability.

2. Study the table below.

NPP account

NPP consumed	Pg/yr	NPP dominated	Pg/yr	NPP lost	Pg/yr
consumed by humans	0.8	croplands	15.0	deforestation for crops	10.0
consumed by domestic animals	2.2	converted pastures	10.0	desertification	4.3
wood used by humans	2.4	tree plantations	2.6	occupation human	2.6
		occupied lands	0.4		
		fires, forage, wood	3.0		
		land clearing	10.0		
Total (4%)	5.2	Total (31%)	41.0	Total (13%)	16.9

One estimate of total Net Primary Production (NPP) of the whole world is 132 Pg/yr. Column 1 gives actual consumption. Column 2 gives the productivity of agricultural systems, and column 3 annual losses to nobody's benefit, but deforestation in column 3 produces land for column 2.

Total NPP sequestered by humans is 58 Pg/yr or 39% of total NPP.

Sources: Olson J S et al, *Carbon in live vegetation of major world ecosystems*. 1983. Oak Ridge Nat Lab. Vitousek P M et al: Human appropriation of the products of photosynthesis. 1986. *Bioscience* 36: 368–373.

▲ **Figure 5.1.11** Data on NPP of the world. (1 Pg (petagram) $= 10^{15}$g)

Note that the terrestrial NPP of 132 Pg carbon/year quoted above differs from most sources which estimate world NPP at 50 Pg carbon per year. The difficulty in calculating NPP is that it is a turnover factor. Satellite images can look at the green colour of chlorophyll and make an estimate of the standing crop in green leaves, but not NPP.

The consequences of the data:

- For 5.2 Pg/yr in direct consumption (col 1), humans are prepared to lose 16.9 Pg/yr permanently (in col 3). The seriousness of the situation is partly hidden by increasing agricultural yields, while not showing the amount of degradation of the seas.

- Net efficiency of human agricultural systems is 5.2/41 = 13%.

- Of the 132 Pg/yr world NPP, humans claim 58 (= 41+16.9) for themselves. The rest is for all other species.

- Of the remaining 80–90 Pg/yr, humans claim by deforestation a large share of 10 Pg/yr.

How much should we leave alone as natural ecosystems for the millions of other species?

What actions can we take to make this more sustainable?

5.2 Terrestrial food production systems and food choices

Significant ideas:

→ The sustainability of terrestrial food production systems is influenced by socio-political, economic and ecological factors.

→ Consumers have a role to play through their support of different terrestrial food production systems.

→ The supply of food is inequitably available and land suitable for food production is unevenly distributed, and this can lead to conflict and concerns.

Applications and skills:

→ **Analyse** tables and graphs that illustrate the differences in inputs and outputs associated with food production systems.

→ **Compare and contrast** the inputs, outputs and system characteristics for two named **food** production systems.

→ **Evaluate** the relative environmental impacts of two named food production systems.

→ **Discuss** the links that exist between socio-cultural systems and food production systems.

→ **Evaluate** strategies to increase sustainability in terrestrial food production systems.

Knowledge and understanding:

→ **The sustainability of terrestrial food production systems** is influenced by factors such as scale, industrialization, mechanization, fossil fuel use, seed/crop/livestock choices, water use, fertilizers, pest control, pollinators, antibiotics, legislation and levels of commercial versus subsistence food production.

→ **Inequalities exist in food production and distribution** around the world.

→ **Food waste** is prevalent in both LEDCs and MEDCs but for different reasons.

→ Socio-economic, cultural, ecological, political and economic factors can be seen to influence choice of food production systems.

→ As the human population grows, along with urbanization and degradation of soil resources the **availability of land** for food production per capita **decreases**.

→ The yield of food per unit area from lower trophic levels is greater in quantity, lower in cost and may require fewer resources.

→ **Cultural choices** may influence societies to harvest food from higher trophic levels.

→ Terrestrial food production systems can be **compared and contrasted** according to inputs, outputs, system characteristics, environmental impact and socio-economic factors.

→ **Increased sustainability** may be achieved by

- Altering human activity to reduce meat consumption and increase consumption of organically grown, locally produced, terrestrial food products.

- Improving the accuracy of food labels to assist consumers in making informed food choices.

- Monitoring and control by governmental and intergovernmental bodies – multinational and national food corporations' standards and practices.

- Planting of buffer zones around land suitable for food production to absorb nutrient run-off.

'Abundance does not spread, famine does.'

Zulu proverb

'This is a sad hoax, for industrial man no longer eats potatoes made from solar energy; now he eats potatoes partly made of oil'.

Howard T. Odum, an American ecologist, referring to the common perception that modern agriculture has freed society from limits imposed by nature, when in fact it is highly dependent on non-renewable fossil fuels (1971).

Key points

- There is enough food in the world but an imbalance in its distribution.
- Terrestrial and aquatic food production systems have different efficiencies.
- Different food production systems have different impacts and make different demands on the environment.
- Food production is closely linked with culture, tradition and politics.
- Our current use of soil to produce food is not sustainable.

▲ **Figure 5.2.1** Wheat and rice – two staple crops for human populations

Sustainability of terrestrial food production systems is influenced by many factors listed in figure 5.2.2.

Types of farming systems

Subsistence farming is the provision of food by farmers for their own families or the local community – there is no surplus. Usually mixed crops are planted and human labour is used a great deal. There are relatively low inputs of energy in the form of fossil fuels or chemicals. With low capital input and low levels of technology, subsistence farmers are unlikely to produce much more than they need. They are vulnerable to food shortages as little is stored. **Cash cropping** is growing crops for the market, not to eat yourself.

Commercial farming takes place on a large, profit-making scale, maximizing yields per hectare. This is often by a monoculture of one crop or one type of animal. High levels of technology, energy and chemical input are usually used with corresponding high outputs.

Farming may also be described as **extensive** or **intensive**. Extensive farming uses more land with a lower density of stocking or planting and lower inputs and corresponding outputs. Intensive farming uses land more intensively with high levels of input and output per unit area. Animal feedlots are intensive.

Pastoral farming is raising animals, usually on grass and on land that is not suitable for crops. **Arable** farming is growing crops on good soils to eat directly or to feed to animals. **Mixed** farming has both crops and animals and is a system in itself where animal waste is used to fertilize the crops and improve soil structure and some crops are fed to the animals.

Definitions

- **LEDC – Less economically developed country:** a country with low to moderate industrialization and low to moderate average GNP per capita.
- **MEDC – More economically developed country:** a highly industrialized country with high average GNP per capita.
- **Agribusiness:** the business of agricultural production including farming, seed supply, breeding, chemicals for agriculture, machinery, food harvesting, distribution, processing and storage.
- **Commercial agriculture (or farming):** large scale production of crops and livestock for sale.
- **Subsistence agriculture (or farming):** farming for self-sufficiency to grow enough for a family.

Factor	Which is more sustainable?	
	Commercial farming: mostly in MEDCs	**Subsistence farming: mostly in LEDCs**
Agribusiness (commercial vs subsistence)	Commercial farming is epitomized by many of the practices in this column but could still function with more environmentally sound practices.	Subsistence agriculture tends to be more associated with this side of the equation.
Scale of farming	Large-scale farming tends to rely heavily on machinery, chemicals and extensive use of fossil fuels.	Small-scale farming tends to be more labour intensive but may still rely on chemicals to boost production.
Industrialization	MEDCs have many people working in industry and so they must be provided with food from large-scale commercial farming.	LEDCs have limited industry so job opportunities are limited and people may have to grow their own food.
Mechanization	Use of lots of heavy machinery can damage the soil and uses a lot of fossil fuels.	Use of draft animals (donkeys, oxen) or human power is less stressful on the soil and can add manure. They are powered by plants so no burning of fossil fuels.
Fossil fuel use	A heavy dependence on fossil fuels is using a finite resource which produces large amounts of pollution.	Use of manual labour or draft animals does not cause these problems.
Seed/crop/livestock choices	Farming systems sometimes grow crops or keep animals that are not indigenous to the area and this can create the need for irrigation, glass houses, imports of feedstuffs.	Selecting organisms that are indigenous is less likely to create these problems.
Water use	Some agricultural systems have very heavy water demands and require large-scale irrigation solutions which divert water from people and may cause localized water supply problems and a drop in the water table.	Also requires water which can be used unsustainably.
Fertilizers, pest control	Growing the same crop continuously on the same land requires chemicals to support the soil and control pests. These often make their way into the local ecosystems – terrestrial and aquatic.	Crop rotation, biological pest control and other environmentally sound practices cause fewer problems but many subsistence farmers use large amounts of pesticides if they can afford them.
Antibiotics	Keeping animals in close quarters (often inside) causes the spread of diseases and this requires large amounts of antibiotics, often used routinely. If these make it into the local ecosystem they can cause super-bugs.	Free range animals tend to be healthier and in less need of antibiotics.
Legislation	Large-scale commercial interests are controlled by legislation so may pollute less.	Small-scale operations often go un-noticed by legal bodies.
Pollinators	Many commercial crops, eg almonds in California, require pollination by bees and other insects. Honey bee hives are brought in to provide this but honey bee colony collapse disorder is killing many bees. Some crops have no pollinators left and humans have to do this by hand.	With a more biodiverse farm, pollinators have different habitats and there are usually enough insects to pollinate crops.

▲ **Figure 5.2.2** Factors influencing sustainability of agriculture

Figure 5.2.2 shows the two extremes of any situation but you should notice that many of these factors are inter-related and difficult to separate.

To read

Some basic food facts (from Myers and Kent)[1]

- Wheat is a staple food for over one third of the world's population (bread and pasta).

- Wheat, rice, maize (coarse grains), potato, barley, sweet potato, cassava, sorghum and millet are the staple carbohydrate foods for most humans.

- Grain production provides half the human population's calories.

- World food production is concentrated in the Northern hemisphere temperate zones.

- There are 23 billion livestock animals on Earth.

- There are 1 billion pigs and 3 billion ruminants.

- Livestock need 3.5 billion hectares of land.

- Crops take up 1.5 billion hectares of land – about 11% of the total land area.

- There are 13.5 billion hectares of land on Earth, but 90% of this is too dry, wet, hot, cold, steep or poor in nutrients for crops.

- In Africa, only 7% of the total land area is cultivated.

- LEDCs have 80% of the world's human population and eat 56% of the world's meat.

- The human population has increased by 70% in the last 30 years but world food production has increased by 17%.

Food facts

Malnutrition is an umbrella term for 'bad' nutrition and it is the result of a diet that is unbalanced. Nutrients may be:

- Lacking – undernourishment, usually a lack of calories.

- Excessive – overnourishment, usually too many calories leading to obesity.

- Unbalanced – the wrong proportion of micro-nutrients.

The Food and Agriculture Organization (FAO)[2] estimates that 925 million people in the world do not get enough energy from their food – they suffer from **undernourishment**. About 2% of these are in MEDCs. Most of the 98% are in Asia, Africa or Oceania. Of these, about 200 million are children and infants. Chronic undernourishment during childhood years leads to permanent damage: stunted growth, mental retardation and social and developmental disorders. Many people are also suffering from **an unbalanced diet** – their food contains enough energy, but lacks essential nutrients like proteins, vitamins and certain minerals.[3] According to the Hunger Site www.thehungersite.com about 10% of the undernourished millions die each year from starvation or malnutrition and three quarters of these are children under the age of five. How can we justify about a sixth of the world (approximately 13%) not having enough food when there are, apparently, large surpluses stored in MEDCs?

Food is potentially one of the most important resource issues facing global society today, alongside potable water (drinking water). As populations increase, as global trade expands and as market choice develops, greater and greater demands are being made on food supplies and food production systems. These are primarily agricultural systems operated on an industrial scale – agribusiness. In the last 50 years technology and science have made huge advances in agricultural practice and agricultural production but then human population has increased too.

In many MEDCs, the cost of food is relatively cheap. Most people purchase foods out of choice and preference rather than basic nutritional need. Seasonality of produce has disappeared in most MEDCs. Exotic foods are freely available all the year round. Modern technology and transport systems mean that New Zealand lamb, beans from Kenya, dates from Morocco and bananas from the Windward Islands can be bought in almost any MEDC supermarket anywhere in the world, all year round.

In LEDCs, many populations struggle to produce enough food to sustain their population. There may be political and economic agendas as well as simple environmental limitations on food production. Cereals can be grown for export and revenue generation rather than to feed indigenous populations – cash cropping. Crops other than food crops can be grown as cash crops: coffee, hemp, flax and biofuels. However they occupy land that could be used for food production and arable land is in finite supply.

[1] Myers N. and Kent J. 2005. *The New Atlas of Planet Management*. University of California Press.
[2] W. P. Cunningham, M. A. Cunningham and B.W. Saigo. *Environmental Science*, 7th Ed. McGraw-Hill Higher Education, New York 2003, 231.
[3] For more information see Cunningham et al., pages 231–235.

The choice of food that we grow and eat is determined by many factors:

1. **Climate** and local ecological conditions determine what will grow where on Earth. We adapt this through irrigation and using greenhouses to artificially alter the climate but most crops are grown without this.

2. **Cultural and religious** – some religions proscribe certain foods, eg Islam and Judaism proscribe eating pork, Hindus do not eat beef. Our traditions determine what foodstuffs we prefer.

3. **Political** – governments can subsidize or put tariffs on some foods to encourage or discourage their production, eg the EU manipulates production in this way.

4. **Socio-economic** – market forces determine supply and demand in a free-market economy, eg if there is a short supply of almonds or beans and prices rise, farmer may go into this crop, supply increases, prices fall and then they may stop growing it.

We all need food – it supplies the fuel that we need to live and breathe and some of us often take for granted the fact that food is readily available. But many of us have nowhere near enough food, let alone nourishment to keep us healthy and energetic. Why is this? Can we do anything about it? Is it anyone's fault?

> **To do**
>
> Consider your own food choices. Which of the factors listed (left) determine what you eat? Are there other factors involved as well?

To read

The rising cost of food

There will be billions more human mouths to feed in the next 50 years so an increase in the demand for food is not going to go away. Human population is now over 7 billion and is predicted by the UN to be 8 billion in 2025 and 9.2 in 2050. Because it is profitable at the moment, there is a massive increase in growing crops for **biofuel** (living plants converted to fuel to replace fossil fuels) and this uses land that would otherwise be used to grow crops for food. We cannot have it both ways as crop land is limited.

In 2008, there were food riots (for example in Burkina Faso, Egypt and Bangladesh) when people protested about the rising price of food. In Haiti, the poorest country, the government was overthrown due to the high cost of food. According to the FAO, 37 countries are facing food crises. Food costs a higher proportion of income for the poor than for the rich. For 100 million people or more, this could mean starvation and many are going without meals because they cannot afford them.

The World Food Programme could only buy half as much food in 2008 as in 2007 with the same amount of money. Wheat and soya prices, rapeseed oil and palm oil prices spiked but have fallen back again (figure 5.2.3). Biofuels are one of the reasons for this as farmers get subsidies for growing them so do not put their crops into the food chain. The increasing wealth of India and China means that more people can afford and want more meat to eat so demand for meat has not just increased

because there are more of us but because many of us eat more meat.

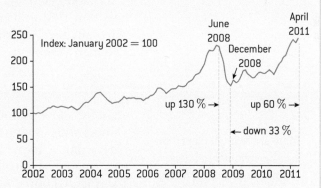

▲ **Figure 5.2.3** World food prices 2002–2011

While there is still enough food, we need to review the strategies on global food production.

1. What has happened to food prices now? Do your own research on this.

2. Do your own research on biofuels and read the section in sub-topic 7.1 on them.

   a. Evaluate the costs and benefits of growing biofuels on agricultural land.

   b. Which countries are most likely to see an increase in the production of biofuels and what impact will that have?

3. Is the country in which you live a net importer or exporter of food?

Food production and distribution around the world

Let's look first at agricultural production.

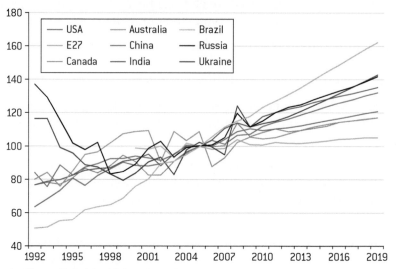

Figure 5.2.4 Agricultural production in key producers 1992–2019 (2004–6 = 100 on y axis)

Source: OECD/FAO

To do

Compare and contrast the agricultural production trends shown in figure 5.2.4.

So why is there such a difference in world agricultural production? If you search the internet or textbooks, many will say that there is plenty of food to go round. The American Association for the Advancement of Science suggests that there is an average of 2,790 calories of food available each day for every human on the planet – 23 percent more than in 1961 and enough to feed everyone.[4] Food production has actually kept up with world population growth, so why are there still so many problems with famine, hunger and malnutrition?

To think about

There are many factors that may be considered but is the main one distribution? It is well known that countries like the USA, Canada and Australia have more food than their indigenous population needs.

Consider who should pay to ship the excess foods from where they are stored to countries like Bangladesh, Ethiopia and Sudan, and is it the type of food they want?

What impact would this have on the local farmers and the markets?

Politics of food supply

If excess food is not paid for, does this put the receiving country forever under the power of the exporting country? What impact does this food have on the local farmers and the economy? What happens when corrupt governments do not distribute to those who need the food, and who decides who needs the food anyway? These are some of the questions that need to be considered where the issues involved in the imbalance in global food supply are concerned.

All people need to eat. However, huge differences exist between the diets of people in MEDCs and those in LEDCs.

	MEDCs	LEDCs
Average energy intake: cal/cap/day (calories per capita per day)	3314 USA 3774	2666 Eritrea 1512
Major types of food (% of total calorie intake)		
meat	12.9	7.3
fish	1.4	0.9
cereals	37.3	56.1

Figure 5.2.5 Average energy intake per day in MEDCs and LEDCS

[4] http://atlas.aaas.org/index.php?part=2&sec=natres&sub=crops

To start answering these issues, it is necessary to have an overview of the global pattern of world food supply. Figure 5.2.6 shows per capita calorie availability in MEDCs and LEDCs – those with low food supplies are called low income food deficient countries (LIFDCs).

To do

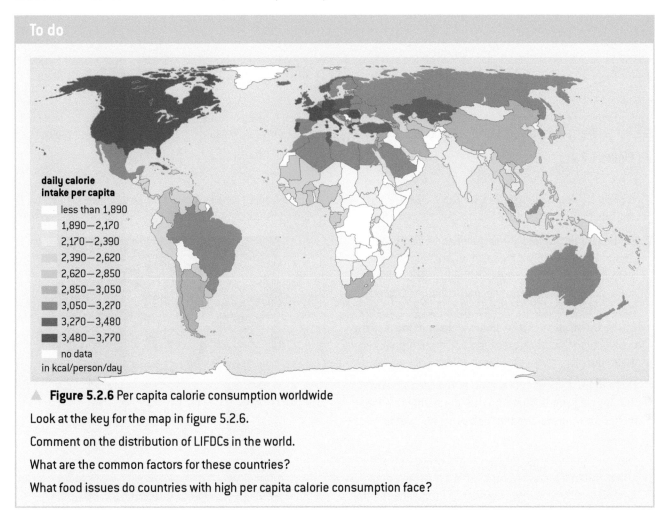

daily calorie intake per capita

less than 1,890
1,890—2,170
2,170—2,390
2,390—2,620
2,620—2,850
2,850—3,050
3,050—3,270
3,270—3,480
3,480—3,770
no data
in kcal/person/day

▲ **Figure 5.2.6** Per capita calorie consumption worldwide

Look at the key for the map in figure 5.2.6.

Comment on the distribution of LIFDCs in the world.

What are the common factors for these countries?

What food issues do countries with high per capita calorie consumption face?

Recent doubts

So far in the history of the human population, food supply has kept pace with population growth, confounding the Malthusians amongst us (see 8.1). But very recently, some are doubting that technology, efficiency and innovation will allow us to feed a world of 9 billion.

As we adapt more and more of the NPP on Earth to human needs, use and degrade more land, demand more meat, we must be reaching the limits of growth.

The 1.1 billion living in poverty appear to be increasing not decreasing and they are getting hungrier.

The global harvest of grain may still be increasing but annual grain yields per hectare have slowed their rate of increase since the Green Revolution as the benefits have been realized and we may be near the limit of productivity. As the human population increases, there is a fall in grain produced per capita (see figure 5.2.7).

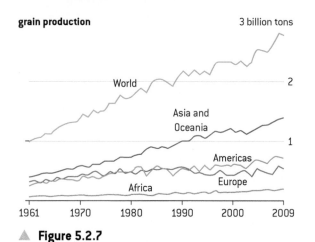

grain production 3 billion tons

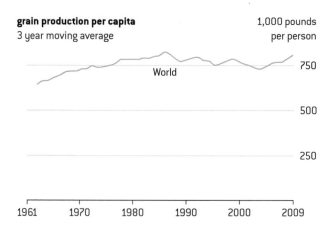

grain production per capita
3 year moving average

1,000 pounds
per person

▲ **Figure 5.2.7**

To think about

Bush meat is any wild animal killed for food. In some countries, this is called game, in others, hunting wild animals. The term bush meat is highly politicized as many see it as illegal hunting, particularly in Central and West Africa and tropical rainforests. Although bush meat can be many species, the emotive element focuses on the killing of the Great Apes for meat, often orphaning their young.

Trade in bush meat is increasing. The reasons are that logging roads built into the forests make access easier for hunters and the high price of meat in markets.

The cane rat is a large rodent pest of crops in West Africa and this is hunted for bush meat. Now, some farmers are successfully starting to farm cane rats for meat in Benin and Togo.

▲ **Figure 5.2.8** A cane rat

TOK

Consumer behaviour plays an important role in food production systems. Are there general laws that can describe human behaviour?

A brief history of agriculture

Livestock farming – why farm animals?

Man has been associated with animals for longer than he has been farming cereals. **Animal domestication** came before crop farming, first the dog, then sheep, goats, pigs and cattle were domesticated and used to fulfil a variety of needs. Dogs were used as hunting companions and ultimately as herding animals. In some cultures the dog was also a source of food. Sheep, goats and pigs probably were hunted as wild prey and then corralled and domesticated to provide a more convenient and more reliable source of food. Cattle, reindeer, horses, donkeys, yaks, camels and llamas were (and are) used as food and as beasts of burden. In addition they provided wool and hides for clothing and even their bones, antlers and teeth had a value as tools and decorations.

Livestock are a useful means of converting plant material unsuitable for human digestion systems (grass) into valued protein. Sheep and goats living in Mediterranean scrub forests browse on woody scrub and trees digesting vegetation that is unpalatable to humans; pigs were often kept on farms as a way of processing waste products and producing valuable protein.

Growing crops

In **arable farming**, seeds of crop plants are deliberately sown into a soil that has been cleared of the natural vegetation. The seeds are usually planted into bare soil that has been previously conditioned (broken up) by plowing (digging over). The plants are a **monoculture** (all of one species) and grown in high density. Conditions that may be limiting are altered so their growth rate is as high as possible.

Fertilizers are added to encourage growth or flowering, soil that is too acidic has lime added, irrigation may be used. The plant varieties are most often 'nutrient hungry' – they extract from the soil quantities of nutrients and minerals often orders of magnitude greater in amount than indigenous species. They are plants that have over time been genetically selected to produce large volumes of biomass whether in tubers, seed heads or the general body of the plant. At the end of the season the crop is harvested as a food supply.

Harvesting requires the removal of the biomass from the field, the soil and the ecosystem. There has been a net loss of biomass, nutrients, minerals and water from the system. The soil is more impoverished than it had been. If another crop is extracted, the soil fertility is reduced again.

Crop rotation is one way of addressing loss of soil fertility. Leguminous crops (soya beans, peas, beans) add nitrogen to the soil so may be grown every fourth year, in a rotation with other crops.

Palm oil – rainforest in your shopping

▲ **Figure 5.2.9** Oil palm

The oil palm is a tropical palm tree indigenous to West Africa and Central America but imported to South East Asia in the early 1900s. Here it is grown for its oil. Half of the large plantations are in Malaysia and the rest are in Indonesia and other South East Asian countries. In Indonesia, the area of land occupied by palm oil plantations has doubled in the last 10 years and is still increasing. According to Friends of the Earth, an NGO, demand for **palm oil** is the most significant cause of **rainforest loss** in Malaysia and Indonesia.

Palm oil is high in saturated fats and semi-solid at room temperature. It is found in 1 in 10 food products:

- cooking oil and margarine
- processed foods: chocolate, bread, crisps
- cosmetics (lipsticks), detergents and soaps (eg Sunlight Soap and Palmolive)
- lubricants
- biofuel.

What do you use that contains palm oil?

The benefits of oil palm plantations are in providing employment and exports. Growing a few oil palms can bring an income for a subsistence farmer and large oil plantations and processing plants provide much needed employment.

However, oil palm plantations often replace tropical rainforest and, in Malaysia and Indonesia, primary rainforest has been cleared for oil palm. Often this forest is on peat bogs which are then drained and habitats lost.

To maintain the monoculture of oil palms, herbicide and pesticides are used on the plantations and these poison other animal species. Animals that were in the rainforest, such as elephants, move into the plantations seeking food and are killed as pests.

Comparing farming systems

Farming system	Shifting cultivation	Cereal growing	Rice growing	Horticulture and dairying
Where	Amazon rainforest	Canadian Prairies	Ganges Valley	Western Netherlands
Type	Extensive Subsistence	Extensive Commercial	Intensive Subsistence	Intensive Commercial
Inputs	Low – labour and hand tools	High use of technology and fertilizers	High labour, low technology	High labour and technology
Outputs	Low – enough to feed the family	Low per hectare but high per farmer	High per hectare, low per farmer	High per hectare and per farmer
Efficiency	High	Medium	High	High
Environmental impact	Low – only if enough land to move to and time for forest to regrow	High – loss of natural ecosystems, soil erosion, loss of biodiversity	Low – padi rice has a polyculture, stocked with fish. Also grow other crops	High – greenhouses for salads and flowers are heated and lit. In dairying, grass is fertilized, cows produce waste

▲ **Figure 5.2.10** Table of types of farming system

Farming's energy budget

The efficiency of a farming system can be measured in a number of ways.

1. Energy contained within the crop of harvested product per unit area.

    We can compare the relative energy returns from cereal and root crops, from wheat and corn, from beef or lamb. However this calculation is problematic. Does the calculation consider biomass harvested or does it consider only the marketable/edible portion of the harvest? In livestock terms do we consider the live weight of the animal or the dressed (just the meat) deadweight?

2. **Efficiency** of agricultural systems

    More scientifically and perhaps more honestly we could look at the **efficiency** – a system with inputs, outputs and storage that operate on a range of spatial and temporal scales. At the end of the system there is the production of a marketable product that is usually sold by weight, eg cereal (maize, wheat, rice etc.). It is therefore possible to calculate the energy contained in a food per unit volume (joules per gram). In order to calculate the energy balance of the farming system all you now need to do is calculate the energy it took to produce that food and deliver it to the market. You would need to consider the fuel, labour and energy that was used to prepare the soil, sow the seed, harvest the crop, prepare and appropriately package it for market, then transport to market. You would also have to factor in the energy cost of dealing with waste products associated with the farming system.

Data exists for efficiency of agriculture (see figure 5.2.11). Shockingly, for all but cereal growing, efficiency is less than one – there is more input than output.

To do

Do you think that cattle fed on grass only (pasture-fed) would have a higher or lower efficiency of conversion? What advantages or disadvantages may there be for pasture-fed instead of grain-fed cattle?

Efficiency expressed as energy output:energy inputs for:	
Dairying	0.38
Cattle rearing – grain fed (beef)	0.59
Sheep (lamb)	0.25
Pigs (pork)	0.32
Cereal growing	1.9
Food products	
White bread	0.525
Chicken	0.10
Battery eggs	0.14
Lettuce produced under glass	0.0017

▲ **Figure 5.2.11**

While total energy out is less than energy in when all factors are considered, remember that the **quality of the energy** is different.

- Fats and protein contain more energy content per gram than carbohydrates.

- You need to eat less meat and fish than cereals to get the same amount of energy.

- A higher energy content food costs less to transport as it has a lower volume.

To equate measures, the **grain equivalent** in kilograms (the quantity of wheat grain that would have to be used to produce one kilogram of that product) is sometimes used.

Rice production in Borneo and California

Two food production systems are discussed here:

- traditional, extensive rice production in Indonesian Borneo (Kalimantan)

- intensive rice production in California.

Traditional extensive rice production is characterized by low inputs of energy and chemicals, high labour intensity and a low productivity. Nearly all energy added is in the form of labour and seeds. The rice yield is rather low. However, the energy efficiency (defined as energy output/energy input) is high. As no fertilizer or pesticides are used (remember it takes energy to make these products and get them to market), the rice yield is the only output (no pollution).

Intensive rice production is characterized by high inputs of energy and chemicals, low labour intensity and a high productivity. The energy inputs are in the form of diesel or petrol, not in the form of labour. Large amounts of fertilizer (N, P) and pesticides (insecticides and herbicides) are used. As a result, high rice yields are obtained. However, because of the large energy inputs, the energy efficiency is much lower than in the traditional extensive rice production example. In some intensive agricultural production systems, the energy inputs are larger than the energy outputs. The intensive rice production system also has extra outputs compared to the extensive system: pollution in the form of excess fertilizer and pesticides. Note that these are not shown in figure 5.2.12: they were not measured.

		Energy (000 kcal ha^{-1})	
		Borneo	California
Inputs			
Direct energy:	labour	0.626	0.008
	axe and hoe	0.016	–
	machinery	–	0.360
	diesel	–	3.264
	petrol	–	0.910
	gas	–	0.354
Indirect energy:	nitrogen	–	4.116
	phosphorus	–	0.201
	seeds	0.392	1.140
	irrigation	–	1.299
	insecticides	–	0.191
	herbicides	–	1.119
	drying	–	1.217
	electricity	–	0.380
	transport	0.051	0.121
Outputs	rice yield	7.318	22.3698
	(protein yield)	(141 kg)	(462 kg)
Energy efficiency		7.08	1.55

▲ **Figure 5.2.12** Rice production systems in Borneo and California[5]

Other typical differences between traditional extensive and intensive agricultural production systems are:

Traditional extensive agriculture	Intensive agriculture
Limited selective breeding	Strong selective breeding
No genetically engineered organisms	Genetically engineered organisms
Polyculture (many different crops)	Monoculture (single crop)
Small effect on biodiversity	Reduction in biodiversity
Little soil erosion	Strong soil erosion

[5] K. Byrne, *Environmental Science,* 2nd Ed. Nelson Thornes, Cheltenham 2001, 180.

1. Copy the table below and do your own research to fill it in.

2. Repeat with another example of two named food production systems local to you.

3. For each system, evaluate the environmental impacts.

Factor		Intensive beef production in South America	Extensive beef production by Masai tribe of East Africa
Inputs	Eg fertilizer, water, pest control, labour, seed (GM or not), breeding stock, livestock growth promoters		
Outputs	Food quality, yield, pollutants, transportation, processing, packaging		
System characteristics	Diversity		
	Sustainability		
Environmental impacts	Pollution		
	Habitat loss		
	Biodiversity loss		
	Soil erosion/degradation		
	Desertification		
	Disease epidemics		
Socio-economic factors	Subsistence or for sale crop		
	Traditional or commercial		
	For export or local consumption		
	For quality or quantity		

Terrestrial versus aquatic food production systems

In terrestrial food production systems food is usually harvested at the first (crops) or second trophic level (meat usually originates from primary consumers like cows, pigs and chickens). This means that these production systems are making a rather efficient use of solar energy. Terrestrial systems do have higher losses when it comes to skeletal waste – land-based animals have more energy tied up in their skeletons as they have to support themselves on land.

In aquatic food production systems, most food comes from higher trophic levels. Typical food fish tend to be carnivorous and are quite often at trophic level 4 or higher. Because of the energy 'losses' at each trophic level, the energy efficiency of aquatic food production systems is lower than that of terrestrial systems. Note that, although the energy conversions along the food chain tend to be more efficient in aquatic ecosystems, the initial intake of solar energy is less efficient than in terrestrial ecosystems because of the absorption and reflection of sunlight by water. Also energy losses in the form of heat are higher in water than on land.

Increasing sustainability of food supplies

Feeding the world's population in the future is a challenge, as the population is growing and the area available to agriculture is decreasing. Not only many new mouths need to be fed, also today's undernourished and malnourished people need more and better quality food. The FAO has stated that by 2050, there will be another 2 billion humans to feed and this needs an increase in food production of 70% of current levels.

Factors that contribute to the decrease in agricultural land are:

- soil erosion
- salinization
- desertification
- urbanization.

We could improve sustainability of food supplies by:

1. **Maximize the yield** of food production systems.

    A problem is of course, how to do this without unsustainable practices. Ways to do this are:

    a. **Improve technology** of agriculture.
       i. Mixed cropping and interplanting conserves water and the soil.
       ii. No plow tillage, drilling seeds into the stubble of the previous crop also conserves water and soil.
       iii. Plant buffer zones around agricultural land to absorb nutrient run-off. This also provides habitat for wildlife.
       iv. Biological control of pests and integrated pest management reduce losses.
       v. Trickle irrigation is less wasteful of water.

    b. **Alter what we grow** and how we grow it.
       i. Genetically modified foods (GM). Inserting into cereals the gene from legumes that allows them to fix nitrogen would save the need for nitrogenous fertilizers. (See box on GM crops on p262).
       ii. Aquaculture and hydroponics.
       iii. Soil conservation measures (5.3).

    c. **A new Green Revolution.**
       i. Agroecology – where nutrients and energy are recycled on farms within closed systems with crops and animals balancing inputs and outputs.
       ii. Techniques of breeding plants more adapted to drought, increasing shade, keeping bare soil covered – drought-proofing farms.

2. Reduce **food waste** by improving storage and distribution.

    a. In LEDCs, food waste is mostly in **production and storage** eg loss through pest infestations, severe weather, lack of good storage such as no refrigeration, no canning factory nearby.

    b. In MEDCs, food waste is mostly in **consumption** eg consumers buying more food than needed and letting it go off, supermarkets having too strict standards (round apples, red strawberries) so rejecting much edible if misshapen food, packaging that may preserve foods but contaminates waste food that could otherwise be recycled for livestock.

3. Monitoring and control.

   a. By governmental and intergovernmental bodies – to regulate imports and exports to reduce unsustainable agricultural practices.

   b. By multinational and national food corporations to raise standards and practices on their supplier farms.

   c. By individuals in NGO pressure groups.

4. Change our attitudes towards food and our diets.

   a. Eat different crops.

   b. Eat less meat.

   c. Improve education about food.

   d. Increase consumption of insects – big protein source that reproduces rapidly and in large numbers.

Changing people's diet can improve the efficiency of food production. If we obtain more food from lower trophic levels (plants), we will greatly increase the amount of food available. People in MEDCs eat more meat than they actually need, so they could simply replace some of their meat by food taken from the first trophic level. However, the trend is the other way. More people in LEDCs are eating more meat. On average, in 2000, each human consumed 38 kg of meat per year. In the USA, it was 122 kg, UK and Brazil 77 kg, China 50 kg, India 5 kg.[6] We cannot all eat more meat. Even though animals graze on some land that would not support crops, it is energetically inefficient to feed our grain to animals which we then eat. In MEDCs, we mostly eat more than we need to sustain ourselves. Obesity has become a problem in some countries. We just do not seem able to get the balance right.

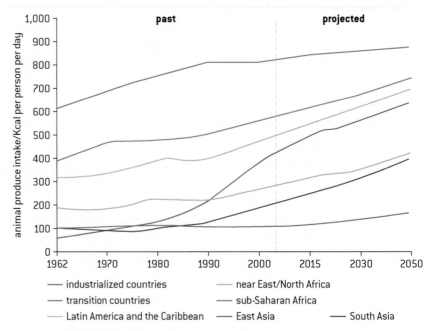

▲ **Figure 5.2.13** Meat consumption past and projected in various regions

6 Myers N. and Kent J. 2005. The *New Atlas of Planet Management*. University of California Press.

5. Reduce food processing, packaging and transport, and be more aware of food production efficiency by improving labelling of foodstuffs and increasing consumer awareness. While local foods may have lower food miles, they may cost more in energy used to produce them. For example, growing tomatoes in a heated greenhouse in a temperate country such as the UK may use more total energy than growing them in the tropics and flying them into the UK.

Predictions for future food supplies

The FAO has reported its predictions on food supply for 2030 and its predictions are as follows:

1. Human population will grow to about 8 billion by 2030, a slowdown in growth rate to 1.1% per year from 1.7% in 2000. Demand for agricultural products will also slow but not in LEDCs where consumption will increase.

2. The human population will be increasingly well-fed with per capita calorie consumption increasing to 3,000 kcal per day from 2,800 now.

3. Numbers of hungry people will decrease to about 440 million but in some areas (sub-Saharan Africa) this will not fall as much.

4. More people will eat more meat.

5. An extra billion tonnes of cereals will need to be grown.

6. LEDCs will have to import cereals.

7. Most increased production will be from higher yields and more irrigation not more land.

8. GM (genetically modified) crops, no tillage planting, soil conservation measures, improved pest control will all increase productivity.

9. Aquaculture will increase.

To do

Organic and local or grown in Kenya and air-freighted?

If all humans alive today adopted the diet of Europeans, we would need 2.5 more planets like the Earth to support us.

If we all ate organic food, we could not produce enough.

Commercial intensive agriculture has done a remarkable job in managing to feed the rapidly increasing human population. Bu t it may not in the future. What has to change?

In MEDCs, there is an increasing supply of organically produced foodstuffs and these are sold at a premium. The premise is that they are better for our health and free of pesticide residues. But there is little scientific evidence to support some claims for organic foods.

Locally grown foods are also marketed as 'better' due to lower food miles and artisanal local producers. But it is the large commercial farms and large supermarket chains that supply most of our food efficiently and at the lowest prices. We could not produce enough food if we all ate organic, local produce.

Food miles is a measure of how far food has moved from grower to your table. The premise is that more food miles means more bad environmental impact for a food. So shop for locally produced foods.

But this is a fallacy as it is efficiency of food production that we should be considering. Growing green beans in a heated glasshouse in Northern Europe uses far more energy than growing them in Kenya even including the carbon emissions of air-freighting them to Europe.

In fact, Kenya's bean industry is very environmentally cost efficient. Beans are grown:

- using manual labour (no fossil fuels)
- using organic fertilizer from cows
- with low-irrigation schemes.

It costs more in carbon emissions to drive 6.5 miles to a supermarket than flying a packet of beans from Kenya to that supermarket.[7]

Check out this article – http://shrinkthatfootprint.com/food-miles. Do you agree with this argument? Will it affect what foods you buy?

To think about

GM crops

GM or **genetically modified crops** have DNA of one species inserted into the crop species to form a **transgenic plant**. GM crops of soya bean, cotton and maize are the most common but the issue of GM crops is surrounded by controversy in some countries over ethics, food security and environmental conservation.

Proponents of GM crops say they are part of the solution to increasing food production. By making a crop disease or pest resistant, fewer chemicals need to be used and less is lost in spoilage. We are only doing what selective breeding has done since farming began but on a molecular scale. **Golden rice** has been made to synthesize beta-carotene, the precursor to vitamin A, so, as a humanitarian tool, it could stop vitamin A deficiency suffered by 124 million people. It has not yet been grown commercially.

▲ **Figure 5.2.14** Golden rice compared with white rice

Opponents of GM crops say we do not know what we are releasing into the environment. Could the GM plants cross-pollinate with other varieties so the introduced DNA escapes into the wild populations? Could the DNA cross the species barrier? If a GM crop can kill a pest species that feeds on it, will that species die out? What will the effect on food chains be? Labeling of GM foods is now demanded and some countries have rejected them totally.

The European Union's (EU) Common Agricultural Policy (CAP)

40% of the EU budget goes towards the CAP, a system of subsidies started in 1962 for agriculture in the EU. In this farmers are guaranteed a price for their produce and a tariff and quota are placed on imports of foodstuffs. The initial aims of the CAP were laudable: to ensure productivity, give farmers a reasonable standard of living, allow food stocks to be secured and provide food in shops at affordable prices. However, it has cost a great deal to implement and it is said that a cow in the EU gets more money in subsidies than a poor person in Africa. The CAP is being reformed with subsidies for individual animals phasing out and a single farm payment being given per hectare of cultivatable land.

One big issue with the CAP is that it is accused of being protectionist – keeping products from other countries out of the EU to the benefit of the 5% of Europeans who work in agriculture – 'Fortress Europe'. It has also been accused of keeping food prices artificially high, causing massive food stocks to build up as producers keep producing even if no consumers want to buy, as the EU will buy the products

7   See http://www.theguardian.com/environment/2008/mar/23/food.ethicalliving for more on food miles

(the wine lake, the butter mountain), and allowing farmers to pollute with high levels of chemicals to increase production. All these issues are being addressed now.

What do you think about protectionism in the trade of food (stopping or adding tariffs to imports from other countries so your own farmers can sell theirs)?

Should small EU farmers have been allowed to go bankrupt and give up?

The Green Revolution of the 20th century

▲ **Figure 5.2.15** Green Revolution poster from Greenpeace

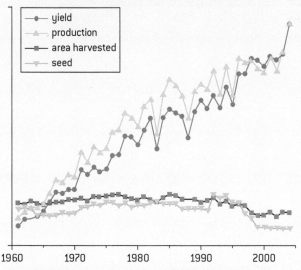

▲ **Figure 5.2.16** Coarse grain production 1961–2004 (coarse grain is wheat, maize, rice, barley)

Here are two very different images about the Green Revolution. Did it save the world from mass famine or has it caused more problems than it solved? From the 1940s to the 1960s, plant breeding of wheat and rice and then other cereals was undertaken to produce varieties that were less prone to disease, had shorter stalks (so they did not lodge or fall over in rain) and gave higher yields. This was done by artificial selection of the varieties and individual plants that had the traits wanted and the results were spectacular in terms of crop yields. (It was before we knew about the possibility of or could carry out genetic engineering which is switching genes from one species to another.)

In Mexico, wheat yields increased such that the country became self-sufficient in wheat and then exported the surplus. In India, the IR8 variety of rice, a HYV or high yielding variety, gave five times the yield of older varieties with no added fertilizer and ten times with. This led to a fall in the cost of rice and to India exporting the surplus. However, in Africa, there was little difference in crop yields.

Some have been critical of the Green Revolution varieties as the result has been that more fertilizer, irrigation and pesticides have been used on them (if farmers can afford them) and these cause eutrophication, salinization and the accumulation of chemicals in food chains. They have also reduced genetic diversity in the crops as most farmers use them. While some say that the poor have become poorer because of this, there is little doubt that the HYVs have allowed us to produce enough food for the population.

1. Explain the relationships between the four graphs in figure 5.2.16.

2. Evaluate the impacts of the Green Revolution on world food supply and the environment.

Practical Work

* Investigate food consumption and/or production patterns.

* Analyse tables and graphs that illustrate the differences in inputs and outputs associated with food production systems.

* Compare and contrast the inputs, outputs and system characteristics for two named food production systems.

* Investigate soil profiles locally.

The FAO

The Food and Agriculture Organization (FAO) is a UN agency. Find its website www.fao.org/worldfoodsituation from where the graph below is taken and answer the questions below.

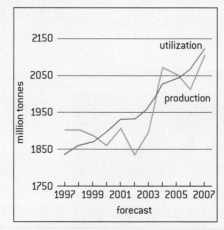

▲ **Figure 5.2.17** World food supply and use 1997–2007

1. Describe the changes in the graph above.

2. Read the 10 FAQs on the world food situation.

3. Answer these questions in your own words:

    a. Why are world food prices rising?

    b. What is the effect of biofuels on food prices?

    c. Who are the winner and losers?

4. Describe the current food situation in your own country and one other. (Select an MEDC and an LEDC.)

5.3 Soil degradation and conservation

Significant ideas:

→ Fertile soils require significant time to develop through the process of succession.

→ Human activities may reduce soil fertility and increase soil erosion.

→ Soil conservation strategies exist and may be used to preserve soil fertility and reduce soil erosion.

Applications and skills:

→ **Explain** the relationship between soil ecosystem succession and soil fertility.

→ **Discuss** the influences of human activities on soil fertility and soil erosion.

→ **Evaluate** soil management strategies in a named commercial farming system and in a named subsistence farming system.

Knowledge and understanding:

→ Soil ecosystems change through **succession**. Fertile soil contains a community of organisms that work to maintain functioning nutrient cycles and that are resistant to soil erosion.

→ **Human activities** which can reduce soil fertility include deforestation, intensive grazing, urbanization, and certain agricultural practices (irrigation, monoculture etc.).

→ **Commercial industrialized food production systems** generally tend to reduce soil fertility

more than small-scale subsistence farming methods.

→ **Reduced soil fertility** may result in soil erosion, toxification, salination and desertification.

→ **Soil conservation** measures exist such as: soil conditioners (for example, organic materials and lime), wind reduction techniques (wind breaks, shelter belts), cultivation techniques (terracing, contour ploughing, strip cultivation) and avoiding the use of marginal lands.

Soil degradation

The problems of our time include climate change, loss of biodiversity, lack of drinking water, poor sanitation and the depletion of fuel wood supplies due to unsustainable rates of use. All of these are significant, but it could be argued that land degradation is the most pressing environmental and social problem facing society today, particularly affecting the world's poor.

It is estimated that an area equal to the size of China and India combined is now classified as having impaired biotic function (damaged ecosystem structure) as a result of poor land management resulting in soil loss. As populations expand, and as social and cultural changes occur, greater and greater demands are being made on larger areas of landscape and soil.

In MEDCs where there has been a relatively long tradition of agriculture (agriculture on an industrial scale) there exists, within the agricultural culture, a knowledge of land management that aims for sustained soil fertility and strives to avoid soil erosion. However even in MEDCs there are occasions when climate and intensive agriculture conspire to bring about unprecedented levels of soil erosion.

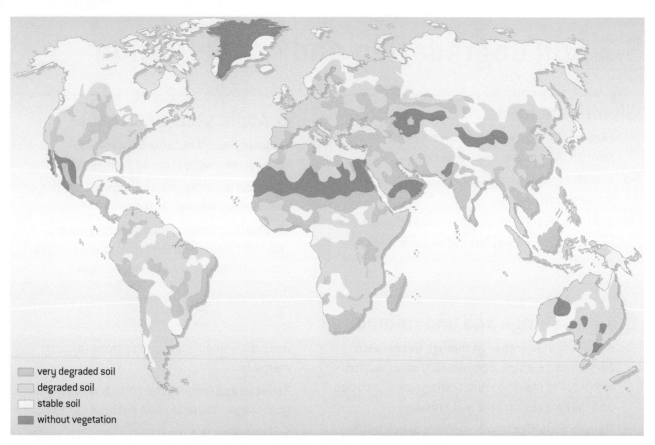

▲ **Figure 5.3.1** An estimate of soil degradation – 15% of ice-free land is affected. This is 11 million km² affected by water erosion and 5.5 million km² by wind erosion

Legend:
- very degraded soil
- degraded soil
- stable soil
- without vegetation

Two types of processes can give rise to soil degradation:

- Processes that **take away the soil** (erosion). This mainly occurs when there is no vegetation on the soil. Wind and water can then simply take the soil away.

- Processes that **make the soil less suitable for use**. In these processes various chemicals end up in the soil and turn the soil useless in the long run.

Examples of human activities that lead to soil degradation are:

- overgrazing
- deforestation
- unsustainable agriculture.

Overgrazing **occurs when too many animals graze in the same area**. Overgrazing of grasslands leaves bare patches where roots no longer hold the soil together. When this is combined with the action of rain and wind the bare patches become bigger and soil is removed from the area.

This happened on a huge scale in the Sahel area in Africa (just south of the Sahara desert) in the 1970s and 1980s. In many African countries the wealth of a man is measured by the number of cattle he has (quantity not quality is important) – this leads to very high stocking levels and overgrazing becomes a problem.

This was then exacerbated in the 70s and 80s when a long dry period strongly reduced the growth of the vegetation which was subsequently

▲ **Figure 5.3.2**

eaten by cattle. The soil particles were no longer kept in place by roots and were blown away by the wind. This resulted in the death of most of the cattle and, later on, in a terrible famine. As soil formation is a very slow process, it will take many years for the Sahel region to recover. In wet climates it is often rain water that takes the soil particles away, especially when the rain water is flowing down slopes.

Overcropping depletes soil nutrients and makes the soil friable (dry and susceptible to wind erosion). This reduces soil fertility as no nutrients are being returned to the soil. If the crop fails then the soil surface again becomes susceptible to erosion. This is especially true in dry regions where crop failure can lead to removal of topsoil by wind. During the 1930s, the American Mid West suffered a major period of wind erosion known as the 'Dust Bowl'. Through overuse of the land an area about twice the size of the United Kingdom, from Nebraska through to Texas, was affected by severe wind erosion. The winds moved soil and dust many thousands of kilometres.

Deforestation is the removal of forest. This can be done in different ways, ranging from careful removal of some of the trees to complete removal of all vegetation. Of course, the more vegetation is removed, the more the soil will be prone to erosion. As most forests are in relatively wet areas, the erosion will mainly be due to water. Deforestation can have a massive effect on soil erosion, especially in tropical regions. The leaves of forest trees both deflect and slow down the progress of rain drops. This helps to stop them explosively removing soil particles. The root systems of forests help to bind the soil together and give it stability, while also absorbing large quantities of water from the soil directly. The absorbed water is eventually returned to the atmosphere via transpiration.

Unsustainable agricultural techniques are techniques that cannot be applied over a long period of time without decrease in productivity or increased inputs of chemicals like fertilizers or energy. Several unsustainable agricultural techniques result in soil degradation:

- (Total) removal of the crops after harvest. This leaves the soil open to erosion.

- Growing crops in rows with uncovered soil in between. Again, erosion will occur, especially if the crops are grown on a slope and the rows are in the direction of the slope.

- Plowing in the direction of the slope. This will leave ready-made channels for rainwater to flow down, taking the soil with it.

- Excessive use of pesticides. This will in the long run make the soil too toxic for further agricultural use. This process is called toxification.

- Irrigation. In many irrigation systems a major part of the water evaporates before reaching the crops. The minerals dissolved in the irrigation water remain in the top layer of the soil and form a hard, salty crust that will make the land unsuitable for growing crops. This process is called salinization (making the soil salty).

- Monocultures. This is where the same crop is grown year after year. This means that the same nutrients are depleted from the soil and the soil loses its fertility.

▲ **Figure 5.3.3** Deforestation for oil palm plantation

Urbanization. For the first time in human history, more of us live in cities than in rural areas. Land in cities is paved and built upon so removing it as a source of agricultural land and increasing run-off which may erode soil elsewhere. Because our major cities expanded from early settlements that were based on agriculture, many of the world's cities have expanded into prime agricultural land.

Soil erosion

If natural vegetation covers a soil, processes that could damage the soil structure are largely eliminated. Leaves deflect heavy rain, roots hold the soil together and humus absorbs large quantities of water. If this natural cover is removed or even when it is replaced by agricultural vegetation soils can become prone to erosion.

Three major processes of soil erosion exist:

1. Sheet wash: large areas of surface soil are washed away during heavy storm periods and in mountainous areas moving as landslides.

2. Gullying: channels develop on hillsides following rainfall. Over time these channels become much deeper.

3. Wind erosion: on drier soils high winds continually remove the surface layer.

Improving the soil – soil conservation

A variety of measures can be taken to conserve soil and soil nutrients.

Addition of soil conditioners

Typical soil conditioners are lime and organic materials. For many centuries farmers have added crushed limestone or chalk to the soil to increase pH and counter soil acidification. Lime has the additional benefit of helping clay particles stick together so that they act more like sand particles. The larger particles created are more free draining than raw clay and they trap more air, helping to improve decomposition by soil microorganisms. Soils become acidic for a number of reasons:

- Acid precipitation. Acid rain or snow means the water the drains through the soil is acidic.

- Some soil processes also make the soil more acidic:

  - The breakdown of organic matter releases carbon dioxide through respiration. This then dissolves into the soil water creating carbonic acid.

  - Nitrification of ammonium ions to nitrates increases acidity.

  - The removal of basic ions through leaching adds to soil acidity.

Organic materials such as straw, or green manure crops that are plowed back into the soil, improve the texture of the soil and act as a supply of nutrients (after decomposition). This is particularly good because the slow decomposition of the organic material means slow release of the nutrients and better absorption by the plant.

▲ **Figure 5.3.4** Gully erosion by water in Virginia, USA.

▲ **Figure 5.3.5** Dust storm of dry soil blown by the wind

▲ **Figure 5.3.6** Shelter belts in Saskatchewan, Canada

Wind reduction

The effect of the wind can be reduced by planting trees or bushes between fields (shelter belts) or by alternating low and high crops in adjacent fields (strip cultivation). An alternative is to build fences.

Soil conserving cultivation techniques

Growing **cover crops** (fast growing crops to cover the soil) between the rows of main crops (for example between rows of corn plants) or between harvest and sowing can keep the soil particles in place.

Terracing is a method to reduce the steepness of slopes by replacing the slope with a series of horizontal terraces, separated by walls. Asian wet rice fields are constructed this way.

Plowing breaks up soil structure and temporarily increases drainage. Traditionally, in the Northern temperate zones, plowing is done in autumn and frost further breaks up clods of soil to make a seed bed for sowing. But there is growing evidence that plowing is bad for soil structure and microbial activity and that no tillage and direct drilling of seed reduces soil erosion and improves soil biodiversity.

Contour farming is plowing and cultivating along the contour lines, ie perpendicular to the slope. By plowing parallel to the slope (contour plowing) instead of up/down hill the furrows and ridges act as small terraces trapping soil and water and the water flow downhill and thereby erosion can be strongly reduced. This method has technical problems however: modern heavy machinery has a tendency to tip over when used parallel to the slope.

Improved irrigation techniques

By careful planning and construction of irrigation systems, evaporation and thereby salinization can be greatly reduced. Covering irrigation canals will prevent evaporation before the water reaches the land. (There are examples known where up to 50% of the water never even reaches the field.) Trickle flow irrigation (also called drip irrigation) systems consist of a network of pipes covering the field. The pipes have small openings next to the plants where water comes out drop-wise and can be taken in by the roots before it evaporates. It is this irrigation system that made it possible to grow roses in desert-like areas in Israel.

Stop plowing marginal lands

Maybe we need to accept that very poor land is simply not suitable for growing crops and could be more suitable for cattle grazing (but with the risk of overgrazing). This may be the case in land at the boundaries of deserts.

Crop rotation

Some crops are more 'hungry' than others. Legumes add nitrogen to the soil as the *Rhizobium* bacteria in their root nodules fix nitrogen from the atmosphere. Cereals take a lot out of the soil. The earliest farmers recognized that growing the same crop on the same land year after year led to pest and disease build-up and impoverished the soil. Shifting cultivators in tropical rainforests move on after a few years to allow the soil to recover. Ancient civilizations practiced crop rotation, sometimes by leaving ground fallow (with no crops), sometimes by growing several

▲ **Figure 5.3.7** Terracing of rice paddies

▲ **Figure 5.3.8** Trickle flow irrigation system

TOK

Our understanding of soil conservation has progressed in recent years. What constitutes progress in different areas of knowledge?

crops in a year to maximize yield. In the Islamic Golden Age, between the 8th and 13th centuries, agricultural improvements such as irrigation, rotations, cash cropping and exporting crops flourished and supported the growth of populations and civilizations. In Europe, a three-year rotation of winter wheat or rye then spring oats or barley then a fallow year was practised until the four-year rotation of the British agricultural revolution introduced turnips and clover so cropping could be continuous and soil fertility maintained.

Sustainability of soils

Saving soils

Human activity can have both positive and negative impacts on soils and on soil fertility. Unmanaged and ignorant exploitation of soils will quickly lead to soil degradation, soil loss and lack of fertility. Soils, like any other part of the global system, exist in a dynamic state. Imbalances will lead to a change in the state of the system and thus the character of the soil both physically and chemically. This will also influence the ecosystem associated with the soil. Our modern global system depends heavily on soil and soil fertility to support a food production system that in turn supports a burgeoning population (90 million extra mouths per year). Soil is therefore an important resource that requires informed conservation. Approximately 24 billion tonnes of soil are eroded from the landscape annually. Croppable land is shrinking in volume annually (though production may be rising).

Soil is formed by a natural set of processes and soil erosion also occurs naturally. However, under natural conditions erosion is often offset by new soil formation, ie the rate of erosion is at or below the rate of accumulation within the soil profile. In natural systems, eroded soil will end up in water courses, estuaries and coastal waters. Here it provides essential minerals, organics and nutrients for the aquatic ecosystems. Soil eroded from one geographical position may also form the basis of soil profiles within another. Eroded upland soils will wash into valleys producing deep fertile floodplain soils.

To think about

Ecocentric or technocentric view of soil fertility management?

As agricultural systems, even those on the relatively small scale, become more and more intensive and more and more commercial, greater and greater demands are made on soil per unit area. The soil is required to 'work harder', produce greater volumes per unit area, and cope with greater demands from genetically modified, nutrient-hungry plants. The soil in its natural state and operating under ambient environmental conditions can no longer fulfil the demands being made upon it, therefore the soil environment must be modified artificially with the addition of fertilizers and additional irrigation water. This is the technocentrist's response. But very quickly the farmers can find themselves on a treadmill with spiralling costs as they attempt to maintain or increase productivity per unit area while at the same time maintaining soil health and thus guaranteeing long-term sustainability of the industry. It costs relatively large amounts to buy synthetic fertilizers and the hardware for irrigation. Therefore the farmer must produce more to get the revenue to pay for this investment in chemicals and hardware. In producing more, more strain is put on the health of the soil system thus more external inputs are required. Potentially a vicious circle is set in motion. What would the ecocentric farmer do?

Danum Valley, Sabah, Malaysia

▲ **Figure 5.3.9** Danum Valley rainforest

The Danum Valley area of Sabah, East Malaysia is an area of lowland tropical forest. Annual rainfall is over 2,500 mm yr^{-1} with little seasonality in the rain pattern. This means that forestry operations occur throughout the year. Large parts of the forest are selectively felled (only certain trees are taken out of the forest rather than just cutting down the entire forest area), but even where this occurs there is an increase in stream flow compared to areas with full canopy cover. This is because of the loss of transpiration returning water back to the atmosphere, though this increase is much less than in clear felled areas.

Because of impermeable geology in the Danum catchment, water tends to flow in the top surfaces of the soil rather than percolating down into bedrock. This means that heavy rainfall can create large flash floods. This makes the soils more sensitive to disturbance, especially from the large vehicles used by foresters.

The Danum catchment is covered by a network of channels, most of which continually flow with water. This makes it very difficult for forestry vehicles to cross the terrain without damaging the banks of stream channels and adding to soil erosion. Also where forestry companies have tried to drain areas, or keep logging roads clear of water, some of these drainage channels have developed into wide gullies that carry eroded soil into the river systems.

Another major factor affecting soil erosion in Danum is the logging roads that have been put in place. Many of these have been built at very steep angles up slopes. The result of this is often mass movement of soil in large landslides. These eroded soils end up in the local streams and rivers and are lost from the forest.

To what extent are our solutions to soil degradation mostly aimed at prevention, limitation or restoration?

To what extent can food production become more sustainable?

BIG QUESTIONS

Soil systems and society

Evaluate the soil system model

In your opinion, discuss whether inequitable food distribution may cause major conflict in the future?

Reflective questions

→ The soil system may be represented by a soil profile. Since a model is strictly speaking not real, how can it lead to knowledge?

→ Consumer behaviour plays an important role in food production systems. Are there general laws that can describe human behaviour?

→ Our understanding of soil conservation has progressed in recent years. What constitutes progress in different areas of knowledge?

→ Why are regional food production systems different?

→ How are food choices influenced by culture and religion?

→ Many LEDCs produce cash crops that are sold to MEDCs; this reduces food availability for the LEDC's population.

  ▪ Why does this happen?

  ▪ What socio-political and factors influence this?

  ▪ What ecological and factors influence this?

  ▪ What are the international issues associated with this?

Quick review

All questions are worth 1 mark

1. Modern commercial agricultural practices tend to

   A. lower species diversity in the community.

   B. reproduce early stages of succession to maximize net productivity.

   C. create conditions favourable for r-selected species.

   D. do all the above.

2. The downward leaching of nutrients within the soil profile is primarily an example of which of the following?

   A. Transformation of energy

   B. Transfer of energy

   C. Transfer of materials

   D. Transformation of materials

3. Which is the best explanation of the advantage of contour-plowing?

   A. Plowing across the contours improves the efficiency of harvesting techniques.

   B. Plowing across the contours improves drainage of the soil.

   C. Plowing parallel with the contours prevents erosion of the soil.

   D. Plowing parallel with the contours improves drainage of the soil.

4. Which of the combinations in the table below correctly completes the following sentence?

   Human activities such as . . . I . . . can cause soil . . . II . . .

	I	II
A.	Overgrazing	Conservation
B.	Deforestation	Degradation
C.	Urbanization	Conservation
D.	Addition of organic material	Degradation

5. Which process reduces storages of soil nutrients?

   A. Addition of well-composted organic matter

   B. Addition of lime

   C. Application of inorganic fertilizers

   D. Heavy rainfall

6. Which of the following is most likely to lead to desertification?

   A. Eutrophication

   B. Use of organic fertilizers

   C. Irrigation

   D. Ozone depletion

7. Which of the following most correctly represents characteristics of each type of soil?

	Sandy soil	Clay soil
A.	Good water-infiltration capacity	Good water-infiltration capacity
B.	Poor nutrient-holding capacity	Good nutrient-holding capacity
C.	Good aeration	Good aeration
D.	Good water-holding capacity	Poor water-holding capacity

6 ATMOSPHERIC SYSTEMS AND SOCIETY

6.1 Introduction to the atmosphere

Significant ideas:
→ The atmosphere is a dynamic system which is essential to life on Earth.
→ The behaviour, structure and composition of the atmosphere influence variations in all ecosystems.

Applications and skills:
→ **Discuss** the role of the albedo effect from clouds in regulating global average temperature.
→ **Outline** the role of the greenhouse effect in regulating temperature on Earth.

Knowledge and understanding:
→ The **atmosphere is a dynamic system** (with inputs, outputs, flows and storages), which has undergone changes through geological time.
→ The atmosphere is predominantly a mixture of nitrogen and oxygen, with smaller amounts of carbon dioxide, argon, water vapour and other trace gases.
→ **Human activities impact** the atmospheric composition through altering inputs and outputs of the system. Changes in the concentrations of atmospheric gases such as ozone, carbon dioxide and water vapour have significant effects on ecosystems.

→ Most reactions connected to living systems occur in the inner layers of the atmosphere, which are the troposphere (0–10 km above Earth) and the stratosphere (10–50 km above Earth).
→ Most clouds form in the troposphere and play an important role in the **albedo** effect for the planet.
→ The **greenhouse effect** of the atmosphere is a natural and necessary phenomenon maintaining suitable temperatures for living systems.

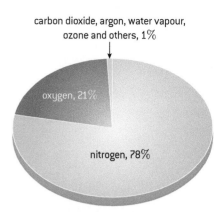

The atmospheric system

The atmosphere is a dynamic system with inputs, outputs, storages and flows. Heat and pollutants are carried across the Earth by air currents in the atmosphere.

Over geological time, the composition of the atmosphere has changed greatly and the changes in the concentrations of atmospheric gases such as ozone, carbon dioxide, methane and water vapour have significant effects on ecosystems (7.2).

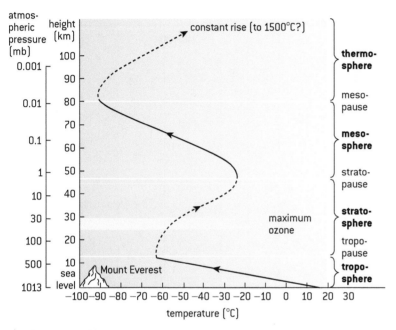

▲ **Figure 6.1.2** The vertical structure of the atmosphere

▲ **Figure 6.1.1** Atmospheric composition

Although the atmosphere is approximately 1,100 km in depth, the **stratosphere** (10–50 km) and the **troposphere** (less than 10 km) are where most reactions affecting life occur, eg ozone and cloud formation.

Human activities and activities by other organisms impact the atmospheric composition through altering inputs and outputs of the system. And the atmospheric composition affects activities of humans and other organisms.

Past atmospheric changes

The atmosphere and climate of Earth have always changed. Climate is unstable and has fluctuated greatly in the past. Factors influencing climate are:

- Abiotic factors – mainly temperature and precipitation.

- Biotic factors – plants and animals.

We cannot measure precipitation in the distant past but we can measure temperature both directly and indirectly by proxy. We can also measure atmospheric gas concentration in bubbles trapped in ice. Various direct and indirect measurements are taken on sediments or fossilized animal shells from the period, but it is hard to say how accurate they are.

Before plants evolved to photosynthesize, there was no free oxygen in the atmosphere on Earth. It gradually increased to about 35% at the

To do

Draw a systems diagram of the atmosphere.

275

end of the Carboniferous period (300 MYA (million years ago)). In the geological timescale, average temperature on Earth was 20 °C in the Early Carboniferous period 350 MYA and this cooled to 12 °C later in the Carboniferous, slightly lower than our 15 °C today. When the temperature was 20 °C, the carbon dioxide concentration in the atmosphere was probably about 1500 ppm (parts per million) and this decreased to about 350 ppm when the average temperature was 12 °C. Today there are about 400 ppm carbon dioxide in the atmosphere (0.04%), less than in the previous 600 million years excepting the Carboniferous.

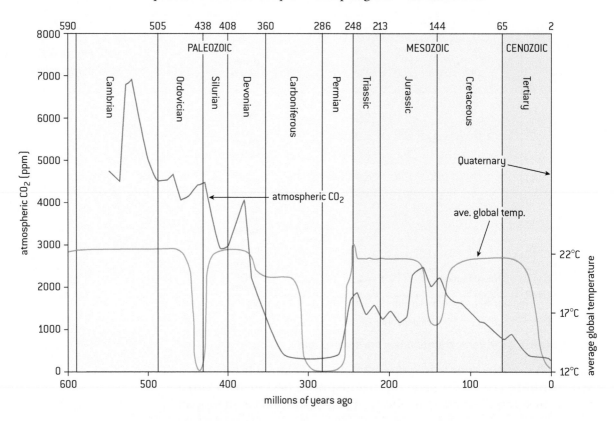

▲ **Figure 6.1.3** Global temperature and atmospheric CO_2 over geological time

1. What is the relationship between carbon dioxide and temperature in the graph?

2. What other factors besides atmospheric carbon may influence Earth temperatures?

3. Can we rely on the data collection methods used?

The greenhouse effect

The greenhouse effect is a natural and necessary phenomenon maintaining suitable temperatures for living systems – a good thing for life on Earth. Indeed, without it, there would be no life on Earth as we know it. The nearest planets to the Earth are Mars with a surface temperature of −53 °C and Venus with a surface temperature of +450 °C. Earth's average surface temperature is a comfortable +15 °C, so just right for life and 33 °C warmer than it would be without the greenhouse effect. A key feature of this temperature is the fact that water is liquid.

(See below.)

The effect is caused by gases in the atmosphere reducing heat losses by radiation back into space. They trap heat energy that is reflected from the Earth's surface and reradiate it – some back to space and some back to the Earth. A common misnomer is that this works in the same way as a glass greenhouse. But greenhouses trap heat by reducing convection which removes the heat more than by radiation. The atmosphere reduces loss of heat by radiation.

Incoming solar radiation is mostly made up of visible light, ultraviolet light and infrared heat. This passes through the atmosphere of Earth mostly unaffected (although the longer wavelength infrared is absorbed by the atmosphere). Nearly 50% of this is absorbed, scattered or reflected by the atmosphere and clouds before it reaches the surface of the Earth. Of the 55% of incoming solar radiation that reaches the Earth's surface, 4% is reflected and 51% absorbed. This is used in several processes including photosynthesis, heating the ground and seas, and evaporation. It is released back to the environment and the atmosphere as longer wavelength infrared energy – heat energy. (See figure 6.1.4.)

Practical Work

* How are the issues addressed in atmospheric systems and society of relevance to sustainability or sustainable development?
* To what extent have the solutions emerging from studying the effect of human activity on the atmosphere been directed at preventing causes of environmental impacts or problems, limiting the extent of impact from these causes, or restoring systems in which impacts have already occurred?
* In what ways might the solutions explored in atmospheric systems and society alter your predictions for the state of human societies and the biosphere some decades from now?

▲ **Figure 6.1.4** Fate of solar radiation reaching Earth

If we had no greenhouse gases in the atmosphere, this heat would go straight back into space and the temperature on Earth would fall drastically every night. But some of this heat is absorbed by gases in the atmosphere which re-emit it as heat energy back to Earth. It's a bit like having a blanket around the Earth.

There are many gases that can do this – greenhouse gases (GHG) – but the main ones are **water vapour**, **carbon dioxide** and **methane**. They cause our atmosphere to be warmer than it would be without them. Only molecules with two or more bonds joining the atoms can absorb and re-emit energy in this way. Nitrogen and oxygen make up most of the atmosphere but are not greenhouse gases.

6.2 Stratospheric ozone

Significant ideas:

→ Stratospheric ozone is a key component of the atmospheric system because it protects living systems from the negative effects of ultraviolet radiation from the Sun.

→ Human activities have disturbed the dynamic equilibrium of stratospheric ozone formation.

→ Pollution management strategies are being employed to conserve stratospheric ozone.

Applications and skills:

→ **Evaluate** the role of national and international organizations in reducing the emissions of ozone-depleting substances.

Knowledge and understanding:

→ Some ultraviolet radiation from the Sun is absorbed by stratospheric ozone causing the ozone molecule to break apart. Under normal conditions the ozone molecule will reform. This ozone destruction and reformation is an example of a dynamic equilibrium.

→ **Ozone-depleting substances** (including halogenated organic gases such as chlorofluorocarbons – CFCs) are used in aerosols, gas-blown plastics, pesticides, flame retardants and refrigerants. Halogen atoms (eg chlorine) from these pollutants increase destruction of ozone in a repetitive cycle so allowing more ultraviolet to reach the Earth.

→ **Ultraviolet radiation** reaching the surface of the Earth damages human living tissues, increasing the incidence of cataracts, mutation during cell division, skin cancer and other subsequent effects on health.

→ The effects of increased ultraviolet radiation on biological productivity include damage to photosynthetic organisms, especially phytoplankton which form the basis of aquatic food webs.

→ Pollution management may be achieved by reducing the manufacture and release of ozone-depleting substances. Methods for this reduction include:

- recycling refrigerants,
- developing alternatives to gas-blown plastics, halogenated pesticides, propellants and aerosols, and
- developing non-propellant alternatives.

→ The **United Nations Environment Programme (UNEP)** has had a key role in providing information and creating and evaluating international agreements for the protection of stratospheric ozone.

→ An **illegal market** for ozone-depleting substances continues and requires consistent monitoring.

→ The **Montreal Protocol** (1987) and subsequent updates is an international agreement on the reduction of use of ozone-depleting substances signed under the direction of UNEP. National governments complying with the agreement made national laws and regulations to decrease consumption and production of halogenated organic gases such as chlorofluorocarbons (CFCs).

Introduction to ozone

Ozone is found in two layers of the atmosphere – the stratosphere where it is 'good' and the troposphere (6.3) where it is considered to be 'bad'.

Ozone is a molecule made up of three oxygen atoms – O_3. Oxygen gas is diatomic, O_2. The stratospheric ozone blocks incoming ultraviolet radiation from the sun and protects life from damaging ultraviolet (UV) radiation.

Ozone is also a greenhouse gas (GHG) (see 7.2), so do not get confused over its environmental effects.

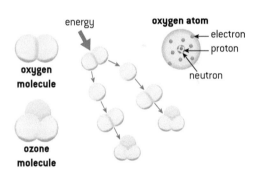

▲ **Figure 6.2.1** Ozone in the atmosphere

The ozone layer

Ozone is a reactive gas mostly found in the so-called ozone layer in the lower stratosphere. The highest ozone concentrations are usually seen at altitudes between 20 and 40 km (at the poles between 15 and 20 km). But it is a very thin layer of about 1–10 ppm (parts per million) ozone.

The ozone layer is an example of a **dynamic equilibrium** because ozone is continuously made from oxygen atoms and is continuously converted back to oxygen. In **both** the formation and the destruction of ozone, UV radiation is absorbed. Under the influence of UV radiation, oxygen molecules are split into oxygen atoms. Oxygen atoms are extremely reactive, and can combine with an oxygen molecule to form ozone. Ozone can also absorb UV radiation and then splits into an oxygen molecule and an oxygen atom. The oxygen atom can react with another ozone molecule, making two oxygen molecules (figure 6.2.2).

The adsorption of UV radiation by the ozone layer is crucial, for without it, life on land would be impossible. UV radiation is usually divided into UV-C, UV-B and UV-A radiation. UV-C radiation has the highest energy (shortest wavelength) and is therefore the most harmful type of UV radiation, while the longer wavelength (low energy) UV-A radiation is relatively harmless. The ozone layer absorbs more than 99% of the UV-C radiation and about half of the UV-B radiation (M. B. Bush, *Ecology of a Changing Planet*, Prentice Hall, Upper Saddle River, 2003, p. 381).

▲ **Figure 6.2.2** Formation of ozone

Effects of ultraviolet radiation

Damaging effects

Increased exposure to UV radiation will have a variety of damaging effects:

- Genetic mutation and subsequent effects on health.

- Damage to living tissues.

- Cataract formation in eyes.

- Skin cancers.

- Suppression of the immune system.

- Damage to photosynthetic organisms, especially phytoplankton.

- Damage to consumers of photosynthetic organisms, especially zooplankton.

Ultraviolet radiation can cause mutations – changes in a species' DNA. This risk is especially high in Australia and New Zealand where the number of cases of skin cancer in humans has increased dramatically. People are advised to wear clothes on the beach and to use sun blocks to protect their skin. In New Zealand the daily weather report in summer includes isolines to show burn times.

It also causes cataracts in the lenses of eyes when the protein of the lens denatures and turns cloudy instead of clear, causing blindness if untreated.

Photosynthetic organisms are sensitive to UV radiation. This can have disruptive effects on food pyramids.

Beneficial effects

- In animals, UV radiation stimulates the production of vitamin D in our bodies. Vitamin D deficiency causes rickets when a child's bones are short of calcium and too soft to support the body.

- It can also be used to treat psoriasis and vitiligo, both skin diseases.

- Use as a sterilizer as it kills pathogenic bacteria and as an air and water purifier.

- Industrial uses in lasers, viewing old scripts, forensic analysis, lighting.

Air pollution and the ozone hole

Since the 1950s, scientists have been measuring the amount of ozone in the stratosphere above Antarctica. They discovered what later would be called the ozone hole: the amount of ozone decreased significantly during the spring (September and October) and increased again in November. Apart from this annual ozone cycle, the scientists discovered that the ozone hole was growing. During the last 30 years, the minimum thickness of the ozone layer has reduced drastically and recovery has been taking longer. These results were later confirmed by NASA satellite data. Reductions in the amount of stratospheric ozone have been observed in other areas as well, including the arctic region (K. Byrne, *Environmental Science*, 2nd ed., Nelson Thornes, Cheltenham, 2001, p. 27).

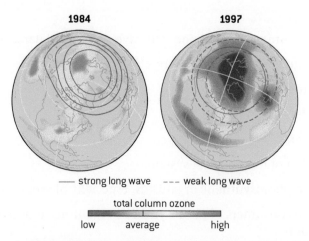

▲ **Figure 6.2.3** Ozone hole in North America 1984 and 1997

Ozone-depleting substances (ODS)

Ozone depletion is the result of air pollution by chemicals that are mostly human-made. The most important ozone-depleting gases are halogenated organic gases, for example chlorofluorocarbons or CFCs, but there are others too.

Substance	Use/source	Remarks
Chlorofluorocarbons (CFCs or freons)	Propellants in spray cans, plastic foam expanders, refrigerants	Release chlorine atoms
Hydrochlorofluoro carbons (HCFCs)	As replacements for CFCs	Release chlorine atoms, but have a shorter lifetime in the atmosphere. Stronger greenhouse gases
Halons	Fire extinguishers	Release bromine atoms
Methyl bromide	Pesticide	Releases bromine atoms
Nitrogen oxides (NO, NO_2, N_2O, often summarized as NO_x)	Bacterial breakdown of nitrites and nitrates in the soil (intensive farming) High-flying supersonic aircraft	The nitrogen oxides are converted to NO, which reacts with ozone

▲ **Figure 6.2.4** The most important ozone-depleting substances (ODS)

The action of ODS

When chlorofluorocarbons (CFCs or freons) were developed during the 1930s, they seemed to be the answer to many technological problems, largely because they are inert (non-reactive) at ground level.

They were used as:

- propellants in aerosols,
- expanders of gas-blown plastics,
- pesticides,
- flame retardants, and
- refrigerants.

Previously used refrigerants were very toxic and flammable, and were therefore quickly replaced by the non-toxic and non-flammable CFCs.

Later on, CFCs were also used as propellants in spray cans and to expand plastic foam. CFCs are extremely stable at ground level and it took a long time before it was discovered that they were not so stable when exposed to UV radiation in the stratosphere. UV radiation releases chlorine atoms

Practical Work

* Investigate changes in tropospheric ozone away from a source of pollution (eg a road) and attitudes towards solving the problem.

* Investigate two examples of when national economic approaches have an impact on international environmental discussions.

from the CFCs. These chlorine atoms can react with ozone, which results in **ozone destruction**. They can also react with oxygen atoms, thereby preventing ozone formation. In both of these processes, the chlorine atoms are formed back and are again able to react with ozone or oxygen atoms. One chlorine atom can thus destroy many molecules of ozone in a **chain reaction** with **positive feedback**.

While replacing CFCs in spray cans and as blowing agents for plastic foam is relatively easy, it is much more difficult to find a suitable refrigerant. The refrigerants used before the introduction of CFCs are not an option because of their dangerous properties. The most suitable CFC replacements are the so-called hydrochlorofluorocarbons (HCFCs). These substances are nearly as good as refrigerants as CFCs and are also non-toxic and inflammable. However, HCFCs also destroy ozone and they contribute to the greenhouse effect. Only their shorter lifetime in the atmosphere makes them less harmful to the ozone layer than CFCs.

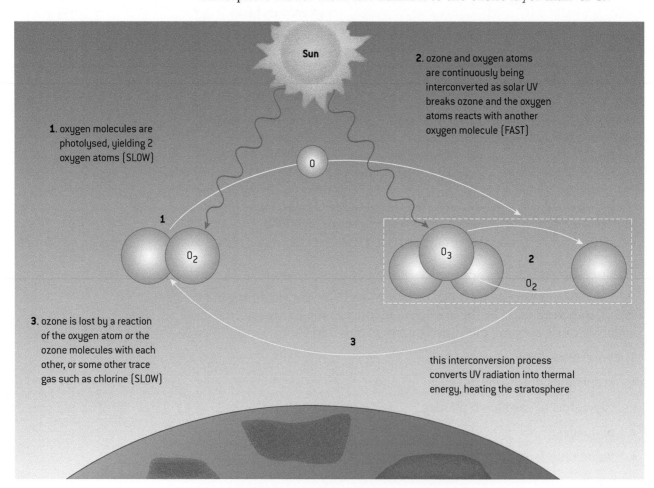

▲ **Figure 6.2.5** The ozone cycle

CFCs are extremely stable, and will therefore persist in the atmosphere for up to 100 years after their release. Measures taken to prevent release of CFCs into the atmosphere will therefore take a long time to result in a thicker ozone layer.

Reducing ODS

Using the 'replace, regulate and restore' model of pollution management strategy, figure 6.2.6 shows some actions for reducing ozone-depleting substances.

Strategy for reducing pollution	Example of action
Altering the human activity producing pollution	Replace gas-blown plastics.
	Replace CFCs with carbon dioxide, propane or air as a propellant.
	Replace aerosols with pump action sprays.
	Replace methyl bromide pesticides. (But most of the gases that can be used to replace CFCs are greenhouse gases).
Regulating and reducing the pollutants at the point of emission	Recover and recycle CFCs from refrigerators and AC units.
	Legislate to have fridges returned to the manufacturer and coolants removed and stored.
	Capture CFCs from scrap car air conditioner units.
Clean up and restoration	Add ozone to or remove chlorine from stratosphere – not practical but it was once suggested that ozone-filled balloons should be released.

▲ **Figure 6.2.6** Table of replace, regulate and restore model of pollution management for ozone depletion

The Montreal Protocol

The discovery of the ozone hole led to a fast response on national and international levels. But even before governments and international organizations took steps, the general public in many developed countries started to boycott products containing CFCs (mainly spray cans). The aerosol industry reacted quickly by changing to ozone-friendly spray cans, and even before CFCs were forbidden by law, hardly any CFC-containing spray cans were produced anymore.

The United Nations organization involved in protecting the environment is UNEP (United Nations Environmental Programme), which:

- forges international agreements
- studies the effectiveness of these agreements and the difficulties in implementing and enforcing them
- gives information to states, organizations and the public.

One of the treaties signed under the direction of UNEP is the **Montreal Protocol** (1987). It is an international agreement to phase out the production of ozone-depleting substances. The signatories agreed to freeze consumption and production of many CFCs and halons to 1986 levels by 1990 and to strongly reduce the consumption and production of these substances by 2000. Since 1987, the original Montreal Protocol has been strengthened in a series of seven amendments. In the protocol a distinction was made between MEDCs and LEDCs. The LEDCs got more time to implement the treaty.

> **TOK**
>
> The Montreal Protocol was an international agreement made by the UN. Can one group or organization know what is best for the rest of the world?

Some 197 countries ratified the agreement – that is all countries in the world at the time, so it was the first universally ratified UN agreement. Most countries followed the rules of the Montreal Protocol and made national laws and regulations accordingly. China and India however, were still producing and using huge amounts of CFCs. These countries' need for refrigerators and air conditioners is quickly growing because of the fast economic growth they are experiencing. But China has now agreed to phase out ODS production two years before schedule and India also has a plan to do so.

The reduction of emissions of CFCs and other ozone-depleting substances has been one of the most successful international cooperative ventures to date.

Timeline in CFC reduction

1970s   Ozone-depleting properties of CFCs recognized. In 1974 USA and Sweden banned them from non-essential aerosol uses. Concerns continue to mount through 1980s.

1985   British Antarctic Survey reports the ozone hole.

1987   Montreal Protocol organized by the United Nations Environmental Program (UNEP). Over 30 countries agree to cut CFC emissions by half by 2000.

1990   London Amendment to strengthen Montreal Protocol: phase-out dates and rates inadequate so Montreal Protocol amended. Industrialized countries would eliminate CFC production by 2000 and developing countries by 2010.

1992   Further measures to accelerate phasing out of ODS and replacement by substitute chemicals.

1995   Nobel Prize for Chemistry awarded to Molina, Sherwood Rowland and Crutzen for their work on solving the ozone depletion issue.

2006   NASA and NOAA record the Antarctic ozone hole as the largest ever measured.

2012   25th anniversary of signing of the protocol.

Significance of the Montreal Protocol

There are several significant facts about it.

1   The best example of international cooperation on an environmental issue.

2   An example of the precautionary principle in science-based decision making.

3   An example of many experts in their different fields coming together to research the problem and find solutions.

4   The first to recognize that different countries could phase-out ODS chemicals at different times depending on their economic status.

5   The first with regulations that were carefully monitored.

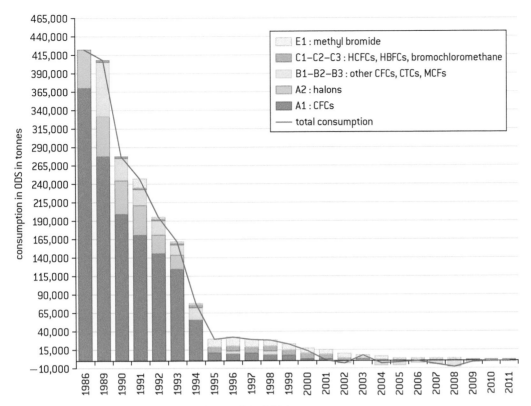

▲ **Figure 6.2.7** EU consumption of ODS 1986–2011

However this is not the end of the story. Due to the long life of CFCs in the atmosphere, chlorine did not reach its peak in the stratosphere until 2005 nor will it return to pre-ODS levels much before 2050. LEDCs are still allowed to make and use some HCFCs until 2030. There is also an illegal market in ODS chemicals.

To do

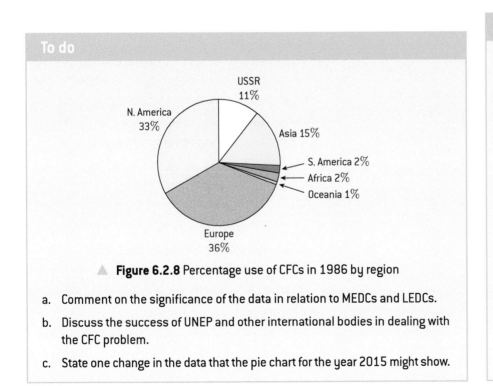

▲ **Figure 6.2.8** Percentage use of CFCs in 1986 by region

a. Comment on the significance of the data in relation to MEDCs and LEDCs.

b. Discuss the success of UNEP and other international bodies in dealing with the CFC problem.

c. State one change in the data that the pie chart for the year 2015 might show.

To do

1. What steps is your national government taking to comply with these agreements?

2. Name local or non-governmental organizations concerned with ODS reduction.

3. What actions are they taking to persuade governments to comply?

4. What can you do to help protect the ozone layer?

5. **Evaluate** the role of national and international organizations in reducing the emissions of ozone-depleting substances.

6.3 Photochemical smog

Significant ideas:

→ The combustion of fossil fuels produces primary pollutants which may generate secondary pollutants and lead to photochemical smog, whose levels can vary by topography, population density and climate.

→ Photochemical smog has significant impacts on societies and living systems.

→ Photochemical smog can be reduced by decreasing human reliance on fossil fuels.

Applications and skills:

→ **Evaluate** pollution management strategies for reducing photochemical smog.

Knowledge and understanding:

→ **Primary pollutants** from the combustion of fossil fuels include carbon monoxide, carbon dioxide, black carbon/soot, unburned hydrocarbons, oxides of nitrogen, and oxides of sulfur.

→ In the presence of sunlight, **secondary pollutants** are formed when primary pollutants undergo a variety of reactions with other chemicals already present in the atmosphere.

→ **Tropospheric ozone** is an example of a secondary pollutant, formed when oxygen molecules react with oxygen atoms that are released from nitrogen dioxide in the presence of sunlight.

→ Tropospheric ozone is highly reactive and damages crops/forests/plants, irritates eyes, creates respiratory illnesses and damages fabrics and rubber materials. **Smog** is a complex mixture of primary and secondary pollutants, of which tropospheric ozone is the main pollutant.

→ The frequency and severity of smog in an area depends on local topography, climate, population density and fossil fuel use.

→ **Thermal inversions** occur due to lack of air movement when a layer of dense, cool air is trapped beneath a layer of less dense, warm air. This causes concentrations of air pollutants to build up near the ground instead of being dissipated by 'normal' air movements.

→ Deforestation and burning may also contribute to smog.

→ Economic losses caused by urban air pollution can be significant.

→ **Pollution management strategies** include:

- **Altering human activity** to consume less. Example activities include the purchase of energy efficient technologies, the use of public/shared transit, and walking or cycling.

- **Regulating and reducing pollutants at point of emission** via government regulation/taxation.

- Using catalytic converters to clean exhaust of primary pollutants from car exhaust.

- Regulating fuel quality by governments.

- Adopting **clean-up measures** such as reforestation, re-greening, and conservation of areas to sequester carbon dioxide.

Urban air pollution

UNEP estimates that:

- 1 billion people are exposed to outdoor air pollution per year

- 1 million people die prematurely due to air pollution

- 2% of GDP is lost by air pollution in MEDCs, 5% in LEDCs

- over 90% of urban air pollution in LEDCs comes from old motor vehicles which are poorly maintained.

Air pollution is caused by chemicals, particulates and biological materials. These may be:

1. **Primary pollutants** – emitted directly from a process. The process may be natural, eg volcanic eruptions or anthropogenic (human-made), eg industry, motor vehicle exhausts. A major source of anthropogenic air pollution is from the combustion of fossil fuels and this produces:

   - carbon monoxide – from incomplete combustion of fossil fuels

   - carbon dioxide

   - unburned hydrocarbons

   - nitrogen oxides – especially nitrogen dioxide, a brown gas, but also nitrous oxide and nitric oxide

   - sulphur dioxide – from coal with high sulphur content

   - particulates or particulate matter (PM) – eg black carbon or soot – fine particles of solid or liquid suspended in air, dangerous as they are taken into the lungs when breathing and cause lung diseases and cancers.

   Other sources of primary pollutants include:

   - building sites

   - forest fires.

2. **Secondary pollutants** – formed when primary pollutants undergo a variety of reactions with other chemicals already present in the atmosphere. Sometimes this is a **photochemical reaction** in the presence of sunlight.

   Examples of secondary pollutants are:

   - tropospheric ozone

   - particulates produced from gaseous primary pollutants

   - peroxyacetyl nitrate (PAN).

Tropospheric ozone

Only 10% of atmospheric ozone is in the troposphere (fig 6.1.2). Here it is in concentrations of only 0.02–0.3 ppm. Ozone is also a greenhouse gas with a global warming potential 2,000 times that of carbon dioxide.

Formation of tropospheric ozone

Among the pollutants emitted during the combustion of fossil fuels are hydrocarbons and nitrogen oxides. Hydrocarbons are emitted because not all the fuel is combusted. Nitrogen oxides are formed when oxygen and nitrogen (both originating from air) react as a result of the high temperature during combustion reactions.

Nitrous oxide contributes to the formation of tropospheric ozone (O_3). First nitric oxide (NO) reacts with oxygen to form nitrogen dioxide (NO_2). This is a brown gas that contributes to urban haze. Other pollutants like hydrocarbons and carbon monoxide accelerate the formation of nitrogen dioxide. When this absorbs sunlight, it breaks up into nitric oxide and oxygen atoms. These oxygen atoms subsequently react with oxygen molecules, forming ozone.

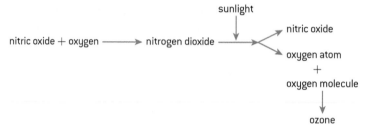

▲ **Figure 6.3.1**

Under normal conditions, most ozone molecules oxidize nitric oxide back into nitrogen dioxide, creating a virtual cycle that leads to only a very slight build-up of ozone near ground level.

Possible effects of ozone

- Ozone is a toxic gas and an oxidizing agent – oxygen causes oxidation, ozone has that extra O so is worse.

- Damage to plants – tropospheric ozone is absorbed by plant leaves. In the leaves, ozone degrades chlorophyll so photosynthesis and productivity are reduced.

- Damage to humans – at low concentrations, photochemical smog can reduce the actions of the lungs, so causing breathing difficulties, and may increase susceptibility to infection. It also causes eye, nose and throat irritation.

- Damage to materials and products:
  - Ozone attacks natural rubber, cellulose and some plastics.
  - It reduces the lifetime of car tyres.
  - It also bleaches fabrics.

Formation of particulates

Burning almost any organic material or fossil fuel releases small particles of carbon and other substances, referred to as particulates. Poorly maintained diesel engines, in particular, release large amounts of

particulates (small solid particles) in exhaust fumes. They are called PM10 or particulate materials and are smaller than 10 micrometres in diameter.

Dangers of particulates:

1. The problem with them is that our respiratory filters (nose and hairs lining the passages of the bronchi and lungs) cannot filter them out so they enter our bodies and stay there causing asthma, lung cancer, respiratory problems and even premature death.

2. Many particulates are carcinogenic (cancer-causing).

3. In areas close to industrial or dense urban areas and especially in developing regions crops become covered with particulates, which reduces their productivity because less sunlight reaches the leaf.

The formation of photochemical smog

On warm, sunny days with lots of traffic, photochemical smog can be formed over cities. Although usually associated with the combustion of fossil fuels, forest burning can also contribute to photochemical smog. In Kalimantan, Indonesia, forest fires in 1997 caused smog over much of S E Asia. The fires burned 8 million hectares, cost the government nearly US$ 5 billion and released a huge amount of carbon dioxide and other gases into the atmosphere.

▲ **Figure 6.3.2** Indonesian forest fires in 1997 causing air pollution over a wide area

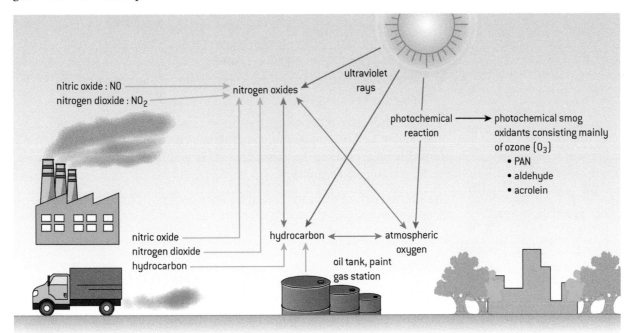

▲ **Figure 6.3.3** Formation of photochemical smog

Photochemical smog is mainly nitrogen dioxide and ozone but is a complex mixture of about one hundred different primary and secondary air pollutants. The biggest contribution to photochemical smog is from motor vehicle exhausts in cities. It is formed when ozone, nitrogen oxides and gaseous hydrocarbons from vehicle exhausts interact with strong sunlight.

Complex reactions create many chemicals in photochemical smog including VOCs (volatile organic compounds), PANs (peroxyacyl nitrates), ozone, aldehydes, carbon monoxide and nitrous oxides. Highly reactive VOCs oxidize nitrogen oxide into nitrogen dioxide without breaking down any ozone molecules in the process. This leads to a build-up of ozone near ground level and smog formation.

Because nitrogen dioxide is an important component of smog, smog can be seen as a brown hue above the city. All these chemicals are strongly oxidizing and affect materials and living things. At higher concentrations, smog can cause coughs and decreased ability to concentrate.

Even though the main primary pollutants – nitrous oxides and hydrocarbons reach a maximum concentration during the morning and evening rush hours, photochemical smog is at its maximum in the early afternoon. This is due to the fact that the important smog-causing reaction is a photochemical reaction, so it reaches its peak in the afternoon sun.

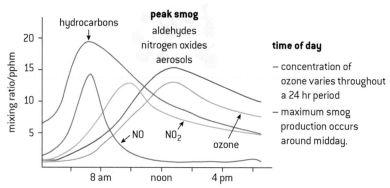

▲ **Figure 6.3.4** Concentrations of main components of photochemical smog in a day

Because photochemical smog first caused problems in Los Angeles, it is often called Los Angeles-type smog. Other cities that frequently suffer from this type of smog are Santiago, Mexico City, Rio de Janeiro, Sao Paulo, Beijing and Athens.

The occurrence of photochemical smog is governed by a large number of factors, including the

- local topography
- climate
- population density
- fossil fuel use.

Smog is most often formed over large cities that are low-lying and/or in valleys. The hills or mountains surrounding these cities take away most of the wind and on warm, calm days severe smog can occur.

Thermal inversion makes things worse. Normally, air over cities is relatively warm and has a tendency to rise. On warm days however, an even warmer layer of air on top of the warm polluted air can prevent this air rising, trapping the pollution at ground level. This occurs most often in warm, dry climates. Weather plays an important role in the disappearance of smog: rain cleans the air of pollutants while winds can disperse the smog.

▲ **Figure 6.3.5** Beijing on a smoggy day. A rainstorm would clear the air temporarily.

Practical Work

* Investigate changes in tropospheric ozone away from a source of pollution (eg a road) and attitudes towards solving the problem

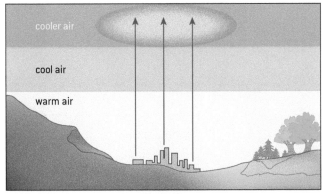

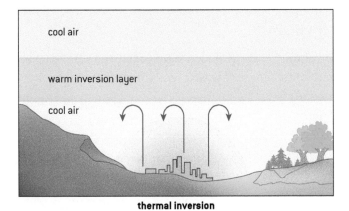

▲ **Figure 6.3.6** Thermal inversion trapping smog over a city

London-type smog

This was the first time the phrase "smog" was coined. London-type smog (smog = smoke + fog) is a completely different type of smog. It occurs at low temperatures. The main primary pollutants are also different: sulphur dioxide and smoke particles produced by burning coal. These smogs killed many people until clean air acts in 1956 legislated to stop coal-burning (unless smokeless fuels) in cities.

Explain why London smogs were not photochemical smogs. (Hint – what was being burned in homes?)

Under the above conditions the concentration of pollutants can reach harmful or even lethal levels. Smog is not only affecting life in the city itself. Often, smog is blown out of the city by the wind and causes damage in the countryside, sometimes up to 150 km away from the city where the smog was formed.

Complete the table below to evaluate the pollution management strategies for reducing photochemical smog.

Strategy for reducing urban air pollution	Example of action	Evaluation
Altering the human activity producing pollution	Consume less, burn less fossil fuel – especially in the internal combustion engine.	
	Act as informed consumers for purchase of energy efficient technologies.	
	Use public/shared transit, walking and cycle paths.	
	Lobby governments to increase renewable energy use.	
Regulating and reducing the pollutants at the point of emission	Government regulation/taxation.	
	Catalytic converters to clean exhaust of primary pollutants from car exhaust.	
	Fuel quality may be regulated by government.	
Clean up and restoration	Afforestation to increase carbon sinks and filter air. But remember this does not reduce emissions.	
	Re-greening of cities – more trees, more parks – absorb carbon dioxide.	

▲ **Figure 6.3.7**

6.4 Acid deposition

Significant ideas:

→ Acid deposition can impact living systems and the built environment.

→ The pollution management of acid deposition often involves cross-border issues.

Applications and skills:

→ **Evaluate** pollution management strategies for acid deposition.

🌐 Knowledge and understanding:

→ The **combustion of fossil fuels** produces sulphur dioxide and nitrous oxides as primary pollutants. These gases may be converted into secondary pollutants of dry deposition (eg ash and dry particles) or wet deposition (eg rain and snow).

→ The possible **effects** of acid deposition on soil, water and living organisms include:

- **direct effect**, for example, acid on aquatic organisms and coniferous forests

- **indirect toxic effect**, for example, increased solubility of metal (eg aluminium ions) on fish

- **indirect nutrient effect**, for example, leaching of plant nutrients.

→ The impacts of acid deposition may be limited to areas **downwind** of major industrial regions but these areas may not be in the same country as the source of emissions.

→ **Pollution management strategies** for acid deposition could include the following measures:

- Altering human activity eg through reducing use, or using alternatives to fossil fuels. International agreements and national governments may work to reduce pollutant production through lobbying.

- Regulating and monitoring release of pollutants eg through the use of scrubbers or catalytic converters that may remove sulphur dioxide and nitrous oxides from coal burning power plants and cars.

- Clean-up and restoration measures may include spreading ground limestone in acidified lakes or recolonization of damaged systems but these are limited.

Acid deposition is the general term for acid coming down from the air. Often, the acid comes down in the form of rain (or snow), this is called **wet deposition**. **Dry deposition** is when the acid comes down as ash or dry particles.

Acidity

Acids are chemicals that are able to give a hydrogen ion (H^+) away. The acidity of solutions is measured using the pH scale. On this scale, a pH value of 7 is neutral (pure water). Values below 7 indicate acidic solutions whereas values above 7 indicate basic (alkaline) solutions. The pH scale is not a linear scale, it is logarithmic. A solution with pH 2 is ten times more acidic than a solution with pH 3.

Normal unpolluted rain is slightly acidic and has a pH of about 5.6. This is caused by the presence of carbon dioxide in the atmosphere. Precipitation is called acidic when its pH is well below pH 5.6. Certain pollutants increase the acidification of rain, which can sometimes fall lower than pH 2.

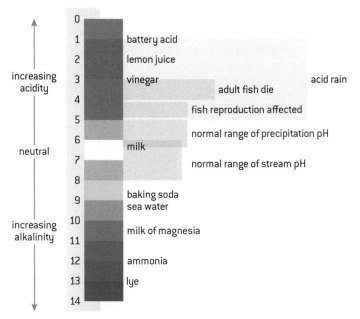

▲ **Figure 6.4.1** The pH scale

Main acid deposition pollutants and sources

Pollutants

Air pollutants can be divided into **primary** and **secondary air pollutants** (6.3). Primary pollutants are those pollutants that are emitted directly, for example those pollutants leaving the chimney of a factory or the exhaust pipe of a car. Primary pollutants can react to form secondary pollutants (1.5). So, secondary pollutants are made after the pollutants leave the chimney. The main primary pollutants leading to acid deposition are sulphur dioxide (SO_2) and nitrogen oxides (NO_x). They react with water to form strong acids – sulphuric and nitric acids. Carbon dioxide (CO_2) is also acidic but forms a weak acid – carbonic acid.

Sources

Naturally, sulphur dioxide is produced by volcanic eruptions and nitrogen oxides by lightning.

The most important human activities that lead to the emission of these pollutants are the combustion of **fossil fuels** in motor cars, industry and thermal power stations which use fossil fuels (coal, oil, gas) to produce steam to drive the turbines.

Sulphur dioxide is formed when sulphur-containing fuels are combusted. Sulphur is common in coal and oil, but is usually absent in natural gas.

▲ **Figure 6.4.2** Industrial air pollution

nitrogen dioxide + water ⟶ nitric acid

sulphur dioxide + water ⟶ sulphuric acid

carbon dioxide + water ⟶ carbonic acid

▲ **Figure 6.4.3**

Nitrogen oxides are formed by reaction of oxygen and nitrogen from the air, which readily takes place at the high temperature during combustion of fossil fuel. The nitrogen is not part of the fuel.

If the primary air pollutants remain in the atmosphere for a sufficiently long time, a variety of secondary air pollutants can be formed. Sulphur dioxide can react with oxygen from the atmosphere to form sulphur trioxide (SO_3). Both sulphur dioxide and sulphur trioxide can react with water and form sulphurous acid (H_2SO_3) and sulphuric acid (H_2SO_4) respectively. The nitrogen oxides can also react with water and form nitric acid (HNO_3). These secondary pollutants are very soluble in water, and are removed from the air by precipitation in the form of rain, hail and snow (wet deposition).

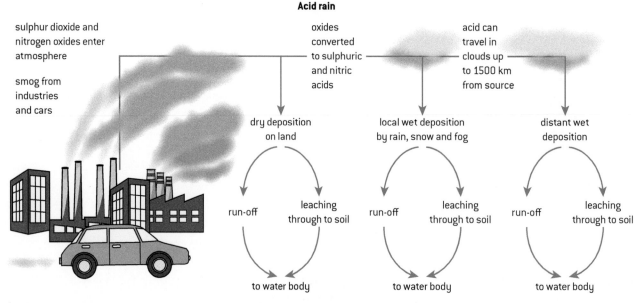

▲ **Figure 6.4.4** The effects of acid deposition on soil, water and living organisms

Acid deposition damage became a focus of environmental attention in the early 1970s when Germany's Black Forest showed a dieback or 'waldsterben' when trees of all ages, both coniferous and deciduous, showed signs of physical damage. But before that, in the Industrial Revolution (which started in 1750), people noticed acidic rain and acid air in cities and the term 'acid rain' was first used in 1872.

Effects of acid deposition

Acid deposition can have direct and indirect effects on soil, plants and water:

- direct effects, eg by weakening tree growth in coniferous forests, acid falling on lakes and ponds decreasing the pH of the water and affecting aquatic organisms

- indirect effects

  - toxic effects, eg increased solubility of metal ions such as aluminium which is toxic to fish and plant roots

  - nutrient effects, eg leaching of nutrients.

▲ **Figure 6.4.5** Forest death from acid deposition

Effect of acid deposition on coniferous forests

Acid deposition affects forests in several ways:

1. Leaves and buds show yellowing (loss of chlorophyll) and damage in the form of lesions, thinning of wax cuticles.

2. These and other changes reduce growth, and allow nutrients to be leached out and washed away and pathogens and insects to gain entry.

3. Symbiotic root microbes are killed and this greatly reduces the availability of nutrients, further reducing tree growth.

4. It reduces the ability of soil particles to hold on to nutrients, such as calcium, magnesium and potassium ions which are then leached out.

5. It releases toxic aluminium ions from soil particles which then damage root hairs.

Trees are weakened and may die.

Toxic effects of acid deposition

Aluminium ions – effect on fish and other aquatic organisms

Aluminium is a common element in the soil. Acid precipitation decreases the pH of the soil, making aluminium more soluble. The aluminium released from the soil eventually ends up in streams and rivers. Fish are particularly sensitive to aluminium in water. At low concentrations, aluminium disturbs the fish's ability to regulate the amount of salt and water in its body. This inhibits the normal intake of oxygen and salt. Fish gasp for breath and the salt content of their bodies is slowly lost, leading to death. At higher concentrations, a solid is formed on the fishes' gills, leading to death by suffocation. Apart from aluminium, other toxic metals can dissolve because of increased acidity.

Lichens

Lichens, which are a symbiotic pairing of an alga and a fungus (2.1), are found growing on trees and buildings. They are particularly sensitive to gaseous pollutants like sulphur dioxide and are used as indirect measures of pollution. Immediately downwind of a heavily polluting industrial region only a few tolerant species are found. These are **indicator species** of high levels of air pollution. As the distance from the source of pollutants increases more and more species are able to survive. Tables of lichen indicator species are used to estimate pollution levels.

Nutrient removal effect on soil fertility

As described above acid rain affects the soil by reducing the ability of soil particles to hold on to nutrients, such as calcium, magnesium and potassium ions which are then leached out. Acid rain also inhibits nitrogen-fixing bacteria and thus their ability to add nitrate ions to the soil.

Buildings

Acid deposition also effects human constructions. Limestone buildings and statues (including many with great archaeological and historical value) react with acid and simply dissolve.

▲ **Figure 6.4.6** Effect of acid deposition on a limestone sculpture – 1908 (top), 1969 (bottom)

Peat bogs

Recent research has also found that peat bogs affected by acid rain produce up to 40% less methane than before. This is because the bacteria that use the sulphates as a food source outcompete the ones that produce methane. This reduction in methane production reduces methane, a GHG, in the atmosphere (7.2).

Human health

Dry deposition is in the form of small particles of sulphates and nitrates and these penetrate into houses and our lungs. Premature deaths from lung disease such as asthma and bronchitis can result from this.

Regional effect of acid deposition

The effects of acid deposition are regional in contrast to climate change or ozone depletion which affect all the Earth. This is because, before the pollutants can spread over long distances, they return to the surface as dry or wet precipitation. The acids seldom travel further than a few thousand kilometres.

- Dry deposition usually occurs quite close to the source of the acidic substances. It consists of sulphur dioxide, sulphur trioxide and the nitrogen oxides.

- Wet deposition occurs at slightly longer distances from the sources of the primary pollutants. It consists of sulphurous acid, sulphuric acid and nitric acid.

It is therefore mainly the downwind areas of major industrial regions that are strongly affected. For example, Scandinavian forests and lakes were mainly affected by acid rain originating in Britain, Poland and Germany brought over by prevailing southwestern winds. Industrial

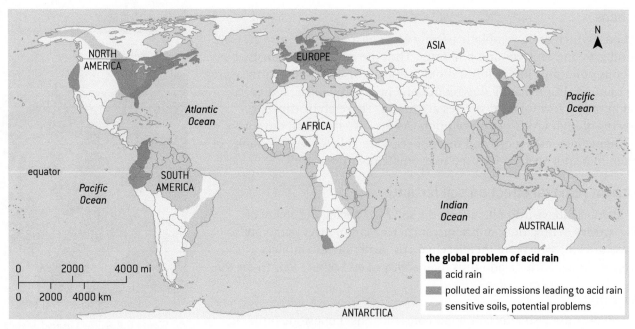

▲ **Figure 6.4.7** Regions of the world with most acid deposition impacts

pollution from the USA is blown by prevailing winds towards Canadian forests. Industrialization in China affects S E Asia.

Soils or bodies of water are most often affected by acid rain. However the impact of acid rain will depend very much on the geology of the area on which it falls. Acid rain does little harm to soils derived from calcium carbonate rocks, ie limestone and chalk. These are alkaline soils, which neutralize (or buffer) the acids. However acid (non-alkaline) rocks produce soils that are very sensitive to acid rain. Acid rain leaches out minerals from these soils. This reduces biodiversity and run-off affects nearby lakes.

Pollution management strategies for acid deposition

Using the 'replace, regulate and restore' model of pollution management strategy, the table below gives some actions for reducing acid deposition and their evaluations.

To do

Research the effect of acid deposition and intergovernmental agreements or legislation and their effectiveness for one of:

a) Canada affected by acid deposition from the USA

b) Sweden and Norway affected by acid deposition from Poland, Germany, UK

c) China.

Draw a poster to show the impact and pollution management strategies.

Strategy for reducing pollution	Example of action	Evaluation
Altering the human activity producing pollution	Replace fossil fuel use by using alternatives: • Ethanol to run cars. • Renewable energy sources for electricity.	Also reduce CO_2 emissions but we live in a fossil fuel reliant economy. Demand for power is ever increasing, particularly in India and China as they industrialize.
	Reduce overall demand for electricity – education campaigns in the population to turn lights off, insulate houses, etc. Use less private transport – car pool, public transport, walk, cycle.	
	Use low sulphur fuels, remove sulphur before burning, or burn mixed with limestone.	
Regulating and reducing the pollutants at the point of emission	Clean-up technologies at 'end of pipe' locations (points of emission), eg scrubbing in chimneys to remove the sulphur dioxide.	Expensive and costs passed on to consumer. Catalysers are cost effective if well maintained but are expensive to buy.
	Catalytic converters convert nitrous oxides back to nitrogen gas.	
Clean up and restoration	Liming acidified lakes and rivers (see below).	Effective in restoring pH but has to be repeated regularly. Costly.
	Recolonize damaged areas.	Affects biodiversity in other ways.
	Liming forestry plantations. Trees acidify soils as they remove nutrients.	Treats symptoms and not the cause.
	International agreements	Agreements are difficult to establish and to monitor.

▲ **Figure 6.4.8** Table of replace, regulate and restore model of pollution management for acid deposition

Practical Work

* Investigate acid deposition in a selection of countries.

The role of international agreements in reducing acid deposition

Acid rain is an international issue because emission of a particular pollutant from one country does not equal the deposition of that pollutant in the same country.

Timeline for political solutions

In Europe

1970s   Evidence for acid rain accumulates including death of vast tracts of German forests, loss of lake biodiversity and accelerated weathering of buildings.

1979   UN Convention on Long Range Transboundary Air Pollution.

1983   Convention modified and fifteen European countries, USA and Canada agreed to cut sulphur emissions by 30% of 1980 levels by 1993.

1985   30 countries adopted the Protocol on the Reduction of Sulphur Emissions, agreeing to reduce sulphur dioxide emissions by 30% (from 1980 levels) by 1993. This was called the 30% club. All of the countries that signed the Protocol achieved this reduction, and many of those that did not sign have also met these reductions.

1988   Sofia convention for reducing nitrogen oxides by 30% of 1987 levels by 1998. Most countries missed this target due to increase in motor vehicles.

1994   UN Convention further modified to cut emissions by 80% of 1980 levels by 2003.

1999   27 countries signed up to new protocol on Long Range Transboundary Air Pollution to reduce and prevent air pollution.

In North America

1995   Clean Air Act target to reduce sulphur dioxide emissions to below 1980 levels. Mostly affected coal-burning power stations in Eastern states. Also target of reducing nitrogen oxides.

Allowance trading scheme – producers of acid deposition gases allocated allowances to emit an amount of sulphur dioxide. These allowances can be bought, sold and traded.

Evaluation of success

In Europe

Average of 50% reductions achieved in Europe by 2000. However developing countries are rapidly industrializing and emissions are set to increase rapidly unless developed countries can help developing countries 'leapfrog' to clean technologies and lifestyles with reduced emissions of these gases. Due to the regional character of acid deposition, international agreements concerning acid deposition are more of

bilateral or regional character. One example is the EU Large Combustion Plants Directive. This directive regulates the emissions by large installations that have a thermal capacity of more than 50 MW, basically electricity plants and large industries. Though the Kyoto protocol aimed at reducing CO_2 emissions will also help reduce acid deposition.

Reducing the effects of acid deposition
Liming lakes to neutralize acidity

Starting in the 1950s, the loss of many fish and invertebrate species in large numbers of Scandinavian lakes was linked to high levels of lake acidity. By 1990, over 400 lakes were virtually lifeless. In the 1980s, Sweden experimented with adding powdered limestone to their lakes and rivers. However results were mixed. The pH of treated lakes is quickly raised but short-lived because of the acidic nature of the inflow of water; liming only treats the symptoms and not the cause. Biodiversity was not immediately restored; the lime seemed to affect nutrient balance as nutrients other than calcium were absent.

Reducing emissions

One way to reduce the emission of sulphur dioxide and nitrogen oxides is to reduce the combustion of fossil fuels. Reducing the need for electricity, reducing the use of cars, developing more efficient or electric-powered cars and switching to alternative energy sources like waterpower and wind and solar energy can do this. Also biofuels (though this may add to the loss of food crops and increase the malnutrition problems discussed in sub-topic 5.2) and nuclear power generation will reduce SO_2 and NO_x emissions.

Precombustion techniques

Precombustion techniques aim at reducing SO_2 emissions by removing the sulphur from the fuel before combustion. The sulphur removed from the fuel can be obtained in several useful forms: as the element sulphur which can be used in the chemical industry, as gypsum, which can be used in construction, or as sulphur dioxide which can be used in the production of sulphuric acid, one of the most used chemicals.

End of pipe measures

End of pipe measures remove the sulphur dioxide and nitrogen oxides from the waste gases. Examples include waste gas scrubbers in electricity plants and the motor car catalytic converter. Waste gas scrubbers are intended to remove sulphur dioxide. The catalytic converter removes nitrogen oxides, together with other pollutants.

In USA

Sulphate wet deposition

1983–85

1992–94

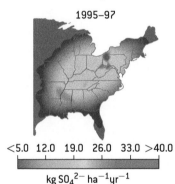

1995–97

<5.0 12.0 19.0 26.0 33.0 >40.0

kg SO_4^{2-} ha^{-1}yr^{-1}

▲ **Figure 6.4.9** Wet deposition in USA 1983–1997

To do

Consider Figure 6.4.9 and research the most recent findings on acid deposition.

To do

Evaluate the use of catalytic converters in cars.

REVIEW

To what extent are our solutions to atmospheric pollution mostly aimed at prevention, limitation or restoration?

Human society depends on oil but this is not sustainable. Discuss your predictions for the state of a world without oil.

BIG QUESTIONS

Atmospheric systems and society

Discuss whether economic development can ever be sustainable development.

Reflective questions

→ The Montreal Protocol was an international agreement made by the UN. Can one group or organization know what is best for the rest of the world?

→ Environmental problems are often emotive. Under what circumstances should we maintain a detached relationship with the subject matter under investigation?

→ Explain why ozone depletion is a global issue.

→ Explain why acid deposition is a regional problem.

→ Explain why national economic decisions have an impact on international environmental discussions.

Quick review

All questions are worth 1 mark

1. Which impact(s) on ecosystems may be associated with the release of sulphur oxides into the atmosphere?

    I. Increased uptake of aluminium ions by living organisms.

    II. A decrease in certain mineral storages in the soil.

    III. Reduced leaf surface area in coniferous forests.

    A. I, II and III **C.** I only

    B. I and III only **D.** III only

2. Which of the following is **not** a result of acid deposition from burning of fossil fuels?

    A. Leaching of calcium from soils.

    B. Death of coniferous trees in forests.

    C. Killing of fish due to high levels of aluminium in lakes.

    D. Thermal expansion of oceans.

3. Which of these human activities both increases global warming and depletes the ozone layer?

    A. Emission of carbon dioxide from vehicle exhausts.

    B. Emission of sulphur dioxide from power stations.

    C. Leakage of methane from gas pipelines.

    D. Release of CFCs from old refrigerators.

4. Which statement about acid rain is correct?

    A. Acid rain is a problem of developed countries because it is always deposited onto industrial areas.

    B. Rain with pH above 7 is considered to be acid rain.

    C. Lime can be used to restore acidified lakes.

    D. Corrosion of buildings and rising sea levels are two of the main effects of acid rain.

5. Which of the following statements is correct?

    A. Ozone gas is increasing in the upper atmosphere through the action of CFCs.

    B. Ozone gas is increasing in the upper atmosphere because of global warming.

    C. Ozone gas is decreasing in the upper atmosphere because of the increase in the amount of nitrogen oxides produced by the combustion of fossil fuels.

    D. None of the above statements is correct.

6. Which of these statements is correct?

    A. The formation of ozone involves the absorption of ultraviolet radiation.

    B. Ozone is destroyed by carbon dioxide released by burning fossil fuels.

    C. The type of ultraviolet radiation absorbed by the ozone layer does not affect living organisms.

    D. Chlorofluorocarbons in the stratosphere are rapidly broken down allowing them to escape into the outer atmosphere.

7. This question refers to the graph, right.

    The highest concentration of ozone is in the altitude band:

    A. 0–10 km.

    B. 10–20 km.

    C. 20–40 km.

    D. 40–80 km.

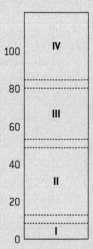

8. The Vienna and Montreal agreements were aimed at reducing the

    A. amount of ozone-depleting substances released into the atmosphere.

    B. loss of biodiversity, particularly in tropical rainforests.

    C. release of all greenhouse gases through burning of fossil fuels.

    D. amount of acid rain affecting Europe and North America.

9. The concentration of ozone in the atmosphere is greatest in the

    A. upper stratosphere.

    B. lower stratosphere.

    C. upper troposphere.

    D. lower troposphere.

7 CLIMATE CHANGE AND ENERGY PRODUCTION

7.1 Energy choices and security

Significant ideas:
→ There is a range of different energy sources available to societies that vary in their sustainability, availability, cost and socio-political implications.
→ The choice of energy source is controversial and complex. Energy security is an important factor in making energy choices.

Applications and skills:
→ **Evaluate** the advantages and disadvantages of different energy **sources**.
→ **Discuss** the factors that affect the choice of energy sources adopted by different societies.
→ **Discuss** the factors which affect energy security.
→ **Evaluate** an energy strategy for a given society.

Knowledge and understanding:
→ **Fossil fuels** contribute to the majority of human energy supply and these vary widely in the impacts of their production and their emissions and their use is expected to increase to meet global energy demand.
→ Sources of energy with lower carbon dioxide emissions than fossil fuels include **renewable energy** (solar, biomass, hydropower, wind, wave, tidal and geothermal) and their use is expected to increase. **Nuclear power** is a low-carbon low emission non-renewable resource but is controversial due to radioactive waste and potential scale of any accident.
→ **Energy security** depends on adequate, reliable and affordable supply of energy that provides a degree of independence. An inequitable availability and uneven distributions of energy sources may lead to conflict.
→ The **energy choices** adopted by a society may be influenced by availability, sustainability, scientific and technological developments, cultural attitudes and political, economic and environmental factors. These in turn affect energy security and independence.
→ Improvements in **energy efficiencies** and **energy conservation** can limit growth in energy demand and contribute to energy security.

1990
so, this climate change thing could be a problem...

1995
climate change: definitely a problem.

2001
yep, we should really be getting on with sorting this out pretty soon...

2007
look, sorry to sound like a broken record here...

2013
we really have checked and we're not making this up.

2019
is this thing on?

TAP TAP TAP

▲ **Figure 7.1.1**

'*I have no doubt that we will be successful in harnessing the sun's energy... If sunbeams were weapons of war, we would have had solar energy centuries ago.*'

Sir George Porter

Energy security

Imagine having no source of energy for a year other than what you yourself can collect. What could you get? Perhaps firewood to make a fire to cook on? Some coal washed up on a beach? Burn your furniture? Could you make electricity? What would not work? Could you have a solar-powered charger for your smart phone and laptop? However enterprising you may be, it would be tough and your lifestyle would change enormously. We rely on energy from electricity, gas and oil to function in our daily lives.

Energy security is an issue that national governments wrestle with so they can make the right choices between the use of resources available to them to produce useful energy and maintain national security. It may be cheaper to buy energy from another country but what happens if you fall out with that country? If fuel prices increase, how long will your people keep paying higher prices without complaint?

Energy price fluctuations and political changes can be risky for governments so they need to spread the risk. This can be achieved by using several energy sources from several places including their own (if they have one).

Energy sources are not equally spread over the world. Fossil fuel deposits are clumped in some areas – there are large deposits of coal in China, oil in the Middle East, gas in Russia. So some countries have to buy fuels from others. Depending on other countries for a nation's energy can work well if there is peace and it is economic to do so. But it increases energy security risks if something goes horribly wrong.

The **energy choices** made by a society depend on many factors:

- Availability of supply – within national borders or imported.

- Technological developments – finding new sources of energy, eg shale oils and harnessing wave power.

- Politics – can lead to conflict over energy supplies or choices to use more expensive domestic supplies for increased security, this may also impact the decision to go nuclear or not.

Key points

- Society gets its energy from a range of energy sources, non-renewable and renewable.

- We demand more and more energy to run the world economy.

- Different energy sources have advantages and disadvantages.

- Societies may choose energy sources for different reasons: political, cultural, economic, environmental and technological.

Key term

Energy security is the ability to secure affordable, reliable and sufficient energy supplies for the needs of a particular country.

303

TOK

The choice of energy source is controversial and complex. How can we distinguish between a scientific claim and a pseudoscience claim when making choices?

- Economics – globalization of economies can make it uneconomic to produce your own power and cheaper to import it.

- Cultural attitudes – our love of the motor car powered by fossil fuels means we are very reluctant to give them up or change to electric cars.

- Sustainability – only renewable energy sources are sustainable yet account for a small percentage of world energy supply.

- Environmental considerations – backlash against nuclear power generation – being phased out in Germany when people felt it was too dangerous after the Fukishima accident (1.1)

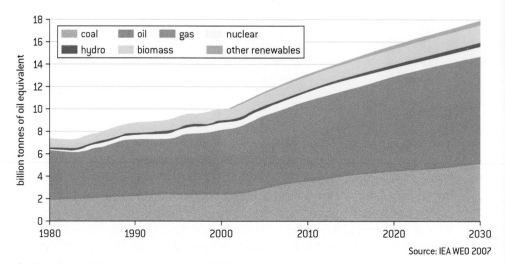

Source: IEA WEO 2007

▲ **Figure 7.1.2** Projected growth in global energy demand

The need for energy to power the global economy is rising and we keep burning non-renewable fossil fuels. Even with improvements in **energy efficiencies,** the inevitable will happen – we will run out of these resources.

Energy conservation can limit growth in energy demand and contribute to energy security but it only has a small impact on total use at the moment. Our need for more and more energy is insatiable.

Examples of energy security choices

1. Ukraine–Russia gas disputes

After the break-up of the USSR, Russia exported gas to Ukraine at below market prices. Some 80% of Russian gas flows through Ukraine to European destinations. In 2006, Russia cut off supplies to Ukraine as they had not paid their debts and were using gas intended for the rest of Europe. The dispute continued until 2010 when an agreement was signed.

Explain how this demonstrates energy security issues.

Investigate recent changes in the relationship between Ukraine and Russia. Have these affected the gas supplies?

2. USA shale oil

Tight oil (figure 7.1.4) is oil that has been discovered but not previously extracted as it was held tightly in rocks and not economic to extract. But with new technologies (fracking) and high oil prices, it has become economic to extract. Since 2008, US tight oil production has increased

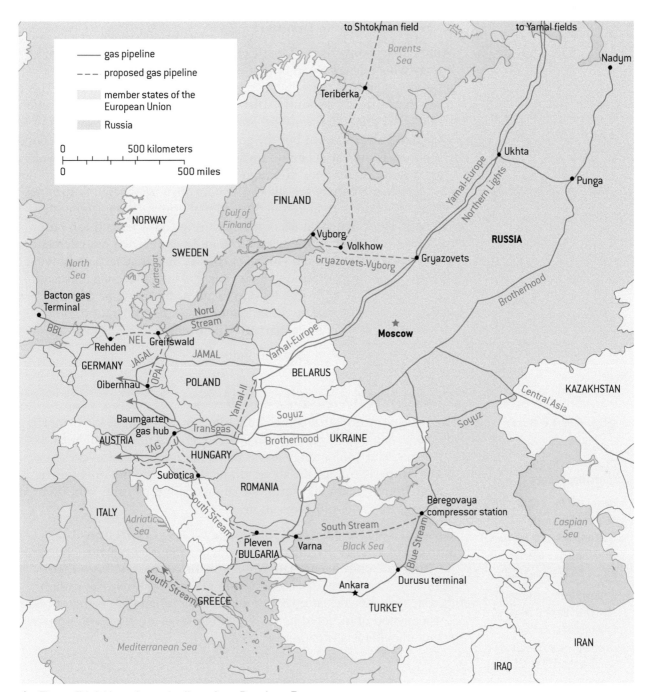

▲ **Figure 7.1.3** Map of gas pipelines from Russia to Europe

from 600,000 to 3.5 million barrels a day. Shale gas and oil is also extracted in the USA now. The result of this is that some predict the USA may change from being a net importer of oil to being the world's largest exporter by 2020 pumping 11.6 million barrels a day of crude oil. Others think this is overestimated.

The result is that now fossil fuel security for the USA has improved dramatically.

Evaluate the production of shale oil in the USA.

Discuss the implications of this for OPEC (Organization of Petroleum Exporting Countries).

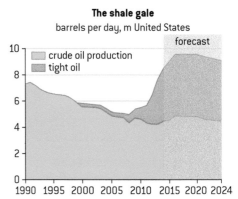

The shale gale
barrels per day, m United States

▲ **Figure 7.1.4** USA production of shale oil

▲ **Figure 7.1.5** Wind turbines in Denmark

To do

Evaluate the energy strategies of your own country and one other named country.

To do

Using figures 7.1.6 and 7.1.7:

1. Calculate the percentage of global energy supply from fossil fuels.

2. Research the population sizes of China and the USA and suggest reasons why they use similar amounts of energy.

3. Research the population size of India and suggest reasons why its use of energy is so much smaller than that of China.

3. Wind turbines in Denmark

Wind energy converts the kinetic energy in the wind to mechanical energy which drives a wind turbine to produce electricity. Denmark produces over 30% of its energy requirements from wind energy. This is more than any other country. The reason for this is that the government drove the change. In the 1970s, much of Denmark's energy was from coal-fired power stations but the government wanted to reduce carbon emissions. There was a ban on nuclear power plants and wind power was seen as the solution. Although the wind speeds in Denmark are not particularly high, there are shallow waters offshore where the turbines can be sited and then linked to the national grid onshore. Denmark is linked to the electricity grids of neighbouring countries and can buy electricity from them if the wind drops and sell it if their own demand is less than that generated. This is, of course, one disadvantage of wind power – you need the wind to blow. There is little evidence that wind turbines kill many birds that fly into them. Although some do, most birds fly over or round the turbines and adapt their migration routes. Power lines kill more birds than do wind turbines.

Evaluate Denmark's wind power industry.

Energy and energy resources

All our energy on Earth comes from our Sun. Without it, the planet would be at absolute zero, which is −273 °C, and there would be no life forms. The Sun's energy drives the climate, geochemical cycles, photosynthesis, animal life – everything. Humans do not only rely on energy from the Sun via plants, we supplement it by using other sources of energy to power our civilization.

Fossil fuels are simply stored solar energy. These **non-renewable sources of energy** are the compressed, decomposed remains of organic life from millions of years ago. Now humans extract and burn them to release that energy. This combustion of fossil fuels releases carbon dioxide that was locked up by photosynthesis when they were formed. This emission of carbon dioxide is having a large effect on atmospheric carbon dioxide levels (sub-topic 7.2).

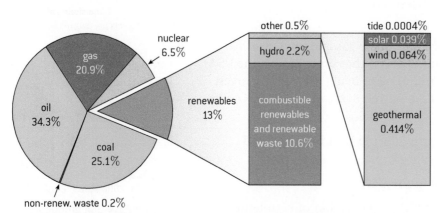

▲ **Figure 7.1.6** Energy sources worldwide (source IEA 2007)

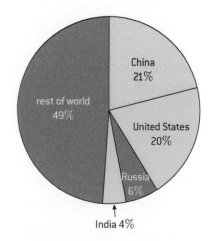

▲ **Figure 7.1.7** World energy use

How much longer for fossil fuels?

There are only estimates of when we shall run out of fossil fuels and we do keep finding more and extracting it. But we shall run out. Ball park figures are that we shall run out of:

- coal in 230 years
- gas in 170 years
- oil in 100 years.

Some say that as the rate at which we use it increases, there will be none left by 2075. Is that likely to be in your lifetime?

Two-thirds of the known oil reserves are in the Middle East. Most of the rest are in Russia and more is now found in the USA and Canada (shale oils and tar sands). The fossil fuel with the most reserves and fastest increase in consumption is coal, mostly because of China's need for power and many power stations there burn coal.

Humans use more and more energy as there are more of us and, as our wealth increases, we each use more energy if we can. But the use is not evenly distributed. Some countries are richer and use more energy than others. Before the industrial revolution, the energy we used was from wood – biomass. Since then, with the discovery of fossil fuels, we use far more than that available to us in biomass. We are using the capital stored as fossil fuels formed 300 million years ago.

Although it is generally agreed that oil will be exhausted in 50–100 years' time, some think it will be as soon as 30–40 years' time – in your lifetimes. We may find more but most experts think that we shall not find more reserves after the next 10–20 years as we shall have looked in most areas by then.

If LEDCs use oil at the rate *per capita* of the United States, it will run out in 17 years or sooner. We have reached or are about to reach 'peak oil', after which the amount produced will decline. There is more natural gas than oil left but we will use all of that in fewer than 80 years. There is far more coal left to mine and some people say that coal is the energy source of the future. We think there are 1,000 billion tonnes of coal left, enough for 150 years. China, in 2007, was building two coal-fired power stations a week and coal's use is growing faster than that of oil and gas. But it is the dirtiest energy source, releasing carbon dioxide and sulphur dioxide, when burned. However, we can build power stations with carbon capture and storage (CCS) technology. Since we are going to burn coal anyway, this should be the future – making coal a clean energy source.

Energy consumption

We are in an **energy crisis** yet continue to use non-renewable fuels at an increasing rate. Oil prices per barrel reached an all-time high in 2008 of well over US$145 per barrel. Are they higher or lower today? We know we use more and more energy and that fossil fuels will run out. We shall have to use energy sources other than fossil fuels. Renewable

> *'Junkies find veins in their toes when the veins in their arms collapse. Developing tar sands is the equivalent.'*
>
> Al Gore, 2008

▲ **Figure 7.1.8** The Airship Hindenberg accident in 1937 when hydrogen that filled the airship ignited

sources as well as nuclear power are the options. The EU, for example, produces less than 7% of its energy from renewable sources but there is a proposal to increase this to 20% by 2020.

In the future, without fossil fuels, humans will have to gain their energy from other sources unless we revert to a much smaller population only dependent on solar energy directly. Possibilities include the **hydrogen economy** where hydrogen is the fuel that provides energy for transport, industry and electrical generation. There are prototype engines and cars that use hydrogen as their fuel, releasing water as the waste product. But hydrogen is a highly flammable gas and difficult to transport and store.

Nuclear fusion involves extracting heavy water (deuterium) from water and fusing two hydrogen atoms to make helium. In theory this works, but in practice it is not feasible yet. (Nuclear fission uses uranium, a non-renewable resource.)

We have to increase our use of **renewable energy sources** for which we have the technology now, to have absolute reductions in energy consumption and to increase its efficiency of use.

To do

Energy consumption

Energy consumption is measured in tonnes of oil equivalent or TOE (the amount of energy released in burning one tonne of crude oil).

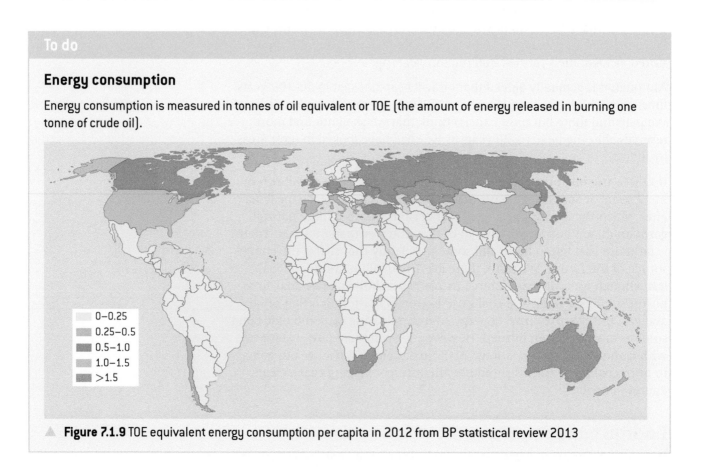

Legend:
- 0–0.25
- 0.25–0.5
- 0.5–1.0
- 1.0–1.5
- >1.5

▲ **Figure 7.1.9** TOE equivalent energy consumption per capita in 2012 from BP statistical review 2013

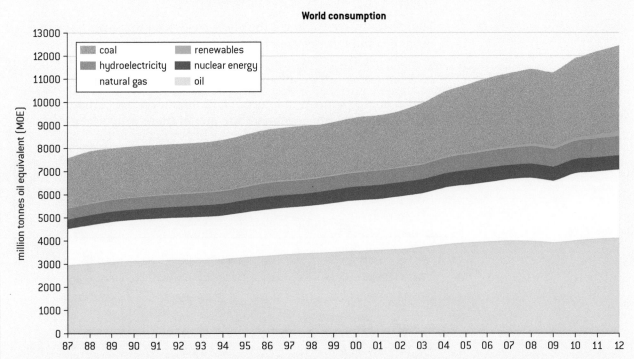

World primary energy consumption grew by a below-average 1.8% in 2012. Growth was below average in all regions except Africa. Oil remains the world's leading fuel, accounting for 33.1% of global energy consumption, but this figure is the lowest share on record and oil has lost market share for 13 years in a row. Hydroelectric output and other renewables in power generation both reached record shares of global primary energy consumption (6.7% and 1.9%, respectively).

▲ **Figure 7.1.10** Consumption of fossil fuels by type from 1987 to 2012

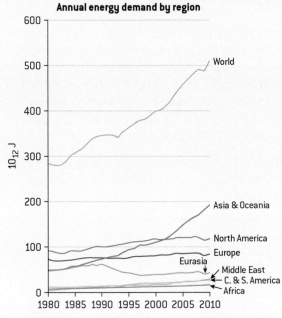

▲ **Figure 7.1.11** World energy demand by region 1980–2010

Study the data in the three figures, 7.1.9, 7.1.10 and 7.1.11, on energy consumption.

1. Describe the pattern of consumption *per capita* in figure 7.1.9.

2. Describe the lifestyle of someone with the highest consumption and someone with the lowest. Name the regions they live in.

3. Suggest reasons for the sharp rise in energy consumption in Asia and Oceania in figure 7.1.11.

4. Suggest reasons for the drop in total energy consumption in 2008.

5. Consider your own use of fossil fuels, list the products you have and activities you do that rely on fossil fuel consumption. (Remember, plastics are derived from oil.)

Renewable energy resources

Theoretically, we could get the energy we need to power the world economy and our own requirements from renewable resources. But, in practice, worldwide we obtain a small percentage at the moment from renewable resources – about 14%.

Which energy sources in figure 7.1.12 do not release carbon dioxide when used to produce energy? Are any of them truly 'clean'?

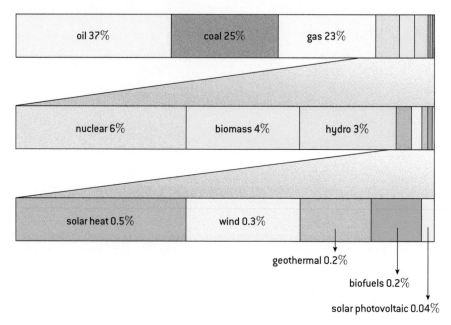

▲ **Figure 7.1.12** Percentage of world energy sources

We know that we must increase the percentage of our energy produced from renewable resources and progress has been and is being made. The EU plans to get 20% of its energy needs from renewables by 2020. The US gains about 12% of domestically generated power from renewables. But investment in research on making renewable energy sources more efficient, for example in wave and tidal power or solar cells, is small compared with research into finding more oil or gas. Reasons for this may be:

- That the TNCs (transnational corporations) and heavy industry are committed to the carbon economy – all machines are made to run on fossil fuels – and the scale of a change is hard to imagine.

- It is cheaper to produce electricity from fossil fuel burning than from most renewable resources at the moment (ignoring the environmental cost).

- Countries are locked into the resource that they currently use – by trade agreements or convenience.

- All renewable sources are location dependent:

  - wave or tidal power are not possible for land-locked countries

  - solar energy requires a sunny climate for maximum efficiency

  - wind power has a range of wind speeds within which it can operate effectively.

- What other reasons can you think of?

Carbon emissions from fossil fuel burning

Although figures vary from source to source, there is general agreement about the amount of carbon dioxide released to the atmosphere by burning fossil fuels. Carbon dioxide is responsible for two-thirds of the anthropogenic (enhanced) greenhouse effect. Most of our efforts are going to reduce these emissions of carbon dioxide which may not look much in parts per million but amount to over four tonnes per year per capita on average worldwide.

Who produces what carbon emissions?

China and the US produce the most and China may now have overtaken the US as the biggest emitter. The rapid growth in emissions in China and India reflect their rapid industrialization and large size. But per capita emissions do not relate to the size of the country.

Ice-core data before 1958. Mauna Loa data after 1958.

▲ **Figure 7.1.13** Atmospheric carbon dioxide levels 1700–2012

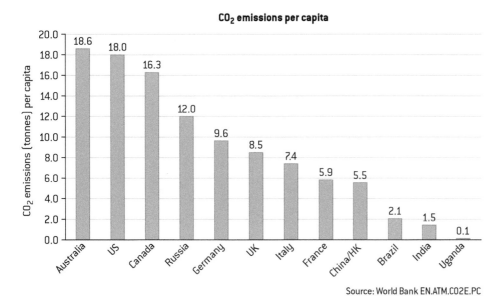

CO$_2$ emissions per capita

Source: World Bank EN.ATM.CO2E.PC

▲ **Figure 7.1.14** Carbon dioxide emissions per capita of selected countries in 2012

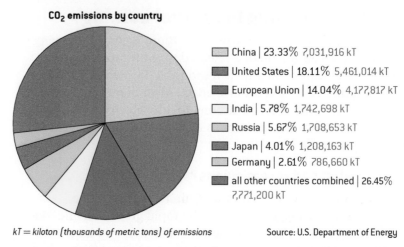

CO₂ emissions by country

- China | 23.33% 7,031,916 kT
- United States | 18.11% 5,461,014 kT
- European Union | 14.04% 4,177,817 kT
- India | 5.78% 1,742,698 kT
- Russia | 5.67% 1,708,653 kT
- Japan | 4.01% 1,208,163 kT
- Germany | 2.61% 786,660 kT
- all other countries combined | 26.45% 7,771,200 kT

kT = kiloton (thousands of metric tons) of emissions   Source: U.S. Department of Energy

▲ **Figure 7.1.15** Carbon dioxide emissions by country or region in 2008

Singapore, Gibraltar and some Gulf states have the highest per capita emissions, followed by the US with emissions several times that of China per capita.

The United Nations calculates that an average air-conditioner in Florida is responsible for more CO_2 every year than a person in Cambodia is in a lifetime, and that a dishwashing machine in Europe annually emits as much as three Ethiopians.

Deforestation also contributes to carbon emissions and Indonesia tops this table due to deforestation and peat and forest fires.

Evaluation of energy sources and their advantages and disadvantages

Energy source	Facts	Advantages	Disadvantages
Non-renewable			
Coal (fossil fuel)	Fossilized plants laid down in the Carboniferous period. Mined from seams of coal which are in strata between other types of rock. May be opencast mined (large pits) or by tunnels underground. Burned to provide heat directly or electricity by burning to create steam-driven turbines in power stations.	• Plentiful supply. • Easy to transport as a solid. • Needs no processing. • Relatively cheap to mine and convert to energy by burning. • Up to 250 years of coal left.	• Non-renewable energy source. • Cannot be replaced once used (same for oil and gas). • Burning releases carbon dioxide which is a greenhouse gas. • Some coals contain up to 10% sulphur. • Burning sulphur forms sulphur dioxide which causes acid deposition. • Particles of soot from burning coal produce smog and lung disease. • Coal mines leave degraded land and pollution. • Lower heat of combustion than other fossil fuels, ie less energy released per unit mass.

Oil (fossil fuel)	Fossilized plants and micro-organisms that are compressed to a liquid and found in porous rocks. Crude oil is refined by fractional distillation to give a variety of products from lighter jet fuels and petrol to heavier diesel and bitumen. Extracted by oil wells. Many oil fields are under the oceans so extraction is dangerous. Pipes are drilled down to the oil-bearing rocks to pump the oil out. Most of the world economy runs on oil either burned directly in transport and industry or to generate electricity.	• High heat of combustion, many uses. • Once found is relatively cheap to mine and to convert into energy.	• Only a limited supply. • May run out in 20–50 years' time. • Like coal, gives off carbon dioxide when burned. • Oil spill danger from tanker accidents. • Risk of terrorism in attacking oil pipelines.
Natural gas (fossil fuel)	Methane gas and other hydrocarbons trapped between seams of rock. Extracted by drilling like crude oil. Often found with crude oil. Used directly in homes to produce domestic heating and cooking.	• Highest heat of combustion. • Lot of energy gained from it. • Ready-made fuel. • Relatively cheap form of energy. • Cleaner fuel than coal and oil.	• Only limited supply of gas but more than oil. • About 70 years' worth at current usage rates. • Also gives off carbon dioxide but less than coal and oil.
Nuclear fission	Uranium is the raw material. This is radioactive and is split in nuclear reactors by bombarding it with neutrons. As it splits into other elements, massive amounts of energy are also released. Uranium is mined. Australia has the most known reserves, Canada exports the most, other countries have smaller amounts. About 80 years' worth left to mine at current rates but could be extracted from sea water.	• Raw materials are relatively cheap once the reactor is built and can last quite a long time. • Small mass of radioactive material produces a huge amount of energy. • No carbon dioxide or other pollutants released (unless there are accidents).	• Extraction costs high. • Nuclear reactors are expensive to build and run. • Nuclear waste is radioactive and highly toxic for a long time. Big question of what to do with it. Needs storage for thousands of years. May be stored in mine shafts or under the sea. Accidental leakage of radiation can be devastating. • Accidents rare but worst nuclear reactor accident at Chernobyl, Ukraine in 1986. • Risk of uranium and plutonium being used to make nuclear weapons. • Terrorist threat.

Renewable			
Hydroelectric power (HEP)	Energy harnessed from the movement of water through rivers, lakes and dams to power turbines to generate electricity. Pumped-storage reservoirs power turbines.	• High quality energy output compared with low quality energy input. • Creates water reserves as well as energy supplies. • Reservoirs used for recreation, amenity. • Safety record good.	• Costly to build. • Can cause the flooding of surrounding communities and landscapes. • Dams have major ecological impacts on local hydrology. • May cause problems with deltas – no sediment means they are lost. • Silting of dams. • Downstream lack of water (eg Nile) and risk of flooding if dam bursts.
Biomass	Decaying organic plant or animal waste is used to produce methane in biogas generators or burned directly as dung/plant material. More processing can give oils (eg oilseed rape, oil palms, sugar cane) which can be used as fuel in vehicles instead of diesel fuel = biofuels	• Cheap and readily available energy source. • If the crops are replanted, biomass can be a long-term, sustainable energy source.	• May be replacing food crops on a finite amount of crop land and lead to starvation. • When burned, it still gives off atmospheric pollutants, including greenhouse gases. • If crops are not replanted, biomass is a non-renewable resource.
Wood	From felling or coppicing trees. Burned to generate heat and light.	• A cheap and readily available source of energy. • If the trees are replaced, wood can be a long-term, sustainable energy source.	• Low heat of combustion, not much energy released for its mass. • When burned it gives off atmospheric pollutants, including greenhouse gases. • If trees are not replanted wood is a non-renewable resource. • High cost of transportation as high volume.
Solar – photovoltaic cells	Conversion of solar radiation into electricity via chemical energy.	• Potentially infinite energy supply. • Single dwellings can have own electricity supply. • Safe to use. • Low quality energy converted to high.	• Manufacture and implementation of solar panels can be costly. • Need sunshine, do not work in the dark. • Need maintenance – must be cleaned regularly.
Concentrated solar power (CSP)	Mirrors are arranged to focus solar energy on one point where heat energy generated drives a steam turbine to make electricity.	• Solar energy renewable source. • Cost of power stations equivalent to fossil fuel power stations.	• Required area of high insolation – so usually in tropics. • Relatively new technology but improving all the time.

Solar - passive	Using buildings or panels to capture and store heat.	• Minimal cost if properly designed.	• Requires architects who can design for solar passive technology.
Wind	Wind turbines (modern windmills) turn wind energy into electricity. Can be found singularly, but usually many together in wind farms.	• Clean energy supply once turbines made. • Little maintenance required.	• Need the wind to blow. • Often windy sites not near highly populated areas. • Manufacture and implementation of wind farms can be costly. • Noise pollution though this is decreasing with new technologies. • Some local people object to on-shore wind farms, arguing that visual pollution spoils countryside. • Question of whether birds are killed or migration routes disturbed by turbines.
Tidal	The movement of sea water in and out drives turbines. A tidal barrage (a kind of dam) is built across estuaries, forcing water through gaps. In future underwater turbines may be possible out at sea and without dam.	• Should be ideal for an island country such as the UK. • Potential to generate a lot of energy this way. • Tidal barrage can double as bridge, and help prevent flooding.	• Construction of barrage is very costly. • Only a few estuaries are suitable. • Opposed by some environmental groups as having a negative impact on wildlife. • May reduce tidal flow and impede flow of sewage out to sea. • May disrupt shipping?
Wave	The movement of sea water in and out of a cavity on the shore compresses trapped air, driving a turbine.	• Should be ideal for an island country. • These are more likely to be small local operations, rather than done on a national scale.	• Construction can be costly. • May be opposed by local or environmental groups. • Storms may damage them.
Geothermal	It is possible to use the heat under the earth in volcanic regions. Cold water is pumped into earth and comes out as steam. Steam can be used for heating or to power turbines creating electricity.	• Potentially infinite energy supply. • Is used successfully in some countries, such as New Zealand.	• Can be expensive to set up. • Only works in areas of volcanic activity. • Geothermal activity might calm down, leaving power station redundant. • Dangerous underground gases have to be disposed of carefully.

▲ **Figure 7.1.16** CSP and photovoltaic solar energy

To do

Select an energy source from the table. Research for information on this source.

Individually or in small groups, present a case to your class about that energy source. Include in your presentation: what it is, where it is mostly used, how much it is used, whether its use is sustainable, its relative cost, advantages and disadvantages of the resource use, future prospects, reasons why some countries may use this resource although it may not be the cheapest.

Peer evaluate the presentations based on:

Quality of spoken presentation – did they present in a clear, interesting way? Did they not read it out? Did they look at the audience and engage them?

Quality of content – did you understand the content? Was it clear?

Quality of associated documentation (handout, slide presentation, data) – did this enhance your understanding of the presentation?

To think about

The Athabasca oil sands

In Alberta, Canada, there is a lot of oil in tar sands. It is a very heavy crude oil, rich in bitumen which is used to make roads. It takes so much energy to refine this oil that it is only economic to do so if oil prices are high. The cost of extracting a barrel of this oil is about US$30. Prices for crude oil are high so it is being extracted – by surface mining from the Athabasca Oil Sands in Alberta. With some smaller deposits, these oil sands cover 54,000 square miles of land. Growing on the land is boreal forest and muskeg ecosystems. Muskeg is a type of peat bog with acidic conditions so vegetation is semi-decomposed. It can be up to 30m deep, is 1,000 years old and is a habitat to beavers, pitcher plants and mosses. Caribou deer (reindeer) also live there. Few humans live there. The amount of oil in these tar sands is roughly equal to the rest of the world's reserves put together. It is mined by open-pit mining which destroys the vegetation and changes the shape of the landscape as huge volumes of tar sands are removed. Mining companies do restore the land to pasture or plant trees but do not recreate the original boreal forest or muskeg.

Not only does it cost to extract the oil, it also takes a lot of energy – one barrel of natural gas to extract two of crude oil. It also takes a lot of water – two barrels of water for each barrel of oil and the waste water has to be kept in large tailing (mining waste) ponds. Rather than reducing, the carbon emissions of Canada are increasing, partly due to this mining of tar sands.

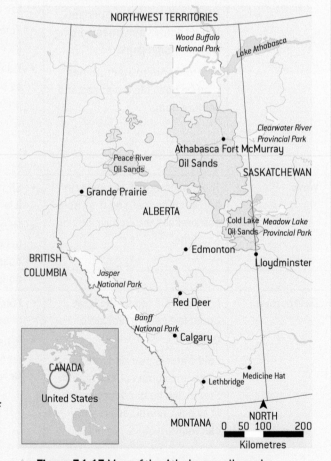

▲ **Figure 7.1.17** Map of the Athabasca oil sands

To think about

Biofuels – the answer or the problem?

Biofuels are fuels made from crops.

Why? In theory, they are greener as there are fewer carbon emissions as they are carbon neutral, meaning that all the emissions of carbon dioxide made from burning them are fixed by growing the plants to replace them. As 25% of carbon emissions are from transport, it is a neat idea but things are not that neat in complex systems.

What crops are used? Sugar cane has been used for decades in Brazil to make ethanol by fermentation. The ethanol (an alcohol) is sold alongside or mixed with petrol (gas) and 80% of cars sold in Brazil now have hybrid engines – they run on petrol, alcohol or a mixture of the two called gasoline. From 2000 to 2005, the world's output of plant ethanol has doubled and biodiesel, made from oily plants like oil palm and soya bean, has increased.

In the USA, maize production for ethanol is heavily subsidized; it has increased five-fold and is set to increase more. In the EU, a regulation was passed that all fuels must contain 2.5% biofuel, rising to 5% and 10% by 2020.

What's the problem?

Amazingly, in calculating the carbon balance of biofuels, no-one considered the additional carbon costs in extracting the ethanol, transporting the crop to the extraction plant and transporting the processed fuel. Some of those costs also apply to conventional fossil fuel extraction, of course, but there is also the cost of (oil-based) fertilizer applied to the crop. Fertilizers release nitrous oxide which is 310 times more powerful a greenhouse gas than carbon dioxide. Some calculate that maize ethanol requires 30% more energy to make than it contains.

The other problem is the land it takes to grow biofuels. It would take 40% of the EU cropland to meet the 10% target. And this means less food is produced.

What are the consequences? Deforestation is happening to plant crops for biofuels and this releases the carbon trapped in the trees. Wetlands are being drained and grassland being planted as well. Indonesia has planted so much oil palm on what was forest land that it is now the third largest emitter of carbon in the world. US farmers are selling 20% of their maize for biofuel instead of food, so soya farmers there are switching to the more profitable maize, so Brazilian farmers grow more soya bean to export, so they plant it on grazing land, so the cattle ranchers cut more forest in the Amazon to turn into grassland.

What should be done?

The calculations need to be checked and all costs included for both fossil fuels and biofuels. Cellulosic ethanol, from wood chips, switchgrass or straw may be the answer as its production will reduce far more carbon emissions than maize or soya bean growing.

There is inertia in the systems so once subsidies and legislation are in place, it will take time to change them if we need to. If biofuels are leading to increased greenhouse gas emissions, not reduced ones, and taking up land needed to grow food, what are we doing?

Practical Work

* Investigate people's attitude to global climate change.
* Evaluate the advantages and disadvantages of different energy sources.
* Evaluate an energy strategy for a named society.
* Investigate the relationship between carbon footprint and wealth / education level.

7.2 Climate change – causes and impacts

Significant ideas:

→ Climate change has been a normal feature of the Earth's history, but human activity has contributed to recent changes.

→ There has been significant debate about the causes of climate change.

→ Climate change causes widespread and significant impacts.

Applications and skills:

→ **Discuss** the feedback mechanisms that would be associated with a change in mean global temperature.

→ **Evaluate** contrasting viewpoints on the issue of climate change.

Knowledge and understanding:

→ **Climate** describes how the atmosphere behaves over relatively long periods of time whereas **weather** describes the conditions in the atmosphere over a short period of time.

→ Weather and climate are affected by **ocean and atmospheric circulatory systems**.

→ **Human activities** are increasing levels of greenhouse gases (eg carbon dioxide, methane and water vapour) in the atmosphere, which leads to:

- an increase in the mean global temperature
- increased frequency and intensity of extreme weather events
- the potential for long-term change in climate and weather patterns
- rise in sea level.

→ The potential **impacts of climate change** may vary from one location to another and may be perceived as either adverse or beneficial. These impacts may include changes in water availability, distribution of biomes and crop growing areas, loss of biodiversity and ecosystem services, coastal inundation, ocean acidification and damage to human health.

→ Both **negative and positive feedback mechanisms** are associated with climate change and may involve very long time lags.

→ There has been **significant debate** due to conflicting EVSs surrounding the issue of climate change.

→ **Global climate models** are complex and there is a degree of uncertainty regarding the accuracy of their predictions.

'The famous quote from a century ago, attributed to Mark Twain and Charles Dudley Warner "everybody talks about the weather but nobody seems to do anything about it" may have been true once. But it isn't true anymore.'

Peter Gleick

Climate and weather

It is important to understand the different between climate and weather with respect to climate change.

Weather is the daily result of changes of temperature, pressure and precipitation in our atmosphere. Weather varies from place to place, sometimes over very short distances. We try to predict weather but can only really do so with some accuracy about five days ahead. There are too many variables that interact in such a complex way. Weather can fluctuate wildly – a very hot day or a very cold one does not mean that average temperatures are changing.

Climate is the average weather pattern over many years for a location on Earth. Climate may show long-term trends and changes if records are kept for long enough.

The difference between weather and climate is the timescale on which they are measured.

The similarity is that both are affected by ocean and atmospheric circulatory systems. Both are also affected by:

- Clouds – clouds may trap heat underneath them or reflect sunlight away from Earth above them.

- Forest fires – release carbon dioxide, a GHG, but regrowth traps it again in carbon stores.

- Volcanic eruptions – release huge quantities of ash which circulate in the atmosphere, cooling the Earth. For example, Mt Pinatubo in the Philippines erupted in 1990 lowering global temperatures for a few years.

- Human activities – we burn fossil fuels and keep livestock, both of which release GHGs.

Climate change is long-term change and has always happened (see later in this sub-topic). Factors that influence climate change:

- Fluctuations in solar insolation affecting temperature.

- Changing proportions of gases in the atmosphere released by organisms.

For climate to change on a global scale, inputs and outputs have to change, eg heat input increases or heat output decreases or both if the climate warms up.

Greenhouse gases reduce heat loss from the atmosphere (see later in this sub-topic). If there are more greenhouse gases, less heat is lost. The system changes in a dynamic equilibrium which may stabilize or reach a new equilibrium if a **tipping point** is passed (1.3).

Long-term records show (figure 7.2.1) that the global average surface temperature of Earth is increasing although there are fluctuations from year to year.

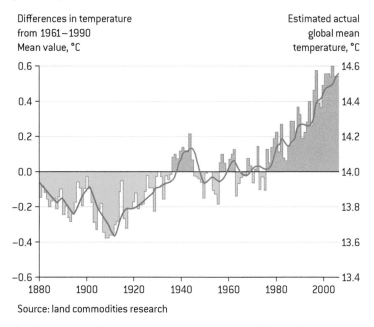

Source: land commodities research

▲ **Figure 7.2.1** Global average temperature 1860–2008

To think about

Stalling of temperature increase?

Since 2000, average global temperature has shown no increase and some climate sceptics (who do not believe humans are causing climate change) use this as evidence to support their hypothesis. But we know that there are variations in climate and it is very complex. Volcanic eruptions, ENSOs, cloud cover and ocean variability affect it so this stalling may be only a 'pause'. But how do we know?

Cherry-picking data is a characteristic of pseudoscience and figure 7.2.2 shows a short period in which average global temperature fell.[1]

In figure 7.2.2:

1. Which was cooler 2011 or 2010?
2. Which was warmer 2011 or 2008?
3. Which was cooler 2011 or 2007?

Taking such snapshots does not give long-term trends as they are minor variations.

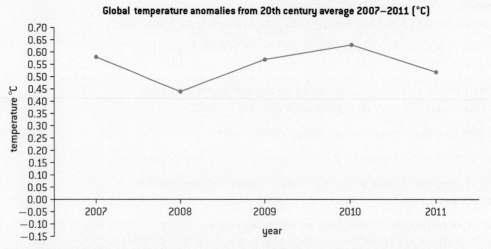

Figure 7.2.2 Global temperature anomalies from 20th century average 2007–2011

4. Compare figure 7.2.3 and figure 7.2.4.
5. What are the trends (blue lines) in these graphs?
6. What conclusions can you draw?

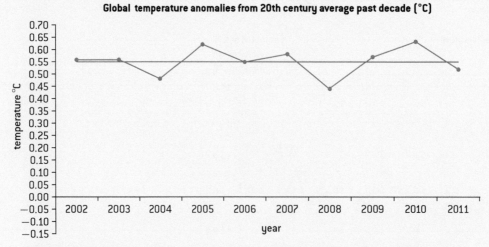

Figure 7.2.3 Global temperature anomalies from 20th century average 2002–2011

---

1   Charts from Peter Gleick's article 'Global Warming Has Stopped'? How to Fool People Using 'Cherry-Picked' Climate Data. Forbes 2012.
    http://www.forbes.com/sites/petergleick/2012/02/05/global-warming-has-stopped-how-to-fool-people-using-cherry-picked-climate-data/

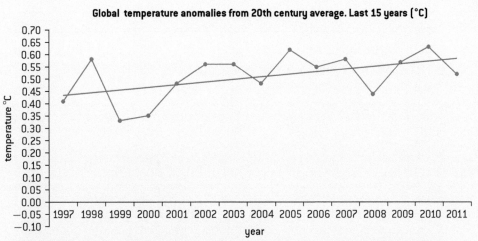

Figure 7.2.4 Global temperature anomalies from 20th century average 1997–2011

We know that the decade of 2000–9 was the warmest on record since records started in 1850 and that the past 15 years are amongst the warmest since 1850.

7. What does that say about figure 7.2.3?

If you take records for the last 130 years, figure 7.2.5 shows the data.

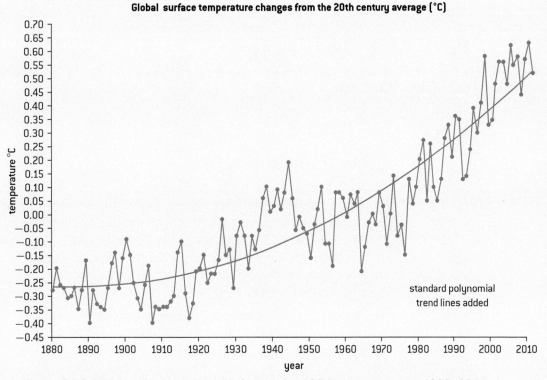

Figure 7.2.5 Global surface temperature changes from 20th century average 1880–2011

8. What does this tell you?

But remember, these are only average global temperatures. They do not record variation across the planet. We know that more extra heat is warming the Arctic and melting Greenland and Arctic ice. Figure 7.2.6

shows that since 1970 far more heat has gone into the oceans than is absorbed by land.

9. So do you now think the Earth is warming up?

10. And what do you think is causing this?

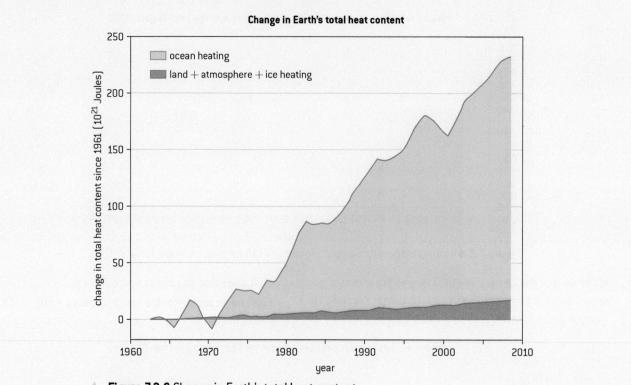

Figure 7.2.6 Change in Earth's total heat content

Taken from Church et al. Sept 2011. Revisiting the Earth's sea-level and energy budgets from 1961 to 2008. *Geophysical research letters*. Article first published online: 16 Sep 2011

Global climate models

Modelling climate change is a complex business requiring huge computing resources. Simple models of the climate system have been developed to predict changes with a range of emissions of greenhouse gases. The models solve complex equations but have to use approximations. They have improved over 30 years. The early ones included rain but not clouds. Now they have interactive clouds, rain, oceans, land and aerosols.

The latest climate models predict similar possible global average temperature changes to those predicted by models 5 or 10 years ago, with increases ranging from 1.6 to 4.3°C. (See 1.2 on models.)

Greenhouse gases

There is very little carbon dioxide in the atmosphere (0.04% of the total gases) but it and other GHGs are increasing through **anthropogenic activities** (activities of humans).

The list of greenhouse gases not only includes carbon dioxide, water vapour and methane but also chlorofluorocarbons (CFCs and HCFCs), nitrous oxide and ozone.

There are three points that may be confusing when reading about or reviewing statistics on climate change.

1. The role of ozone and CFCs.

2. The role of water vapour.

3. Whether figures refer to total GHG effects or the enhanced (anthropogenic) greenhouse effect.

The enhanced greenhouse effect

As humans increase emissions of some greenhouse gases (GHGs), the greenhouse effect is exaggerated or enhanced. Most climate scientists believe that this is causing global warming and climate change.

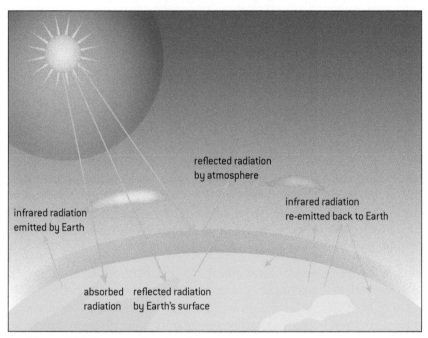

reflected radiation by atmosphere

infrared radiation re-emitted back to Earth

infrared radiation emitted by Earth

absorbed radiation

reflected radiation by Earth's surface

▲ **Figure 7.2.7** The greenhouse effect

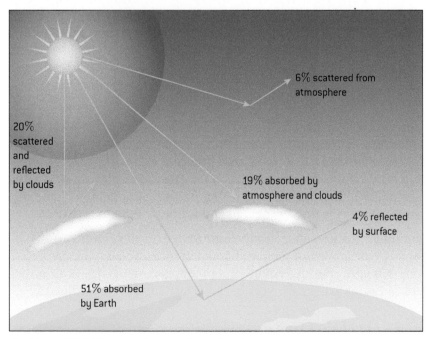

6% scattered from atmosphere

20% scattered and reflected by clouds

19% absorbed by atmosphere and clouds

4% reflected by surface

51% absorbed by Earth

▲ **Figure 7.2.8** The greenhouse effect showing percentages absorbed, reflected or scattered

Global warming potential (GWP)

The greenhouse effect of a molecule of a GHG varies depending which gas it is.

Carbon dioxide has a GWP of 1. Methane has a GWP of 21 so traps 21 times as much heat as the same mass of carbon dioxide.

Ozone occurs in the lower two layers of the atmosphere. In the troposphere it is a GHG but in the stratosphere it forms a layer that absorbs much of the ultraviolet radiation from the sun and so acts as a coolant. There is no direct link between global warming and ozone depletion but, because the climate is so complex, there are indirect links. The thinning of the ozone layer is certainly allowing more ultraviolet radiation to reach the Earth's surface but this amounts to less than 1% of solar radiation reaching Earth and is not significant in causing warming.

CFCs are chemicals made by humans that coincidentally break down ozone when they reach the stratosphere but act as GHGs in the troposphere. CFCs are human-made chemicals so are not present in the atmosphere as a result of natural processes. There are many types eg CFC-11 and CFC-12 as well as HCFCs. Although their concentration in the atmosphere is measured in parts per trillion (10^{-9}), they have a large contribution to the enhanced greenhouse effect because each molecule has a high GWP and a long lifetime in the atmosphere. Their GWPs may be thousands of times that of carbon dioxide. That means a molecule of a CFC is up to 10,000 times more effective at trapping long-wave radiation than a molecule of carbon dioxide which has a GWP of 1.

When data is presented, consider whether the contribution of water vapour is included or excluded. Water vapour has the largest effect on trapping heat energy so is the most potent greenhouse gas but it is not usually listed because it varies so much in concentration and is constantly condensing to water, snow and ice that stops it acting as a GHG. Somewhere in the region of 36–66% of the greenhouse effect is due to water vapour. The IPCC (Intergovernmental Panel on Climate Change), and most scientists, omit water vapour from their calculations but the IPCC work on the figure of a 50% contribution by water vapour. Clouds may contribute up to 25% (depending on the type of cloud and its altitude) and other GHGs cause the rest, with carbon dioxide having the largest effect.

Also remember that most GHGs in the atmosphere are there through natural processes (except CFCs which are human-made) and it is the **increase in these due to anthropogenic activities** that is of concern. Carbon dioxide concentration may be higher now than at any time during the last 160,000 years – the recent rapid rate of increase of 30ppm in 30 years is unprecedented and is due to human activities.

The amount of carbon added to the atmosphere each year due to human activities may not seem much when measured in parts per million (400 ppm) but this equates to an increase of 3.2–4.1 Gt C in the form of carbon dioxide each year over the last 25 years according to the IPCC. A Gt is a gigatonne or one billion tonnes (10^9) so that is up to 4,100,000,000 tonnes above the natural carbon cycle and does not include carbon in methane. Natural sinks (oceans, plants) absorb about half of this carbon each year (see Carbon cycle 2.3).

Greenhouse gas	Pre-industrial concentration (ppm)	Present concentration (ppm)	GWP	% Contribution to enhanced greenhouse effect	Atmospheric life times/years
Carbon dioxide, CO_2	270	400	1	50–60	50–200
Methane, CH_4	0.7	1.774	21	20	12
Nitrous oxide, N_2O	0.27	0.31	206	4–6	140
CFC-11	0	0.00025	3500	14 (all CFCs)	45
Ozone	not known	variable	2000	variable	not known

▲ **Figure 7.2.9** Major GHGs information

Source: adapted from 'Recent Greenhouse Gas Concentrations' by T. J. Blasing, from http://cdiac.ornl.gov

To do

What does this graph suggest about the changes in the relative importance of the GHGs?

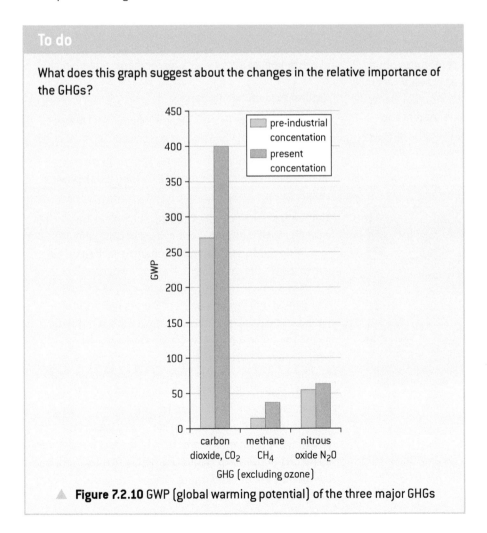

▲ **Figure 7.2.10** GWP (global warming potential) of the three major GHGs

Methane as a greenhouse gas

Methane is a simple hydrocarbon, CH_4. Since about 1950, the concentration of methane in the atmosphere has increased by about 1% per year due to human activities. About 60% of the methane in the air is from human activity and 15% of this from cattle. We also use methane as a fuel. Natural gas is methane, biogas digesters produce methane which is used for cooking or heating.

Sources of methane:

Cattle

There are 1.3 billion cattle in the world and they are ruminants with bacteria in their stomachs that break down the cellulose in the grass that they eat. These bacteria live in anaerobic conditions and release methane as a waste product. It comes out of both ends of the cattle and amounts to 100 million tonnes of methane per year. Each methane molecule is 21 times more effective than one of carbon dioxide at absorbing and radiating heat energy and methane contributes about 20% of the anthropogenic greenhouse gases. If we could capture this methane, we could use it as a fuel. Some scientists are trying to reduce the amount that cows produce by feeding them special diets higher in sugar levels or taking the bacteria in kangaroo stomachs and putting them in cow stomachs.

▲ **Figure 7.2.11** Cows produce methane

Waste tips

Waste tips in richer countries give off methane as waste food decomposes in the tip in anaerobic conditions. There can be so much methane that it is tapped and piped away to be used to generate electricity or for heating. But much is not captured and is released into the atmosphere.

▲ **Figure 7.2.12** Scavenging on a waste tip

Rice paddy fields cover 1.5 million km^2 of land and rice is a staple crop. The fields release up to 100 million tonnes of methane per year due to anaerobic respiration by bacteria in the soil. But they only release methane when flooded which is about one third of the year and the rest of the time may act as a sink for methane, absorbing it.

Natural sources of methane

Swamps and bogs – 5 million km^2 of bogs and marshes release methane.

Termites may produce 5% of atmospheric methane as bacteria in their guts release it as they break down cellulose.

The tundra – the bogs and swamps of the tundra contain much methane produced by decomposition in waterlogged soils. But this methane is locked up as it is frozen in the permafrost. There is evidence that the permafrost is melting and that some methane is being released. In some parts of Siberia or Northern Canada, you can dig a hole and set fire to the methane. As the permafrost melts, it releases more methane which causes more warming – an example of positive feedback. In Arctic seas is another source of methane, locked up as methyl hydrates in clathrates which are molecular cages of water that trap methane molecules within them. These are only stable when frozen and under high pressure at the bottom of the seas. There may be up to 1×10^{10} tons of methane in these structures and companies are already trying to mine them. But it is very dangerous work as the methyl hydrates can bubble up to the surface and sink any ships in the area.

The climate change debate

The only environmental issue to have caused as much debate and discussion as climate change was probably human population growth. But that is, in some ways, clearer to deal with. We can count how many we are, more or less, and can see a direct effect of more people wanting to use more resources from a finite stock.

Climate change and global warming have become very emotive issues where national and international politics, global economics and the fate of national economies are all bound up with scientific debate about the evidence and cause and effect. Added to this are the questions of whether millions or billions will suffer, whether there will be losers and winners if climate shifts to a new equilibrium and whether the power bases of different nations will be affected; you can begin to see what a complex issue this is.

We talked about environmental viewpoints near the start of this book (topic 2). Your viewpoint certainly influences how you interpret the evidence on climate change as technocentrists and ecocentrists clash on the question of what we should do or can do to mitigate the effects that we are seeing.

There are facts that are not in debate:

- there is a greenhouse effect

- GHG emissions are increasing due to human activities and are probably increasing the greenhouse effect

- there has been a recent pattern of increased average global temperature.

There is not total agreement about the cause of the rise in temperature nor over what we should be doing about it.

The vast majority of scientists working in this field accept the correlation between increased GHG emissions and increased temperature, causing climate change and different weather patterns. But there is a minority who question the cause and effect, some citing the earth's rotational wobble, sunspot activity or that increased temperature is causing increased GHG, not the other way round. And there are climate change deniers.

But all agree that the feedback mechanisms are very complex in such a complex system as the Earth and that our models, though much improved, may not exactly model the climate.

Adding the question of what should be done – prevention or cure or no action and the inertia that individuals and nation states have in managing change, you can begin to see why there is so much to discuss and how actions lag behind what we think we know.

Here, we take the view of the IPCC (Intergovernmental Panel on Climate Change) www.ipcc.ch in their fifth assessment report in 2014:

- 'Warming of the climate system is unequivocal, and since the 1950s, many of the observed changes are unprecedented over decades to millennia.

TOK

How should we react when we have evidence that does not fit with an existing theory?

- Atmospheric concentrations of carbon dioxide, methane, and nitrous oxide have increased to levels unprecedented in at least the last 800,000 years.

- Continued emissions of greenhouse gases will cause further [global] warming and changes in all components of the climate system. Limiting climate change will require substantial and sustained reductions of greenhouse gas emissions.

- Human influence on the climate system is clear. It is extremely likely (95–100% probability) that human influence was the dominant cause of global warming between 1951–2010.'

To do

Anthropogenic sources of greenhouse gases

Copy and complete the table with sources due to human activities.

Greenhouse gas	Sources due to human activities
Carbon dioxide	
Methane	
Ozone	
Nitrous oxide	
CFCs	

What is climate change and what will happen?

Changes in the climate can be seen in different ways. It may be in changed temperatures or rainfall patterns, more severe storms, ice sheet thinning or thickening and sea level rises. It may not be a steady process and there was cooling recorded in the 1970s and no rise in average global temperature since 2000. These could be due to sunspot variations and global dimming due to pollution but the general trend is for warming over the last century.

There are five ways in which the climate can change over time due to a change in greenhouse gas levels.

1. There may be a **direct relationship** – more forcing (changes in solar radiation), more change in proportion (Figure 7.2.13a).

2. There may be a **buffering action** in which forcing increases but climate change does not follow in a linear way. It is insensitive to change (Figure 7.2.13b).

3. It may respond slowly at first but then **accelerate** until it reaches a new equilibrium (Figure 7.2.13c).

4. It may reach a tipping point – that is the climate makes no response to changes but then reaches a threshold, at which point it changes rapidly until a new, much higher equilibrium is reached. (Figure 7.2.13d).

5. In addition to the threshold change, it may then get **stuck at the new equilibrium** even when the forcing decreases until it then tips over a new threshold and falls rapidly. These threshold changes could occur in just a few decades (Figure 7.2.13e).

Figure 7.2.13 shows these scenarios. Our problem is that we do not know which one we are living in. So how do we decide what to do?

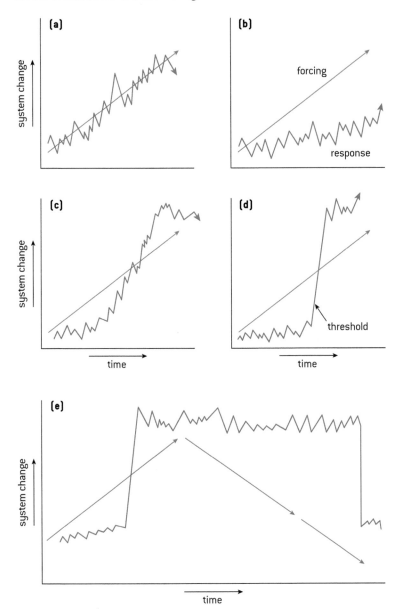

▲ **Figure 7.2.13** Possible climate change system responses to a forcing mechanism

A way of visualizing this is to imagine the climate as a car that you (the forcing mechanism) are pushing uphill. In (a) you push steadily uphill. In (b) you push with the same force but the car moves much more slowly – it has more resistance. In (c) you reach a part of the hill with a shallower gradient and the car moves more easily before the hill gets steeper again. In (d) you push until the car reaches the edge of a cliff and then falls over it. Can you explain (e) in this analogy?

To do

Where is the evidence?

See 6.1 for past climate changes.

More recent climate change

Ice cores have been taken from the Antarctic and Greenland ice caps. The Vostok core in the Antarctic went down as far as ice laid down 420,000 years ago. A more recent one reached 720,000-year-old ice. The bubbles of air trapped in the ice can be analysed to tell us what the climate was like at the time the ice froze. The proportions of different isotopes of hydrogen and oxygen give an indication of the climate then and levels of gases can be measured. The age of the ice can be calculated by ice rings in the top layers (rather like tree rings showing summer and winter) and by the dust from volcanic eruptions lower down. It gets less accurate the deeper you go. From the ice cores, a picture of CO_2 and temperature over time can be built up. See figure 7.2.14.

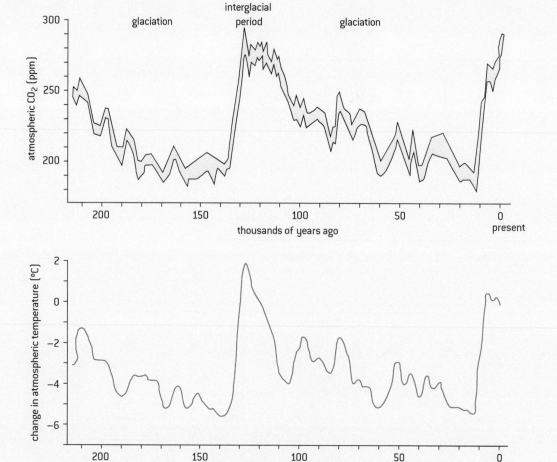

▲ **Figure 7.2.14** CO_2 and temperature levels of the last 240,000 years

1. What conclusion can you draw from the two graphs?

Since we started burning large amounts of fossil fuels, humans have added CO_2 to the atmosphere in addition to the natural amounts in the carbon cycle. Ice core records show CO_2 levels have risen from about 270 ppm before 1750 to 400 ppm by volume today. Although CO_2 levels were far higher than this (see 6.1) in geologic time, the recent increase has been due to human activities and adds 27 billion tonnes a year of carbon to the atmosphere.

In Hawaii in the Pacific, atmospheric CO_2 has been measured since 1958. See figure 7.2.15.

2. Describe and explain this trend.

3. What is the percentage increase in CO_2 levels between 1960 and 2008?

4. Why is there an annual cycle? (Hint: think about which hemisphere has the most land mass and then when plants photosynthesize.)

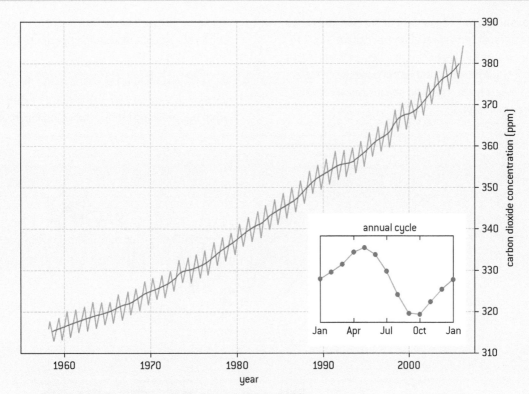

▲ **Figure 7.2.15** Atmospheric CO_2 in ppm 1958–2007

5. The temperature of the Earth since 1850 is shown in figure 7.2.16. Although this has fluctuated, there is a trend. What is it?

6. What would you have said in 1910 or 1950?

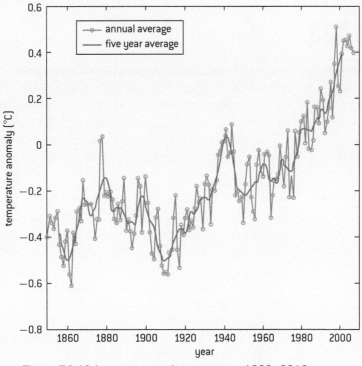

▲ **Figure 7.2.16** Average annual temperature 1880–2010

More recently, over the last 100 years, sea level changes have been recorded as well. But these are difficult to measure as the land is not static but also moves up and down slowly. What measurements do show, however, is that sea levels have risen and sea ice thickness at the poles has decreased.

7. Sea temperature has also been recorded. Figure 7.2.17 shows average surface sea temperature since 1980. What is the trend?

8. What happened to temperature when Pinatubo (a volcano in the Philippines) erupted? Explain this. What happens in El Niño years? Explain this.

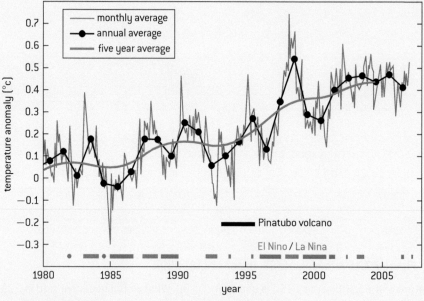

▲ **Figure 7.2.17** Sea surface temperature 1980–2007

TOK

There has been considerable debate about the causes of climate change. Does our interpretation of knowledge from the past allow us to reliably predict the future?

What is the consensus view?

There is evidence that greenhouse gases concentrations have increased since the Industrial Revolution and human activity has caused this. Climate is changing. Some of that may be due to natural climatic variation (sunspot activity, wobble of the earth), some due to atmospheric forcings. But the climate and weather patterns are very, very complex and some parts respond quickly, eg the atmosphere, while others react very slowly, eg the deep oceans.

The IPCC is the Intergovernmental Panel on Climate Change was set up by UNEP (the UN Environment Programme) and the World Meteorological Organization in 1988 to look into this major issue. It is made up of hundreds of scientists from around the world who research and regularly report on climate change. The first four reports used language that qualified some statements as scientists are wary of stating certainty, as further experiment could falsify a theory – for example, 'all swans are white' works as a hypothesis until you see a black swan in Australia. However, the language became more certain over time and in the fourth report they used terms such as 'extremely likely', '>95% confidence of an outcome', 'very high confidence', '9 out of 10 chance of being correct', so went as far as possible in stating their case. The fifth assessment report in 2014 was very clear about 'unequivocal warming of the climate system with 95–100% probability that human influence was the dominant cause'.

Impacts of climate change

Global average temperature has increased by about 0.85 °C since 1880, most of that since 1980 and this has had a number of impacts.

1. On oceans and sea levels

Sea levels are rising. This is because water expands as it heats up and ice melting on land slips off the land and into the sea increasing the volume of sea water. The Greenland and the Antarctic ice sheets are on land and are thinning. This and the thermal expansion of the seas will mean that sea levels rise more. By how much is not clear but predictions are becoming more accurate as climate modelling improves with more variables entered into the programmes. An increase of between 1.5 and 4.5 °C could mean a sea level rise of 15–95cm (IPCC data). But that is assuming a proportional relationship. If there is a threshold and this is exceeded, sea levels could rise by many metres. Up to 40 nations will be affected. Low lying states, eg Bangladesh, the Maldives and the Netherlands would lose land area – some, eg Tuvalu, would disappear completely.

The oceans absorb carbon dioxide and this makes them slightly acidic. They are slightly more acidic by 0.1 pH as they have absorbed about half the carbon produced by anthropogenic activities. This may affect marine organisms, particularly corals. But as they warm, they absorb less CO_2.

2. On polar ice caps

Melting of land ice in Antarctica and Greenland will cause sea levels to rise as it flows into the oceans. Melting of the floating ice cap of the Arctic will not increase the volume of water as ice has the same displacement as liquid water. But glaciers are melting into the seas and these will increase the volume of water. The Greenland ice sheet could melt completely and slow down or stop the North Atlantic Drift current by diluting the salt water. If the North Atlantic Drift current and the Gulf Stream slow or even shut down, the climate of the UK and Scandinavia would be much colder.

Melting in the Arctic could open up trade routes, make travel in the region easier and allow exploitation of undersea minerals and fossil fuel reserves. Methane clathrate is a form of ice under the Arctic Ocean floor that traps methane. If this were to melt and reach the surface, the release of methane may trigger a rapid increase in temperature.

3. On glaciers

In the Little Ice Age between about 1550 and 1850, glaciers increased in size. They then decreased (except for the period 1950–90 when global dimming possibly masked global warming) and have continued to decrease in size. Some have melted completely. Loss of glacier ice leads to flooding and landslides. Glacier summer melt provides a fresh water supply to people living below the glacier and this has provided water to many major Asian rivers (River Ganges, Brahamaputra, Indus, Yellow, Yangtze) which are fed by the Himalayan glaciers. It is causing significant drought problems in Tanzania where the Kilimanjaro glacier has lost over 80% of its volume.

4. On weather patterns

More heat means more energy in the climate and so the weather will be more violent and sporadic with bigger storms and more severe droughts. Global precipitation may increase by up to 15%. This will cause more soil erosion and lack of water will mean more irrigation and consequent salinization. There were more hurricanes in 2007 and some were more violent (eg Hurricane Katrina). There is evidence that severe weather and more extreme rainfall or droughts are occurring. Monsoon rains fail more often than they used to.

5. On food production

Warmer temperatures should increase the rate of biochemical reactions so photosynthesis should increase. But respiration will increase too so there may be no increase in NPP. In Europe, the crop growing season has expanded. But if biomes shift away from the equator, there will be winners and losers. It very much depends on the fertility of the soils as well. If production shifts northwards from Ukraine with its rich black earth soils to Siberia with its thinner less fertile soils, NPP will fall. Various predictions state huge ranges of changes from −70% to +11%. There are just too many variables to be certain. But what is certain is that some crop pests will spread to higher latitudes as they will not be killed by cold winters.

In the seas, a small increase in temperature can kill plankton, the basis of many marine food webs.

Heatwaves and drought kill livestock.

6. On biodiversity and ecosystems

Melting of the tundra permafrost would also release methane which is trapped in the frozen soils. In Alaska, Canada and Russia, permafrost is melting and houses built on it are shifting as it thaws.

Animals can move to cooler regions but plants cannot. The distribution of plants can shift as they disperse seeds which germinate and grow in more favourable habitats. But this happens slowly at about 1 km per year and perhaps too slowly to stop them becoming extinct. Species in alpine or tundra regions have nowhere to go, neither higher up nor towards higher latitudes. Polar species could become extinct in the wild.

Birds and butterflies have already shifted their ranges to higher latitudes.

Plants are breaking their winter dormancy earlier.

Loss of glaciers and decreased salinity of marine waters and changes to ocean currents alter habitats.

If droughts increase, then wildfires are more likely to wipe out other species or at least habitats for animals.

Increase in temperature of fresh and salt water may kill sensitive species and national parks and reserves could find their animals dying and the park boundaries static.

Indonesian forest fires have set fire to the peat bogs which have burned continually for years. The amount of carbon released by these adds significantly to carbon in the atmosphere.

Pine forests in British Columbia are being devastated by pine beetle which is not being killed off by cold winters as they are too mild. Corals are very sensitive to increased sea temperature. An increase of one degree can cause coral bleaching as the mutualistic zooxanthellae algae in the corals are expelled and the coral dies. Corals are the basis for many food webs. If the corals die, the ecosystem dies.

7. On water supplies

Increased evaporation rates may cause some rivers and lakes to dry up. Without a water supply, populations would have to move away. The UN says that 2.4 billion people live in the river basins fed by the Himalayas and their water supply is reducing. In Europe and North America, glaciers are also in retreat.

8. On human health

Heatwaves killed many in Europe in 2006. These may increase. Insect disease vectors will spread to more regions as the less cold winters means they will not be killed off. Malaria, yellow fever and dengue fever could spread to higher latitudes. Algal blooms may be more common as seas and lakes warm and some are toxic (red tides) and can kill humans. In a wetter climate, fungal disease will increase; in drier areas, dust increases leading to asthma and chest infections.

Warmer temperatures in higher latitudes would reduce the number of people dying from the cold each year and reduce heating bills for households. And fewer snow storms and icy roads means lower death tolls on the roads.

9. On human migration

If people cannot grow food or find water, they will move to regions where they can. Global migration of millions of environmental refugees is quite possible and this would have implications for nation states, services and economic and security policies. The IPCC estimates 150 million refugees from climate change by 2050.

10. On national economies

Some would suffer if water supplies decrease or drought occurs. Others would gain if it became easier to exploit mineral reserves (tar sands of Canada and Siberia) that would have been frozen in the permafrost or under ice sheets. If rivers do not freeze, hydroelectric power generation is possible at higher latitudes.

The Northwest Passage is a sea route for shipping from the Atlantic to the Pacific via the seas of the Arctic Ocean north of Canada. Many explorers tried to find this route in the northern summers but were stopped by sea ice. In 2007, the passage was navigable for the first time in recorded history.

Overall, there will be gains and losses for national economies. Agricultural production may rise in higher latitudes but fall in the tropics. Africa will probably lose food production and rainfall. Northern Darfur has seen desertification on a massive scale already and many millions of hectares become desert.

To do

Copy and complete the table below using the data in this section.

Advantages of climate change	Disadvantages of climate change
The Northwest Passage will improve shipping	Africa will lose food production

To put a monetary value on this is difficult but the Stern Report (from the former chief economist of the World Bank) suggested in 2006 that 1% of global GDP should now go to mitigating the effects of climate change to save up to 20% of global GDP in a recession later.

How fast could all this happen? It is happening now and there will be more changes in your lifetime.

Feedback mechanisms and climate change

Feedback is the return of part of the output from a system as input, so as to affect succeeding outputs.

There are two kinds of feedback:

Negative feedback is feedback that tends to dampen down, reduce or counteract any deviation from an equilibrium, and promotes stability. For example, increased evaporation in tropical latitudes leads to increased snowfall on the polar ice caps, which reduces the mean global temperature.

Positive feedback is feedback that amplifies or increases change; it leads to exponential deviation away from an equilibrium. For example, increased thawing of permafrost leading to an increase in methane levels, which increases the mean global temperature.

While modelling climate keeps improving, some feedbacks are so complex that we cannot be sure of the results.

To do

Figure 7.2.18 lists changes that may have positive and negative feedbacks.

Draw feedback cycles for three of the boxes in figure 7.2.18.

Also look at sub-topic 1.3 to see some of these feedbacks.

	Positive feedback or amplified change	Negative feedback or dampened down change
Oceans	Oceans are a carbon sink containing 50 times the amount of carbon as the atmosphere. They release more carbon dioxide to the atmosphere as they warm up as warm liquids hold less gas. Stalling of the North Atlantic Drift could reduce transfer of heat to the north and increase temperatures dramatically. Huge amounts of methane are frozen in methane clathrates in the ocean sediments. If these are released, the volume of methane in the atmosphere will increase dramatically.	Oceans absorb more CO_2 in warmer water as phytoplankton photosynthesize faster, producing more phytoplankton that absorb more CO_2 so dampening global warming.
Clouds	More evaporation leads to more clouds which trap more heat.	More evaporation leads to more clouds which reflect more heat.
It could be either. In the dark, clouds keep heat in, in the light they reflect it. But it depends what type of cloud as well. Cirrus (high, thin) clouds have a warming effect, low, thick ones a cooling one.		

Pollution	At night, cloud formation increased by aerosols acts as insulation, trapping heat. More clouds, more heat trapped. Black soot falling on ice decreases albedo, increasing heat absorption, increasing temperature and melting.	Aerosols from pollution, particularly sulphates, form condensation nuclei and more clouds form. These reflect heat and increase albedo, reducing warming in the day.
Polar ice	Ice has a high albedo – reflects heat and light. When it melts, the sea or land have a lower albedo and absorb more heat and more ice melts.	Warmer air carries more water vapour so more rainfall, some of which will be snow so more snow, more reflection, lower temperatures, more snow and ice. Possibly the next ice age.
Forests	Forests are cut down and burned. Less carbon is absorbed. More CO_2 in the atmosphere so higher temperatures. Forests die due to high temperature and may catch fire, more CO_2 released, temperature rises.	CO_2 absorbed – forests act as a carbon sink, removing CO_2 from the atmosphere, so temperature rise decreases.
Tundra	As temperatures rise, permafrost melts, releasing CO_2 which is trapped in the frozen soil. Methane is also released.	

▲ **Figure 7.2.18** Possible feedback mechanisms in climate change

To do

Look at the effects of climate change again and put them in diagrammatic form or pictorial form to include the effects, positive and negative changes and feedback mechanisms.

To think about

Global dimming

Global dimming is a reduction in solar radiation reaching the surface of the Earth. It was first noticed in the 1950s when scientists in Israel and Australia measured pan evaporation rates (evaporation of water from a salt pan). In the 1980s, more research in Switzerland, Germany and the USSR also found that the incoming radiation was less than it was. At the time, most were sceptical of the results which showed a reduction in the rates because of a reduction in sunlight and so a cooler Earth at a time when it conflicted with evidence that the Earth was getting warmer. When aircraft were grounded for a few days in the US after 9/11, the absence of contrails (vapour trails from aircraft) produced a sharp rise in the range of the Earth's surface temperature.

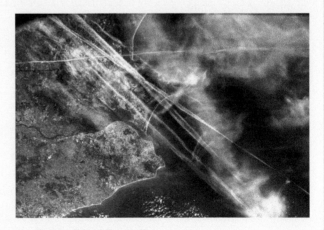

▲ **Figure 7.2.19** Contrails

Other experiments measuring sunlight levels over the Maldives showed that particulates in the atmosphere were causing global dimming. The small pollutant particles of mostly sulphate aerosols in the clouds both block sunlight from reaching the Earth's surface and reflect it back into space. These act as nuclei around which water droplets form and the clouds then reflect more sunlight back into space. Other particles come from volcanic eruptions, dust storms and incomplete burning of fossil fuels which produce black carbon or soot. It appears that global dimming affects weather patterns to the extent of shifting the monsoon rains and the long-term drought in the 1970s and 80s in sub-Saharan Africa.

The drop in temperature caused by global dimming was about 2–3% from 1960 to 1990 but no drop has been recorded since then.[2] The amount varies around the Earth with more in the temperate zones of the Northern hemisphere and may be masking the full increase of global warming. There is some evidence that, as we clean up our atmospheric pollutants, global dimming decreases and we see the full effect of global warming.

While global warming would increase temperature and give more energy to the water cycle, decreased energy input would slow it down and make it more humid with less rain.

It has been suggested that we could control global dimming and so mitigate the effects of global warming. Putting sulphur in jet fuel would cause it to produce sulphates in flight and so cool the earth. But the danger in this is that we would be creating more pollution, not less, and may find ourselves in a cycle of increased warming and increased pollution to counteract it. Perhaps not the best idea?

To do

Select and summarize two opposing views of global climate changes and present your own viewpoint.

To do

Watch the film 'An Inconvenient Truth'.

Watch 'The great global warming swindle'.

Watch 'Global Dimming', a BBC Horizon documentary or the NOVA programme 'Dimming the Sun'.

Read peer-reviewed articles and books on the issue of climate change. (There is a lot of propaganda out there. Make sure you are reading a reliable source.)

Form an opinion based on evidence and decide what actions you will take.

Climate modelling

We all want to know what the weather will be and we base our predictions on past experience (what it was like last June, for instance) and authority (the experts who tell us their informed predictions). GCMs are general circulation models and we have been trying to model the climate with these since the first computers. The early ones had few inputs and were not very accurate predictors but we now have AOGCMs (atmosphere–ocean GCMs) which split the Earth into sub-regions and consider the inputs of the atmosphere, oceans, ice sheets, land and biosphere. This includes the effect that humans are having on forests (deforestation). The latest AOGCM models quite accurately reflect past climate change and have predicted changes with various concentrations of greenhouse gases. These are reported by the IPCC.

What next on climate change?

Humans are resourceful and climate change will not make the human species extinct. But it will alter your lifestyle choices and possibilities and you should understand the science and the politics behind decision-making on these issues.

---

[2] Hegerl, G. C., Zwiers, F. W., Braconnot, P., Gillett, N.P., Luo, Y., Marengo Orsini, J.A., Nicholls, N., Penner, J.E., et al. (2007), 'Chapter 9, Understanding and Attributing Climate Change – Section 9.2.2 Spatial and Temporal Patterns of the Response to Different Forcings and their Uncertainties', in Solomon, S., Qin, D. & Manning, M. et al., *Climate Change 2007: The Physical Science Basis. Contribution of Working Group I to the Fourth Assessment Report of the Intergovernmental Panel on Climate Change*, Intergovernmental Panel on Climate Change, Cambridge, United Kingdom and New York, NY, USA: Cambridge University Press, http://www.ipcc.ch/pdf/assessment-report/ar4/wg1/ar4-wg1-chapter9.pdf.

Your parents' or grandparents' generations were probably concerned about the Cold War, nuclear proliferation and deterrent, World War II or human population growth. Until the 1980s, we were not utilizing more than the Earth could provide and were not all really aware of, or chose to ignore, climate change as an issue affecting all life on Earth. Now and for the rest of your lifetimes, the climate change debate will not go away, as it is a moral as well as an economic and environmental issue that will affect your lifestyle and choices.

The moral dilemmas involve your personal choices (whether to fly or not, to drive or not, what type of car, to live in a city or the countryside) as these all have an effect on your personal carbon emissions. There is an ethical question for governments of MEDCs whose peoples mostly have a high standard of living and why should they deny this to those in LEDCs by asking them to decrease their economic development rate. In the economic model we have, LEDCs aiming to increase their GDP (gross domestic product) and so improve the lives of their people and would be unwilling to reduce growth rates.

The cost of alleviating climate change has been given a value of 2% of the world's GDP so should governments all vote to give 2% of their GDP towards this and so not towards something else? We have and can develop the technology to make cleaner energy production systems but we do not have the collective will to put efforts into this as long as we can burn fossil fuels. Perhaps because democratic politics is short-termist – until the next elections – we do not plan for 50 or 100 years but for less than five.

Practical Work

* Evaluate two different viewpoints on climate change and our reaction to it.

TOK

There is a degree of uncertainty in the extent and effect of climate change. How can we be confident of the ethical responsibilities that may arise from knowledge when that knowledge is often provisional or incomplete?

7.3 Climate change – mitigation and adaptation

Significant ideas:

→ Mitigation attempts to reduce the causes of climate change.

→ Adaptation attempts to manage the impacts of climate change.

Applications and skills:

→ **Discuss** mitigation and adaptation strategies to deal with impacts of climate change.

→ **Evaluate** the effectiveness of international climate change talks.

Knowledge and understanding:

→ **Mitigation** involves reduction and/or stabilization of greenhouse gas (GHG) emissions and their removal from the atmosphere.

→ **Mitigation strategies to reduce GHGs** in general should include:
- reduction of energy consumption
- reduction of emissions of oxides of nitrogen and methane from agriculture
- use of alternatives to fossil fuels
- geo-engineering.

→ **Mitigation strategies for carbon dioxide removal** (CDR techniques) include:
- protecting and enhancing carbon sinks through land management eg United Nations reduction of emissions from deforestation and forest degradation in developing countries (UN REDD) programme
- using biomass as fuel source
- using carbon capture and storage (CCS)
- enhancing carbon dioxide absorption by the oceans through either fertilization of oceans

with N/P/Fe to encourage the biological pump, or increasing upwellings to release nutrients to the surface.

→ Even if mitigation strategies drastically reduce future emissions of GHGs, past emissions will continue to have an effect for some time.

→ **Adaptation strategies** can be used to reduce adverse effects and maximize any positive effects. Examples of adaptations include flood defences, vaccination programmes, desalinization plants and planting of crops in previously unsuitable climates.

→ **Adaptive capacity** varies from place to place and can be dependent on financial and technological resources. MEDCs can provide economical and technological support to LEDCs.

→ There are **international efforts** and conferences to address mitigation and adaptation strategies for climate change (eg Intergovernmental Panel on Climate Change (IPCC), National Adaptation Programmes of Action (NAPAs), United Nations Framework Convention on Climate Change (UNFCCC), etc.

Strategies to alleviate climate change

We can either try to reduce the impact of climate change or adapt to it or do both.

There are three routes we can take on this issue: do nothing, wait and see, take precautions now. Science cannot give us 100% certainty on the issue of global warming nor predict with total accuracy what will happen. What it can do is collect data and provide evidence. How that evidence is interpreted and extrapolated will depend on individual viewpoints, scientific consensus, economics and politics.

There is a minority of scientists and others who do not accept that global warming and climate change is a problem for human activity and development on Earth. They may take the 'do nothing' and business as usual approach, saying that we may forfeit economic development and so progress out of poverty for many by reacting to a non-threat. Alternatively, they say that warming is a good thing, brings benefits and technology can manage the effects.

The danger in the 'wait and see' strategy is that it takes a long time for actions to have results. To move the global economy away from a fossil fuel base is a long, slow process and the possible disruption of national economies in the process may not be necessary. But it is possible that we will reach the **tipping point** when our actions will have little effect as positive feedback mechanisms change the climate to a new equilibrium which could be 8 °C warmer than it is now. So better safe than sorry, perhaps?

The **precautionary strategy** is the majority choice. Act now in case. Even if we find out that burning fossil fuels is not causing global warming, we know they will run out and it makes sense to clean up the pollution caused by burning fossil fuels and find alternative fuel sources now, before we run out. What we are seeing in national policies and international targets are precautions – carbon emission reduction, carbon offset and lifestyle changes – against increased climate change.

These precautions can be divided into three categories:

- international commitments,
- national actions, and
- personal lifestyle changes.

Mitigation strategies

A. Stabilize or reduce GHG emissions
   a. reduction of energy consumption
   b. reduction of emissions of nitrogen oxides and methane from agriculture
   c. use of alternatives to fossil fuels

B. Remove carbon dioxide from the atmosphere

C. Geo-engineering

In mitigation, we use technology and substitutions that reduce or stabilize GHG emissions and remove GHGs from the atmosphere. With

> **Key term**
>
> **Mitigation** involves reduction and/or stabilization of greenhouse gas (GHG) emissions and their removal from the atmosphere. It is anthropogenic intervention to reduce the anthropogenic forcing of the climate system; it includes strategies to reduce greenhouse gas sources and emissions and enhancing greenhouse gas sinks.
>
> IPCC, 2007. http://www.ipcc.ch/publications_and_data/ar4/wg2/en/annexessglossary-e-o.html

> **Key term**
>
> **Adaptation** is the adjustment in natural or human systems in response to actual or expected climatic stimuli or their effects, which moderates harm or exploits beneficial opportunities.
>
> IPCC, 2007. http://www.ipcc.ch/publications_and_data/ar4/wg2/en/annexessglossary-a-d.html

respect to climate change, mitigation means implementing policies to reduce GHG emissions and to enhance carbon sinks.

But even if mitigation strategies drastically reduce future emissions of GHGs, past emissions will continue to have an effect for some time as the gases are circulating in the atmosphere.

There is a difference between stabilizing GHG emissions and stabilizing concentrations of GHGs in the atmosphere. If tomorrow we could stabilize GHG emissions at today's levels, their concentrations in the atmosphere would continue to rise. This is because our human activities are adding GHGs to the atmosphere faster than natural processes can remove them (see carbon cycle 2.3).

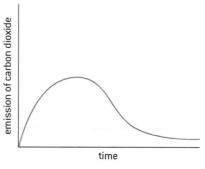 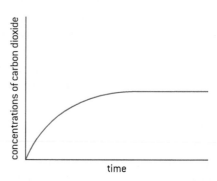

▲ **Figure 7.3.1**

To stabilize GHG concentrations in the next 100 years, we would need to reduce emissions by 80% of peak emission levels. This is extremely unlikely to happen.

A. Stabilize or reduce GHGs

- Reduction of energy consumption.

  - Reduce energy waste by using it more efficiently, eg improving fuel economy in motor car engines, hybrid or electric vehicles, insulate and cool buildings more efficiently, education in schools, energy efficient light bulbs and appliances.

  - Reduce overall demand for energy and electricity by being more efficient and using less – by changing lifestyles and business practices, eg using less private transport, cycling or walking not driving, eat less meat, circular economy, fly less often (7.1).

  - Adopt carbon taxes and remove fossil fuel subsidy.

  - Set national limits on GHG production and a carbon credit system.

  - Personal carbon credits which can be traded and encourage people to reduce their carbon footprint.

  - Change development pathways and socio-economic choices – change priorities in government and educate to change social attitudes, eg London has a toll charge for cars driving into the city and low emission zones for trucks.

  - Improve efficiency of energy production.

- Reduction of emissions of nitrous oxides and methane from agriculture.

  - Reduce methane production, eg methane from cows can be reduced by changing their diets.

  - Capture more methane produced from landfill sites.

  - Sustainable agriculture.

- Use of alternatives to fossil fuels.

  - Replace high GHG emission energy sources with low GHG emission ones, eg hydroelectric and other renewables and nuclear power generation instead of burning fossil fuels.

B. Remove carbon dioxide from the atmosphere

This is termed carbon dioxide removal (CDR).

1. **Increase amount of photosynthesis** so increasing the rate at which atmospheric carbon dioxide is converted into a biomass carbon sink (2.3) by reforesting and decreasing deforestation rates, restore grasslands. This does not reduce emissions so do not get confused. For example, the UN-REDD programme – (**UN R**educing **E**missions from **D**eforestation and forest **D**egradation) in LEDCs. This is a collaboration coming out of the Bali road map aiming to allow member countries to pool resources in reducing GHG emissions from deforestation and degradation.

2. **Carbon capture and storage** (**CCS**)**.** Removal of carbon dioxide is more easily done before it is released to the atmosphere. This means capturing it in emissions from power stations, oil refineries and other industries which emit large amounts of carbon dioxide. But this will increase the cost of energy and products. To store it, suitable rocks have to be found and the captured carbon dioxide transported there. It is then pumped into the rocks under pressure and stays there. An alternative is to store it in mineral carbonates by reacting carbon dioxide with metal oxides using high temperatures. Limestone is calcium carbonate so it would be like making limestone but the energy needed is a huge amount. A few pilot plants have carried out CCS but there are no large-scale CCS power stations.

3. **Use more biomass** as a source of fuel. If the same crop is planted in the following year, an equal amount of carbon dioxide to that released by burning the fuel is then captured by photosynthesis when biomass is replanted ready for the next year's harvest. This should then be a carbon neutral fuel.

   a. Directly by burning it to generate heat or electricity

   b. Indirectly to produce biofuels (7.1) eg

      i. biogas from animal waste in Indian villages or on a larger scale in fermenters

      ii. biodiesel and ethanol from waste organic matter or waste vegetable oils or from planting crops such as sugar cane.

343

C. Geo-engineering

Geo-engineering or climate engineering – large-scale intervention projects. This is somewhat different from the other mitigation strategies because so far they are hypothetical or computer models, they have not been tried and they have ethical questions around them. For example:

1. Scatter iron, nitrates or phosphates on oceans to increase algal blooms which take up more carbon and act as a carbon sink.

2. Release sulphur dioxide from airplanes to increase global dimming (7.2).

3. Send mirrors into space between the Earth and the Sun to deflect solar radiation.

4. Build with light-coloured roofs to increase albedo and reflect more sunlight.

Adaptation strategies

Adaptation aims to reduce adverse effects and maximize any positive effects. Examples of adaptations include flood defences, vaccination programmes, desalinization plants and planting of crops in previously unsuitable climates.

Adaptation initiatives and measures aim to reduce the vulnerability of natural and human systems against actual or expected climate change effect.

But who pays? This depends on the technological and economic resources available and on the will of a country, industry, company or individual. This is called the adaptive capacity.

Having adaptive capacity is a necessary condition for the design and implementation of effective adaptation strategies to reduce the damaging outcomes resulting from climate change. MEDC nations can provide support to LEDC nations.

1. Change land use through planning legislation.
   a. Do not allow building on flood plains – localized flooding in the UK in 2014, see 4.2.

2. Build to resist flooding.
   a. Plan water catchment and run-off to minimize flooding.
   b. Build houses on stilts or with garages which can be flooded underneath.

3. Change agricultural production.
   a. Irrigate more effectively in drought areas.
   b. Store rainwater for times of water shortage.
   c. Breed drought tolerant crops.
   d. Grow different crops.

4. Manage the weather.
   a. Seed clouds to encourage rainfall.
   b. Plant trees to encourage more rainfall.

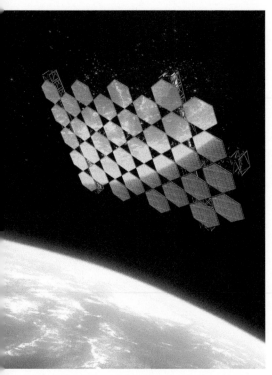

▲ **Figure 7.3.2** Mirrors in space to deflect sunlight from the Earth. Could this really work?

Key term

Adaptive capacity is the ability or potential of a system to respond successfully to climate variability and change, and includes adjustments in both behaviour and in resources and technologies.
From IPCC report 2007

5. Migrate to other areas.

6. Vaccinate against waterborne diseases eg typhoid.

7. Manage water supplies.

    a. Desalination plants.

    b. Increase reservoirs.

    c. Harvest run-off more effectively.

    d. Use water harvesting from clouds in higher areas.

International action: a timeline of agreements and commitments for action

1979   First World Climate Conference. Climate change officially recognized as a serious problem needing an international response when evidence of increasing carbon dioxide levels established.

1988   **Intergovernmental Panel on Climate Change (IPCC)** established by United Nations Environment Programme (UNEP) and the World Meteorological Organization. The IPCC is a collaborative body comprising over 2,000 climate scientists worldwide. Its main activity is to provide at regular intervals an assessment of the state of knowledge on climate change.

1990   **First IPCC Report** on Climate Change. The Report confirmed that *climate change was a reality and was supported by scientific data.*

1992   Rio Earth Summit (United Nations Conference on Environment and Development). **United Nations Framework Convention on Climate Change (UNFCCC)** signed by 154 governments. The objective of the Convention is to stabilize greenhouse gas concentrations. The governments of developed or annex I nations were voluntarily committed to developing national strategies for reducing greenhouse gas emissions to 1990 levels by the year 2000.
NAPA (a **National Programme of Action**) is one form of reporting proposed by the UNFCCC for LEDCs to decide how to meet their most urgent needs to adapt to climate change.

1995   First UNFCCC conference. Governments recognized that voluntary commitments were inadequate and work started to draft a protocol for adoption at the third Conference of Parties in 1997. **Second IPCC report** concludes that *the balance of evidence suggests a discernible human influence on the global climate.*

1997   **The Kyoto Protocol** signed by some 160 nations at third UNFCCC conference. The Protocol calls for the first ever legally binding commitments to reduce carbon dioxide and five other greenhouse gas emissions to 5.2% below 1990 levels before 2012. The US signed but has not ratified the protocol.

2001   **Third IPCC Report** states that *anthropogenic emissions will raise global mean temperature by 5.8 °C by 2050.*

2004   **The Kyoto protocol is still ineffective.** For the Kyoto Protocol to be effective at least 55 countries have to ratify (fully adopt the commitments) and there must be enough annex I (developed) countries which together are accountable for more than 55% of the emissions according to the 1990 levels. However the percentage of annex I countries is only 37.5%.

2005   Kyoto treaty goes into effect, signed by major industrial nations except US. Work to retard emissions accelerates in Japan, Western Europe, US regional governments and corporations.

2007   **Fourth IPCC report** warns that serious effects of warming have become evident; cost of reducing emissions would be far less than the damage they will cause. Dec 2007 UN climate conference in Bali agreed on a **Bali road map** to have a global treaty by end.

2008   **Global economic crisis** – reduction in international will to negotiate on carbon emissions reductions as national economies fall but this fall results in lower GHG emissions from industry.

2009   China overtook the USA as the country with the largest greenhouse gas emissions. 192 governments at the Copenhagen UN climate summit.

2013   Milestone of 400 ppm carbon dioxide in atmosphere reached. Apparent pause in warming explained as oceans have continued to warm.

2014   **Fifth IPCC report** was the strongest warning yet that global warming is happening, human activities are mostly causing it through burning fossil fuels and increasing carbon dioxide levels in the atmosphere.

Climate change management strategies

Management strategies can be looked at using the pollution management model in figure 1.5.6.

Use the section on mitigation and adaptive strategies and your own research to complete the table below.

Strategy for reducing global emissions	Example of action	Evaluation
Altering the human activity producing pollution		
Regulating and reducing the pollutants at the point of emission		
Clean up and restoration		

▲ **Figure 7.3.3** Clean up actions on climate change

To research

1. Find out what major climate change meetings there have been since 2007 and what the outcomes are.

2. Find out:

   a. What your own government policy is on climate change and carbon emissions.

   b. What alternative energy sources the country in which you live is developing.

   c. What are the advantages and disadvantages of these?

3. List the possible ways that countries could reduce their carbon emissions.

4. Research your own carbon footprint. www.carbonfootprint.com is a good place to start.

   a. Calculate your carbon footprint. (This measures your carbon use in tons of CO_2 not hectares as in ecological footprints.)

   b. List as many ways as you can of reducing your own carbon footprint size.

   c. How many of these will you do?

5. Which of the strategies you have listed would be the most effective in reducing CO_2 emissions and why? Consider whether they need people to cooperate, if they reduce your quality of life, if the technology is available, how easy they are to do. Are the strategies ecocentric or technocentric?

6. What do you think are the ethical issues surrounding the geo-engineering strategies?

To do

Carbon offset and carbon emissions trading

As part of the Kyoto Protocol in which 163 countries agreed to aim to limit their GHG production, the concept of **carbon emissions trading** evolved. In this scheme, countries that go over their quota (set by international agencies) on carbon dioxide emissions can buy carbon credits from countries which do not meet their quotas. In this way they still produce carbon dioxide but globally the limits are still met. You can imagine how complex this system is to operate. Monitoring carbon emissions for an industry is hard enough but for a country is very difficult.

1. Which country owns the emissions from an international flight or container ship – where it started or ends up or the country where the airline is based?

2. Who sets the quotas?

While a market has grown up for trading carbon emission permits, it is a volatile one. The EU emissions trading scheme, for example, has seen the value of carbon credits fall due to an overestimate of the allocation required when it started. The scheme does not encourage industries or countries to reduce their emissions either if they can buy permission to continue emitting. Carbon emissions trading is an alternative to a Carbon Tax where organizations are taxed for polluting – for releasing carbon dioxide. Also many TNC (transnational corporations) locate their production in LEDCs.

3. To which country do the carbon emissions belong?

4. Which do you think is the better option?

There are voluntary schemes now to **offset carbon emissions** for individuals and companies. Book a plane ticket now and you will be asked by many airlines if you want to pay to offset the carbon emissions that you are causing by flying on a plane. If you agree to this, the money should go to a company that invests it in a scheme that reduces carbon emissions. This is usually in a renewable energy scheme such as wind turbines, tree planting or hydroelectric power generation. Although small the market is increasing as environmentally aware individuals invest. But some schemes have minimal impact on global carbon emissions as they must invest in a scheme that would otherwise not have happened. Just taking the money and planting trees which would still have been planted is not recapturing any more carbon dioxide.

Personal carbon allowances (PCAs) are talked about. In this idea, we are all issued with an allowance for carbon emissions. If we travel a lot or live in energy inefficient homes, we would either have to change our lifestyles or buy more PCAs on the open market from people who produce less carbon dioxide.

To become **carbon neutral** is a goal that some talk about. Being carbon neutral means that you have no net carbon emissions – all carbon you release is balanced by an equivalent amount that is taken up or offset. This could be by planting trees, buying carbon credits or through projects that reduce future GHG emissions.

5. Do you think this is possible for an individual? What would you do? You may want to make a list of all the things that you do that affect your carbon emissions – researching your carbon footprint may help.

6. How do you think different societies would react to becoming carbon neutral?

Evaluate climate models in predicting climate change.

To what extent can society mitigate the effects of climate change?

Comment on the ways in which your EVS influences your attitude towards your lifestyle choice with respect to climate change.

BIG QUESTIONS

Climate change and energy production

Carbon dioxide levels are now 400ppm in the atmosphere. Discuss whether you think we have reached a tipping point on climate change.

To what extent can nations achieve sustainable development and energy security?

Examine the factors that may make your own EVS differ from that of others with respect to climate change.

Reflective questions

→ How should we react when we have evidence that does not fit with an existing theory?

→ The choice of energy source is controversial and complex. How can we distinguish between a scientific claim and a pseudoscience claim when making choices?

→ There has been considerable debate about the causes of climate change. Does our interpretation of knowledge from the past allow us to reliably predict the future?

→ There is a degree of uncertainty in the extent and effect of climate change. How can we be confident of the ethical responsibilities that may arise from knowledge when that knowledge is often provisional or incomplete?

→ Why does a country's choice of energy source impact another country?

→ What factors determine energy choices for a country?

→ The impacts of climate change are global. Who decides what action is necessary for mitigation?

Quick review

All questions are worth 1 mark

1. The major greenhouse gases are

   A. nitrous oxide and chlorofluorocarbons.

   B. water vapour, carbon dioxide, ozone and methane.

   C. carbon dioxide, nitrogen and ozone.

   D. ozone, water vapour and chlorine.

2. The rise in the Earth's mean surface temperature between 1860 and the present is considered to be caused by

   I. release of methane from wetlands.

   II. deforestation.

   III. burning of fossil fuels.

   A. I, II and III   B. I and II only

   C. II and III only   D. I and III only

3. Which column in the table correctly shows the effects of the pollutant gas?

	A.	B.	C.	D.
	Sulphur dioxide	Halogenated gases	Methane	Carbon dioxide
increases the greenhouse effect	Yes	No	Yes	Yes
depletes stratospheric ozone	Yes	Yes	No	Yes
increases acidity of rain	Yes	No	No	Yes

4. If part of the cost to the environment of fossil fuel use were added to the price of the fuel, the most likely effect would be that

   A. global warming would increase.

   B. use of renewable energy would decrease.

   C. more fossil fuels would be produced.

   D. consumption of fossil fuels would decrease.

5. Which of the following greenhouse gases are produced only by human activities?

   A. Methane and chlorofluorocarbons (CFCs)

   B. Carbon dioxide and water vapour

   C. CFCs

   D. Methane and water vapour

6. Methane is produced by

   I. bacterial activity.

   II. decomposition in landfill sites.

   III. digestive systems of cattle.

   A. I and II only   B. I and III only

   C. II and III only   D. I, II and III

7. Which of these human activities both increases global warming and depletes the ozone layer?

   A. Emission of carbon dioxide from vehicle exhausts.

   B. Emission of sulphur dioxide from power stations.

   C. Leakage of methane from gas pipelines.

   D. Release of CFCs from old refrigerators.

8. What might be a consequence of a significant decrease in the amount of the CO_2 in the atmosphere?

   A. The Earth becoming warmer.

   B. A decrease in CFC levels in the atmosphere.

   C. A rise in sea levels.

   D. The Earth becoming cooler.

9. Which pair of statements about the greenhouse effect is correct?

A.	It is caused by carbon dioxide and methane.	It increases acid rain.
B.	It occurs in the troposphere.	It may cause a rise in sea levels.
C.	It accelerates ozone depletion.	It is caused by CFCs.
D.	It blocks UV light.	It occurs in the asthenosphere.

10. Which list contains only greenhouse gases?

    A. Carbon dioxide, water and methane

    B. Methane, CFCs and sulphur dioxide

    C. Carbon dioxide, lead and methane

    D. Nitrogen, water and CFCs

8 HUMAN SYSTEMS AND RESOURCE USE

8.1 Human population dynamics

Significant ideas:
→ A variety of models and indicators are employed to quantify human population dynamics.
→ Human population growth rates are impacted by a complex range of changing factors.

Applications and skills:
→ **Calculate** values of CBR, CDR, TFR, DT and NIR.
→ **Explain** the relative values of CBR, CDR, TFR, DT and NIR.
→ **Analyse** age/sex pyramids and diagrams showing demographic transition models.
→ **Discuss** the use of models in predicting the growth of human populations.
→ **Explain** the nature and implications of exponential growth in human populations.
→ **Analyse** the impact that national and international development policies can have on human population dynamics and growth.
→ **Discuss** the cultural, historical, religious, social, political and economic factors that influence human population dynamics.

Knowledge and understanding:
→ Demographic tools for quantifying human population include **crude birth rate (CBR), crude death rate (CDR), total fertility rate (TFR), doubling time (DT) and natural increase rate (NIR).**
→ **Global human population** has followed a rapid growth curve but there is uncertainty as to how this may be changing.
→ As the human population grows, increased stress is placed on all of Earth's systems.
→ **Age/sex pyramids and demographic transition models (DTM)** can be useful in the prediction of human population growth. The DTM is a model which shows how a population transitions from a pre-industrial stage with high CBR and CDR to an economically advanced stage with low or declining CBR and low CDR.
→ **Influences on human population dynamics** include cultural, historical, religious, social, political and economic factors.
→ **National and international development policies** may also have an impact on human population dynamics.

Human population statistics

Are there too many humans alive on the Earth today? Have we exceeded the carrying capacity of the Earth? Are we heading for a population crash? The difficulty we have in trying to answer these questions is that humans are able to manipulate the environment. We can increase carrying capacity locally, live in large cities, live in regions that cannot grow enough food for the population and use technology. Here we look at **demographics** – the study of the dynamics of population change.

Type 'World population clock' into a search engine and you will find a number of websites that give an estimate of the human population on Earth. At the time of writing this, the US census bureau figure is **7,150,229,812**. What is it when you are reading this?

World birth and death rates

Estimated 2011

Birth rate	Death rate
· 19 births/1,000 population	· 8 deaths/1,000 population
· 131.4 million births per year	· 55.3 million people die each year
· 360,000 births per day	· 151,600 people die each day
· 15,000 births each hour	· 6,316 people die each hour
· 250 births each minute	· 105 people die each minute
· Four births each second of every day	· Nearly two people die each second

Average life expectancy at birth is approximately 67 years. Sources: Population Reference Bureau & the World Factbook (Central Intelligence Agency)

▲ **Figure 8.1.1** World birth and death rates 2011

Global human population growth rate has, until now, followed an **exponential curve**. This is when population follows an accelerating rate of growth which is proportional to the population size. For example, a population increases in each generation from 2 to 4, 4 to 8, 8 to 16 etc.

Our current population growth rate is phenomenal – each year about 90 million people are born. Predictions are that, even with slowing growth rates, it will double again within another 100 years.

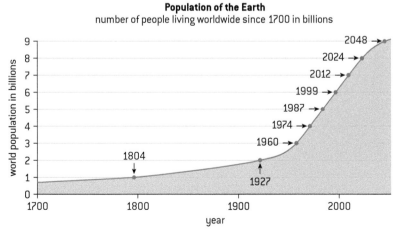

Population of the Earth
number of people living worldwide since 1700 in billions

Source: United Nations world population prospects, Deutsche Stiftung Weltbevölkerung

▲ **Figure 8.1.2** World human population growth since 1700

From figure 8.1.2, calculate the doubling time in years for the human population to increase:

a. 1 to 2 billion

b. 2 to 4 billion

c. 4 to 8 billion

Then complete the table below.

Years	Population	Doubling time
1804–1927	1–2 billion	123 years

Of the 7.1 billion humans alive today, about half of us live in poverty.

We know we are living unsustainably but we can only estimate what future human population growth rate will be, and when and at what number the exponential curve will start to level out and even decrease (see figure 8.1.3).

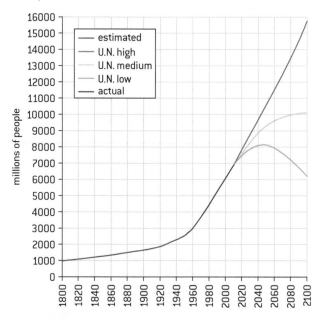

▲ **Figure 8.1.3** Global human population growth rate over time and three predictions for the future

If you look at any statistics or any graph of projected human population growth estimates vary enormously. This is because they are based on past and current trends. We can apply mathematical formulas to current figures but they assume human behaviour is predictable. It is also very hard to build in the impact of the demographic structure of the population. In a population with a high proportion of young people a CBR of 10/1000 will have a significant impact on population growth. But if you have an ageing population with fewer people in the child-bearing years the impact will be far less.

In figure 8.1.3, there are three estimates:

- The high variant – assumes that CDR will continue to fall rapidly but that the CBR will continue to fall slowly.

- The medium variant is the middle ground and a straightforward projection of the curve.

- The low variant assumes we will not find a cure to AIDS or any of the other big killers and that the CBR will fall.

Measuring population changes

The four main factors that affect population size of organisms are birth rate, death rate, immigration and emigration.

Fertility rates higher than 2.0 result in population increase, while lower than 2.0 results in population decrease, because the two parents should be replaced by two children in order to maintain a stable population. Migration is not taken into consideration.

Fertility rate is the number of births per thousand women of child-bearing age. In reality, **replacement fertility** is from 2.03 in MEDCs and 2.16 in LEDCs because of infant and childhood mortality.

Fertility is sometimes considered a synonym to birth rate. However, there is a difference in expression. Birth rate is expressed as a percentage: births per 1000, or hundred (%) of the total population, not of each woman. However, in practice, **crude birth rate (CBR)** is used, which is per thousand individuals, male and female, young and old.

To do

1. Calculate the population density, crude birth rate, crude death rate, and natural increase rate from the data provided. Put the data in a table.

2. **Population density** is the number of people per unit area of land. Calculate the population density in number per km².

Region	Popn 10⁶	Land area km² × 10⁶	Births 10⁶	Deaths 10⁶	Crude birth rate	Crude death rate	Natural increase rate	Popn density
World	6,000	131	121.0	55.8				
Asia	3,500	31	88.2	29.4				
India	1,000	3	29.0	10.0				
Africa	730	29	30.7	10.0				
Tanzania	30	0.9	1.3	0.4				
Europe	730	22.7	8.5	8.2				
Switzerland	7	0.04	0.09	0.07				
N America	460	21.8	9.3	3.6				
USA	270	9.6	4.3	2.4				

3. Describe and explain the differences in the data for the three regions Asia, India and Africa.

Key terms

Measures of total human population change are:

- **Crude birth rate (CBR)** is the number of births per thousand individuals in a population per year.

- **Crude death rate (CDR)** is the number of deaths per thousand individuals in a population per year.

Crude birth and death rates are calculated by dividing the number of births or deaths by the population size and multiplying by 1000. Write these out as a formula.

- **Natural increase rate (NIR)** is the rate of human growth expressed as a percentage change per year.

Natural increase rate = (Crude birth rate − crude death rate) / 10 (migration is ignored)

- **Doubling time (DT)** is the time in years that it takes for a population to double in size.

A NIR of 1% will make a population double in size in 70 years. This is worth remembering.

- The doubling time for a population is 70 / NIR.

Another way to measure births is the:

- **Total fertility rate (TFR)** is the average number of children each woman has over her lifetime.

To think about

1. Research, sketch and describe the shape of the exponential growth curve in the human population over the last 500 years.

2. Exponential growth is characterized by increasingly short doubling times. Doubling time is the number of years it would take to double the size of a population at a particular rate (%) of growth. For example with a 2% growth rate or **natural increase rate**, the population doubling time would be about 35 years, with 4% natural increase rate a population will double in about 17 years. How long would it take for a population to double if the natural increase rate was 1%?

3. Using figure 8.1.2, copy and complete the table with doubling times for the global population (in billions).

Date	Population	Doubling time (yrs)	Date	Population	Doubling time (yrs)
1500	0.5	1,500		5.0	
1800	1.0	300		6.0	
1927	2.0			7.0	
	3.0			8.0	
	4.0			9.0	

4. What do you notice about the changing doubling times?

To do

Inequalities of life

If we could reduce the world's population to a village of precisely 100 people, with all existing human ratios remaining the same, the demographics would look something like this:

The village would have 60 Asians, 14 Africans, 12 Europeans, 8 Latin Americans, 5 from the USA and Canada, and 1 from the South Pacific

51 would be male; 49 would be female

82 would be non-white; 18 white

67 would be non-Christian; 33 would be Christian

80 would live in substandard housing

67 would be unable to read

50 would be malnourished and 1 dying of starvation

33 would be without access to a safe water supply

39 would lack access to improved sanitation

24 would not have any electricity (and of the 76 that do have electricity, most would only use it for light at night)

7 people would have access to the internet

1 would have a college education

1 would have HIV

2 would be near birth; 1 near death

5 would control 32% of the entire world's wealth; all 5 would be US citizens

33 would be receiving —and attempting to live on— only 3% of the income of 'the village'.

How would you illustrate this information?

The *State of The Village Report* by Donella H. Meadows was published in 1990 as *Who lives in the Global Village?* and updated in 2005. It is controversial as some people think she was biased in the use of statistics and some of these are inaccurate. Examples of this are:

Male:female ratio is 1.05:1

Almost 80% of the world's population is now literate

There is less than 1/6th of the world's population malnourished

About 3% of the world's population will have a college education

About 9% will now own a computer

The US controls no more than 30% of the world's wealth.

However, it is a stark demonstration of the haves and have-nots.

Do you think this is an example of propaganda in favour of a particular viewpoint or does it make valid points about the unequal distribution of wealth and goods on earth – or is it both?

LEDCs and MEDCs

The **Human Development Index (HDI)** has been adopted by the UN Development Programme as a measure of the 'well-being' of a country. It combines measures of health (life expectancy), wealth (gross domestic product (GDP) per capita) and education into one value. It is used to rank countries. Iceland, Norway and Canada have been at the top of this list in recent years.

Countries are also grouped into more and less economically developed, based on their industrial development and GDP.

> **TOK**
>
> A variety of models and indicators are employed to quantify human population dynamics. How can we know which indicators to use?
>
> How can we judge which are the most accurate models?

More economically developed countries (MEDCs)	Less economically developed countries (LEDCs)
Most countries in Europe and North America, and South Africa, Israel and Japan	Most of the countries in sub-Saharan Africa, large areas of Asia and South America
Industrialized nations with high GDPs	Less industrialized or have hardly any industry at all
Population is relatively rich	May have raw materials (natural capital) but this tends to be exported and processed in MEDCs.
Individuals are unlikely to starve through poverty	Population has a lower GDP and higher poverty rates
Relatively high level of resource use per capita (per person)	More people are poor with low standards of living
Relatively low population growth rates largely due to low CBR but rising CDRs	High population growth rates largely due to rapidly falling CDRs
Have very high carbon and ecological footprints	Have lower carbon and ecological footprints

▲ **Figure 8.1.4** Comparison of MEDCs and LEDCs

It is easy to put these characteristics into a table of the extremes but in reality this division is a gradually changing continuum with many countries being very hard to categorize.

Various other terms are used to describe the differences in industrialization and wealth of countries. Developed and developing have been used but MEDC and LEDC has replaced these as some LEDCs are in economic decline and are called failed states or least developed countries.

The terms First, Second, Third and Fourth World used to refer to technologically advanced democracies, Communist states within the Soviet bloc, economically underdeveloped countries and stateless nations respectively. Newly industrialized countries (NICs) have accelerated their industrial development and increased GDP, often accompanied by massive foreign investment, population migration to the cities to provide a workforce, free trade and increased civil rights. At the moment, the following countries are considered to be NICs:

China, India, South Africa, Malaysia, Thailand, Philippines, Turkey, Mexico and Brazil. And it would be very hard to define them as LEDC or MEDC.

Human population growth and resources

Human population causing environmental impact appears to be underpinned by a set of simple facts:

- more people require more resources;
- more people produce more waste;
- people usually want to improve their standard of living;
- so the more people there are, the greater the impact they have.

If we can control population increase **and** control resource demand, levels of sustainability should increase.

Demography is the study of the statistical characteristics of human populations, eg total size, age and sex composition, and changes over time with variations in birth and death rates.

Populations remain stable when the **death rate** and the **birth rate** are equal and so there is no net gain in population size.

There are numerous examples of the impact of resource failure and the consequences on human population. Modern history is littered with examples of the direct and indirect impact of famines and droughts across the Earth.

Size of population alone is not the only factor responsible for our impact on our resource base and our impact on the environment in which we live. We need to also consider the wealth of a population, resource desire and resource need (or use). Many population impact models function on the assumption that all individuals (or all populations of a similar size) have the same resource needs and thus have the same impact environmentally (based on resource use and waste associated with exploiting a resource). However, individual resource use (and population resource use) is a dynamic principle. Resource use varies in time and space.

- MEDCs and LEDCs demonstrate contrasting resource use per capita.
- Urban and rural populations demonstrate varying resource use profiles.
- Young people have different resource needs to the elderly.
- Amazonian Indians have different resource needs than Parisians.

Yet all these groups may have an impact, though the impact will vary in scale, type and severity. And the impact may not necessarily be linearly related to population size.

About 20% of us live in MEDCs, 80% in LEDCs. The proportion in MEDCs is falling as birth rates are higher in LEDCs and sometimes negative in some MEDC countries (eg Italy, Germany).

To do

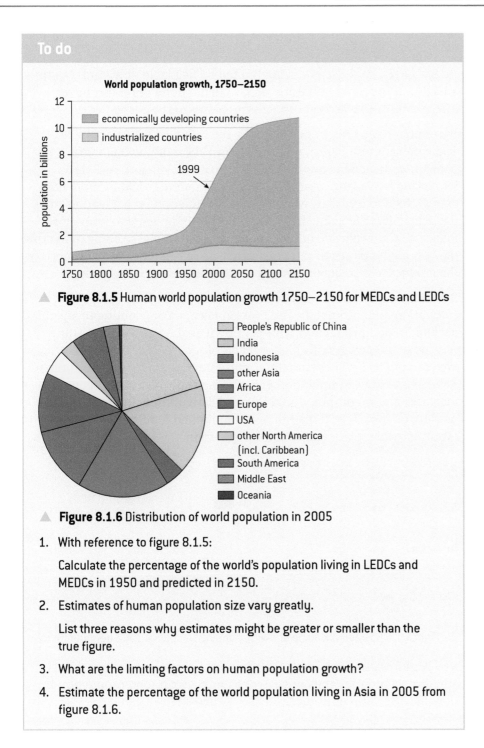

World population growth, 1750–2150

economically developing countries

industrialized countries

1999

Figure 8.1.5 Human world population growth 1750–2150 for MEDCs and LEDCs

People's Republic of China
India
Indonesia
other Asia
Africa
Europe
USA
other North America (incl. Caribbean)
South America
Middle East
Oceania

Figure 8.1.6 Distribution of world population in 2005

1. With reference to figure 8.1.5:

   Calculate the percentage of the world's population living in LEDCs and MEDCs in 1950 and predicted in 2150.

2. Estimates of human population size vary greatly.

   List three reasons why estimates might be greater or smaller than the true figure.

3. What are the limiting factors on human population growth?

4. Estimate the percentage of the world population living in Asia in 2005 from figure 8.1.6.

Population growth and food shortages

There are two main theories relating to population growth and food supply, from Malthus and Boserup.

Malthusian theory

Thomas Malthus was an English clergyman and economist who lived from 1766 to 1834. In his text *An essay on the principle of population*, 1798, Malthus expressed a pessimistic view over the dangers of overpopulation and claimed that food supply was the main limit to population growth.

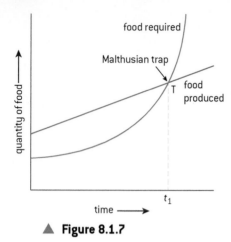

food required

Malthusian trap

T   food produced

quantity of food

time

t_1

▲ **Figure 8.1.7**

Malthus believed that the human population increases geometrically (ie 2, 4, 8, 16, 32, etc.) whereas food supplies can grow only arithmetically (ie 2, 4, 6, 8, 10, 12, etc.), being limited by available new land. Malthus added that the 'laws of nature' dictate that a population can never increase beyond the food supplies necessary to support it.

According to Malthus, population increase is limited by certain 'checks'. These prevent numbers of people increasing beyond the optimum population, which the available resources cannot support. As long as fertile land is available, Malthus believed that there would be more than enough food to feed a growing population. However, as population and the demands for food increase, there is a greater pressure to farm more intensively and cultivate poorer, more marginal land. According to Malthus, though, food production can only increase to a certain level determined by the productive capacity of the land and existing levels of technology.

Beyond the ceiling where land is used to its fullest extent, over-cultivation and, ultimately, soil erosion occurs, contributing to a general decline in food production. This is known as the law of diminishing returns where, even with higher levels of technology, only a small increase in yield will eventually occur. These marginal returns ultimately serve as a check to population growth. Malthus did acknowledge that increases in food output would be possible with new methods in food production, but he still maintained that limited food supply would eventually take place and so limit population.

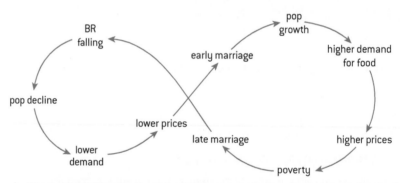

BR falling

pop decline

lower demand

lower prices

early marriage

late marriage

pop growth

higher demand for food

higher prices

poverty

▲ **Figure 8.1.8** Feedback cycle showing population changes and demand for food

Neo-Malthusians agree with Malthus' arguments and believe that we are now seeing the limits of growth as increase in food production is slowing. The Club of Rome, an NGO, is neo-Malthusian.

Limitations of Malthusian theory

Anti-Malthusians criticize the theory as being too simplistic. A shortage of food is just one possible explanation for Malthus' reasoning. This ignores the reality that it is actually only the poor who go hungry. Poverty results from the poor distribution of resources, not physical limits on production. Except on a global scale, the world's community is not 'closed' and so does not enjoy a fair and even distribution of food supplies. Even so, Malthus could not possibly have foreseen the spectacular changes in farming technology which mean we can produce enough food from an area the size of a football pitch to supply 1,000 people for a year, ie there is enough land to feed the whole human population. Thus evidence of the last two centuries contradicts the Malthusian notion of food supply increasing only arithmetically. Rather than starvation, food surpluses exist and agricultural production increases. In 1992, European surpluses reached 26 million tonnes and there are indications that this trend will continue, contrary to Malthusian theory. There were 7 million people in Britain when Malthus lived. Now there are 60 million and most have a high standard of living and enough food – though some is imported. This model is repeated in MEDCs which import food from across the world. Globalization is something Malthus could not have expected.

Boserup's theory

In 1965, Esther Boserup, a Danish economist, asserted that an increase in population would stimulate technologists to increase food production (the optimistic and technocentric view). Boserup suggested that any rise in population will increase the demand for food and so act as an incentive to change agrarian technology and produce more food. We can sum up Boserup's theory by the sentence 'necessity is the mother of invention'.

Boserup's ideas were based on her research into various land use systems, ranging from extensive shifting cultivation in the tropical rainforests to more intensive multiple cropping, as in South East Asia. Her theory suggests that, as population increases, agriculture moves into higher stages of intensity through innovation and the introduction of new farming methods. The conclusion arising from Boserup's theory is that population growth naturally leads to development.

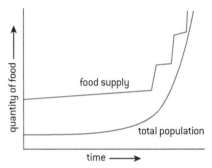

▲ **Figure 8.1.9** Food supply and population curves

Limitations of Boserup's theory

Like Malthus, Boserup's idea is based on the assumption of a 'closed' community. In reality, except at a global scale, communities are not 'closed' because constant in- and out-migration are common features. It has therefore been very difficult to test Boserup's ideas. This is because migration usually occurs in areas of over-population to relieve the population pressure, which, according to Boserup's theory, then leads to technological innovation.

Overpopulation can lead to unsuitable farming practices that may degrade the land so population pressure may be responsible for desertification in the Sahel. From this it is clear that certain types of fragile environment cannot support excessive numbers of people. In such cases, population pressure does not always lead to technological innovation and development.

Application of theories of Malthus and Boserup

There is evidence to suggest that the ideas of both Boserup and Malthus may be appropriate at different scales. On a global level the growing suffering and famine in some LEDCs today may reinforce Malthusian ideas. On the other hand, at a national scale, some governments have been motivated by increasing population to develop their resources and so meet growing demands.

Both Malthus and Boserup can be right because Malthus refers to the environmental limits while Boserup refers to cultural and technological issues.

1. Read the description of the theories of Malthus and Boserup and summarize their models in a table like the one below.

	Malthus	Boserup
Model diagram		
Main ideas		
Limitations		
Applications		

2. Look at the graph of food supply in India (figure 8.1.10).

    i. According to the Indian National Commission on Population, the population of India was about 846 million in 1991, 1,012 million in 2001 and estimated to be 1,179 million in 2009. As the population of India has increased what happens to the per capita food supply?

    ii. Add a third line to show increase in human population.

    iii. Whose theory is represented by this data? Explain your reasoning.

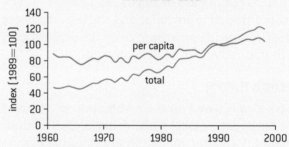

Index of total and per capita food production, India. 1961–1998

Figure 8.1.10 Total and per capita food production in India 1961–1998

3. **Global population growth**

The pattern of human growth is not uniform with most growth currently taking place in LEDCs. Use the following data to construct population growth curves for MEDCs and LEDCs of the world from 1800 until 2100 AD. Plot the LEDCs above the MEDCs (units = 10^9).

Date	MEDC	LEDC	Date	MEDC	LEDC
1800	0.3	0.7	1980	1.1	3.3
1850	0.4	0.8	1990	1.2	4.1
1900	0.6	1.1	2000	1.3	5.0
1950	0.8	1.7	2025	1.4	7.2
1960	0.9	2.1	2050	1.4	8.0
1970	1.0	2.8	2100	1.4	8.0

4. The values for the next century are only estimates. What will be the most important social factor that will determine human population size?

What will be the future world population?

What do fertility rates mean?

Fertility rate	Predicted world population (billions) in 2100
1.5	3.6
2.0	10.1
2.5	15.8

A fertility rate of 2.0 means that a couple replace themselves, and do not add to the population. In this scenario the population will increase from what it is now to 10.1 billion in 2100. If every second woman decides to have three rather than two children, a fertility rate of 2.5, the population will rise to 15.8 billion by 2100. If, however, every second woman decides to have only one child instead of two, a fertility rate of 1.5, the world population will sink to 3.6 billion. Total world fertility is now about 3.0, 1.7 in MEDCs, and averaging 3.4 (but up to 6.0) in LEDCs. Fertility rate is falling although population size continues to increase. The UN has calculated estimates for population change based on fertility rates stabilizing at 2.6 (high), 2.1 (medium/replacement level) and 1.6 (low).

When is a country overpopulated?

If the optimum population is when the population produces the highest economic return per capita, using all available resources, then some countries may have a higher optimum population density than others. The UK and Netherlands have high population densities but can support this population with a high living standard. Brazil with two people per km^2 in the north is overpopulated as resources are much scarcer. The snag is that the richer countries have to import goods and services from elsewhere.

Why do people have large families?

It appears that the decision to have children is not correlated with GNP of a country nor personal wealth. Some reasons may be:

1. **High infant and childhood mortality:** according to UNICEF one child dies every three seconds (26,500 per day) due to malnutrition and disease. It is an insurance to have more than you may need so that some of them reach adulthood.

2. **Security in old age:** the tradition in the family is that children will take care of their parents. The more children the more secure the parents, and the less the burden for each child. If there is no social welfare network, children look after their parents.

3. **Children are an economic asset** in agricultural societies. They work on the land as soon as they are able. More children mean more help but more children need feeding. In MEDCs, children are dependent on their parents during their education and take longer to contribute to society.

4. **Status of women:** the traditional position of women is that they are subordinate to men. In many countries, they are deprived of

many rights, like owning property, having their own career, getting an education. Instead they do most of the agricultural work and are considered worthy only for making children, and their social status depends on the number of children they produce, particularly boys. Breaking down such barriers of discrimination (social or religious), allowing girls to get an education and be capable of gaining status outside the context of bearing children has probably contributed more than anything toward the very low fertility rate in MEDCs.

5. **Unavailability of contraceptives:** in MEDCs this is the prime way of reducing fertility. In LEDCs, many women would like to have them but they are too poor to pay for them or they cannot get them.

The ways to reduce family size are to:

1. **Provide education** in the form of basic literacy to children and adults.

2. **Improve health** by preventing the spread of diseases through simple measures of hygiene (boiling water), by improving nutrition, and by providing some simple medication and vaccines.

3. Make **contraceptives** and family counselling available.

4. **Enhance income** by small-scale projects focusing on the family level. Microlending, as in the Gramin Bank, is a practice that has had high success. Small loans are given for a peasant to buy some seed and fertilizer to grow tomatoes, for a woman to buy pans to bake bread, for a weaver to buy yarn, for an auto mechanic to get some tools. Thus, small enterprises may start that will feed the whole family (at least). Return of the loan is guaranteed through credit associations formed by the members of the community.

5. **Improve resource management.** Local people may grow tree seedlings for transplanting in reforestation projects, prevent erosion through soil conservation measures. We have realized that large projects in LEDCs often do not work. Major projects like building dams for HEP or roads cost an LEDC which is then in debt (Third World debt) and force the population into cash cropping (eg tobacco, oil palm).

To think about

The status of women

According to the UN, women's rights are the key to reducing the population growth rate. Fertility rates remain high where women's status is low. Less than 20% of the world's countries will account for nearly all of the world's population growth this century. Not coincidentally, those countries — the least developed nations in sub-Saharan Africa, south Asia, and elsewhere — are also where girls are less likely to attend school, where child marriage is common, and where women often lack basic rights.

a. Make a list of issues which may maintain the low status of women.

b. Suggest proposals which might lead to a lowering of fertility rate.

c. List four reasons why educating women will reduce fertility/birth rate.

To do

Population pyramids

All pyramids are taken from the US census website at http://www.census.gov/population/international/data/idb/ informationGateway.php. Have a look at this site as it has dynamic pyramids which change over time.

Population or age/sex pyramids show the distribution of individuals in a population, by sex and age. They contain a lot of information.

1. List the pieces of information that you can find in the population pyramid of Afghanistan in 2000.

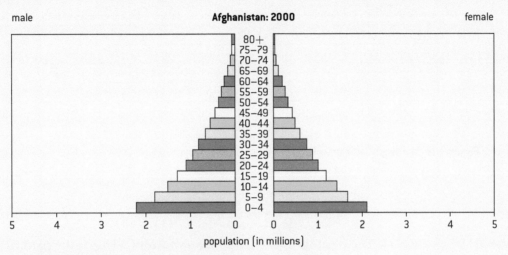

▲ **Figure 8.1.11** Population pyramid of Afghanistan in 2000

2. What changes are there in this predicted pyramid of 2025?

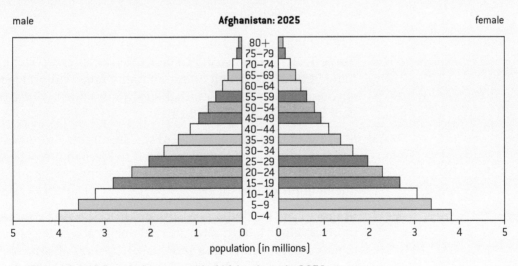

▲ **Figure 8.1.12** Population pyramid of Afghanistan in 2050

3. Draw two horizontal bands (at 15 and 65 years). What do these bands represent?

Population pyramids can indicate political and social changes too. China used the concept of optimum population to try to stabilize its population at 1.2 billion by the year 2000 and reduce the population to a government-set level of 700 million by the end of the century.

4. Explain the decrease in population younger than 35–45 in China in 2007. Think of at least two reasons.

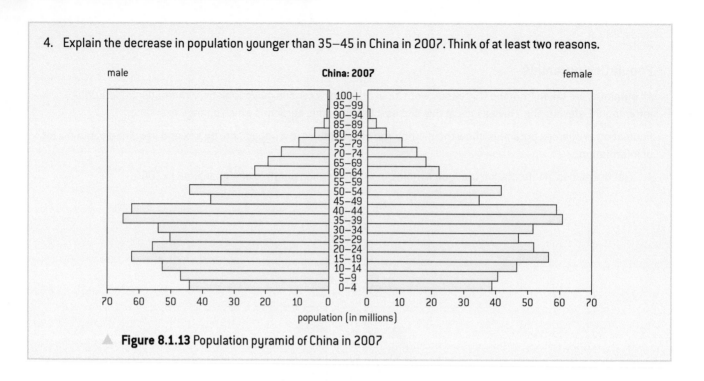

▲ **Figure 8.1.13** Population pyramid of China in 2007

Demographic transition model

The **demographic transition model (DTM)** is the pattern of decline in mortality and fertility (natality) of a country as a result of social and economic development. Demographic transition can be described as a five-stage population model, which can be linked to the stages of the sigmoid growth curve.

The stages are:

1. **Stage 1**: High stationary (Pre-industrial societies) – High birth due to no birth control, high infant mortality rates, cultural factors encouraging large families. High death rates due to disease, famine, poor hygiene and little medicine.

2. **Stage 2**: Early expanding (LEDC's) – Death rate drops as sanitation and food improve, disease is reduced so lifespan increases. Birth rate is still high so population expands rapidly and child mortality falls due to improved medicine.

3. **Stage 3**: Late expanding (Wealthier LEDC's) – As a country becomes more developed, birth rates also fall due to access to contraception, improved healthcare, education, emancipation of women. Population begins to level off and desire for material goods and low infant death rates mean that people have smaller families.

4. **Stage 4**: Low stationary (MEDC's) – Low birth and death rates, industrialized countries. Stable population sizes.

5. **Stage 5**: Declining (MEDC's) – Population may not be replaced as fertility rate is low. Problems of ageing workforce.

As a model, the DTM explains changes in some countries but not others. China and Brazil have passed through the stages very quickly. Some sub-Saharan countries or those affected by war or civil unrest do not follow the model. It has been criticized as extrapolating the European model worldwide.

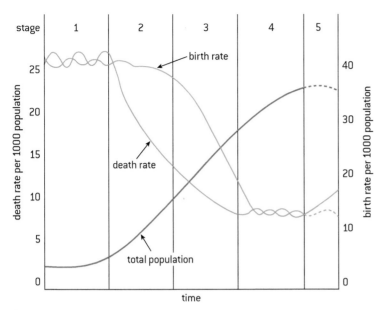

▲ **Figure 8.1.14** The demographic transition model

Population pyramids take one of basic four shapes which reflect the DTM stages:

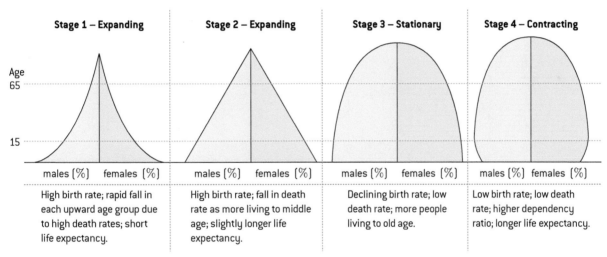

▲ **Figure 8.1.15** The four shapes of population pyramids

	Stage 1 – Expanding	Stage 2 – Expanding	Stage 3 – Stationary	Stage 4 – Contracting
	High birth rate; rapid fall in each upward age group due to high death rates; short life expectancy.	High birth rate; fall in death rate as more living to middle age; slightly longer life expectancy.	Declining birth rate; low death rate; more people living to old age.	Low birth rate; low death rate; higher dependency ratio; longer life expectancy.

To do

1. Copy and complete the table with the characteristics of each pyramid

Stage	1. Expanding	2. Expanding	3. Stationary	4. Contracting
Birth rate				
Death rate				
Life expectancy				
Population growth rate				
Stage of DTM				
Example				

2. For each pyramid below, identify the stage. You might like to look up your own country, if not included, and do the same. Comment on the birth rate, death rate, life expectancy, gender differences and stage of development of the country.

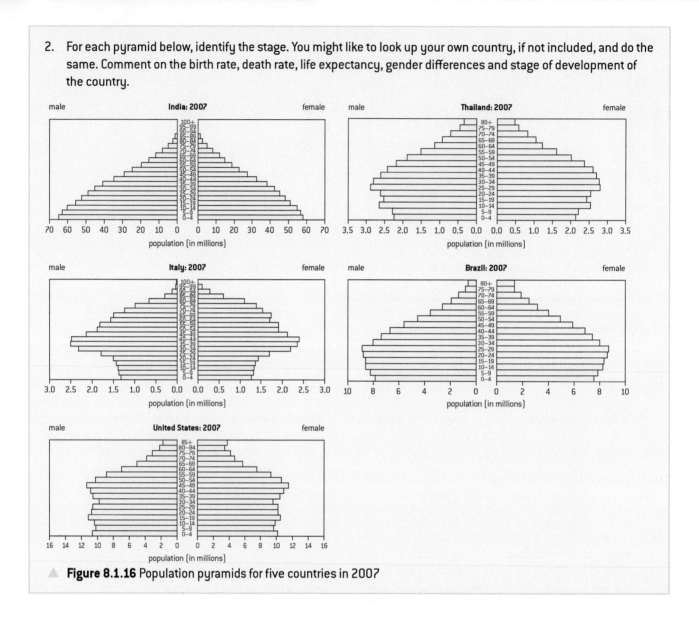

▲ **Figure 8.1.16** Population pyramids for five countries in 2007

The controversial thing about this model is that it is based on change in several industrialized countries yet it suggests that all countries go through these stages. Like all models, it has limitations. These are:

- The initial model was without the fifth stage which has only become clear in recent years when countries such as Germany and Sweden have fallen into population decline.

- The fall in the death rate has not always been as steep as this suggests as movement from the countryside to cities has created large urban slum areas which have poor or no sanitation and consequent high death rates of the young and infirm.

- Deaths from AIDS-related diseases may also affect this.

- The fall in the birth rate assumes availability of contraception and that religious practices allow for this. It also assumes increasing education of and increased literacy rates for women. This is not always the case.

- Some countries have compressed the timescale of these changes. The Asian 'Tiger economies' of Malaysia, Singapore and Hong Kong, for example, have leapt to industrialized status without going through this sequence in the same time period as others.

- This is a eurocentric model and assumes that all countries will become industrialized. This may not be the case in some 'failed states', for example.

To do

Using computer models to predict population change

Go to the UN Economic and Social Affairs office website http://esa.un.org/unpd/wpp/unpp/panel_population.htm

1. Using the basic data and median variant, find the data for the following:

    a. World population in 2050

    b. More developed regions population in 2050

    c. Less developed regions population in 2050

    d. Calculate the percentage of total world population that each region makes in 2050

    e. Repeat these steps for the year 2015

    f. What does this tell you?

2. Now go to the detailed indicators in the left-hand column

    a. Look up and make a note of fertility rates for the same regions and same years.

    b. Which region has the highest fertility rate and why?

    c. What has happened to birth rates of these regions over time and what does this mean?

    d. What has happened to death rates of these regions over time and what does this mean?

    e. What has happened to life expectancy in these regions?

3. Using the same website, pick two variables that you think are important when trying to explain population growth, justify why they are important and explain what they show over the period 1950–2050.

4. What are the problems with using computer models to predict population expansion? How valid are they and what are they useful for?

5. For the country in which you live or where you hold nationality, examine the change in population between 1950 and 2050.

To think about

Models may be very generalized and simple to use or so complex that they are difficult to use. They should present the significant factors without extra detail that may confuse us. They should be useful in helping us to make predictions and make sense of the real world. How far does the DTM help us or does it hinder our understanding of population change which is far more complex than this model suggests?

To think about

The AIDS epidemic – Africa, an orphaned continent

Information taken from UNAIDS, a joint UN agency on AIDS research and relief.

Worldwide, 35 million people live with the HIV virus. 10% of these are aged 15 or under. In 2010, 1.8 million people died from AIDS related illnesses. About 2/3rds of these were in sub-Saharan Africa which is the worst affected region.

South Africa is the worst affected country with more than 15% of the people infected.

Most people who die from AIDS related illnesses are the wage-earners of a family and those who are poor cannot afford anti-retroviral drug treatments. Although the multinational drug companies have agreed to reduce the prices of these drugs in Africa, they are still out of the reach of all but a few. The effect is to leave a generation of orphan children, looked after by their grandparents.

1. Describe the population pyramid for South Africa in 2000.

2. What type of pyramid is this?

3. Explain why it changes for 2025.

4. Explain the factors that influence the pyramid in 2050. (Remember economic factors as well.)

▲ **Figure 8.1.17** The World Aids Day ribbon

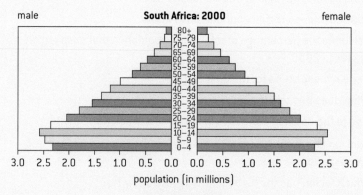

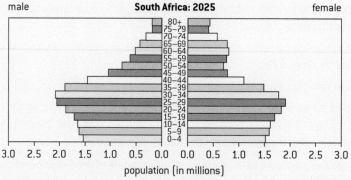

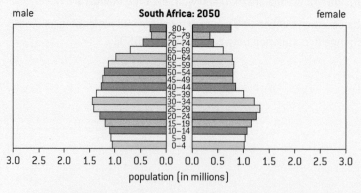

▲ **Figure 8.1.18** Population pyramid for South Africa 2000, 2025 and 2050

Influencing human population growth

Policies that may reduce population growth rates	Policies that may increase population growth rates
Parents in subsistence communities may be dependent on their children for support in their later years and this may create an incentive to have many children. So if the government introduces pension schemes the CBR comes down.	Agricultural development, improved public health and sanitation, etc. may lower death rates and stimulate rapid population growth without significantly affecting fertility.
If you pay more tax to have more children or even lose your job, you may decide to have a smaller family.	Lowering income tax or giving incentives and free education and health care may increase birth rates, eg Australia baby bonus.
Policies that stimulate economic growth may reduce birth rates as a result of increased access to education about methods of birth control.	Encouraging immigration, particularly of workers – for example Russia allows migrants to work who do not have qualifications to fill the gap in manual labour.
Urbanization may also be a factor in reducing crude birth rates as fewer people can live in the smaller urban accommodation.	
Policies directed toward the education of women, and enabling women to have greater personal and economic independence may be the most effective in reducing fertility and therefore population pressures.	

▲ **Figure 8.1.19** National and international policies influence human population growth

While we may be able to count how many people are alive, what age they are and where they live, and even predict changes in the future, we do not really know how many people and other species the Earth can support. All evidence we have at the moment is that we are using the Earth's resources unsustainably but are we inventive enough to either live within our means or find ways to increase productivity?

The greying of Europe

Europe's population growth rate is falling and its population is ageing (going grey). Who will be the workers who support the older population? Immigration may help but immigrants also grow old and need state support.

So the combination of decreased replacement rate and fewer workers means it is not looking good for Europe. Within the geographical boundaries of Europe are some 731 million people (with 499 million of these in the 28 EU member states). In 1900, Europeans made up 25% of the world population; by 2050, they will be 7% according to UN projections. This is because the growth rate in other countries is far higher than that of Europe. By 2050, the median age of Europeans will be 52 years. The total fertility rate across the EU is about 1.59 children but, in some countries, fertility rates are lower than in others. Italy has a fertility rate of 1.2 children per woman. This is not enough to replace the population yet some see immigration as bringing major social problems as well.

Services such as education, medical care and social support are not increasing to meet demand. Immigration is across the EU, mostly from east to west but also into the EU from LEDCs. Debate about immigration soon becomes political with terms like 'illegal immigrant', 'economic migrant' and 'bogus asylum seeker' being used, yet migration has always occurred for a host of economic, social and political reasons. It is inevitable. The skill for politicians will be to intervene to change retirement age and pension systems, improve productivity and stimulate worker mobility.

Practical Work

* To what extent does a country's development depend on its economy and its demographics that influence its development policies?

* Explain the nature and implications of exponential growth in human populations.

* Discuss the cultural, historical, religious, social, political and economic factors that influence human population dynamics.

* Discuss the use of models in predicting the growth of human populations.

* A variety of models and indicators are employed to quantify human population dynamics. To what extent are the methods of the human sciences 'scientific'?

Some European countries are trying to increase their birth rates to alleviate the problems of an ageing population but once women have control over their own fertility they rarely wish to go back to having large families.

Research population policies in one of India, China, Iran, Colombia, Brazil, Singapore or your own country. Write a short case study on this and exchange it with your classmates.

8.2 Resource use in society

Significant ideas:

→ The renewability of natural capital has implications for its sustainable use.

→ The status and economic value of natural capital is dynamic.

Applications and skills:

→ **Outline** an example of how renewable and non-renewable natural capital has been mismanaged.

→ **Explain** the dynamic nature of the concept of natural capital.

Knowledge and understanding:

→ **Renewable natural capital** can be generated and/or replaced as fast as it is being used. It includes living species and ecosystems that use solar energy and photosynthesis. It also includes non-living items, such as groundwater and the ozone layer.

→ **Non-renewable natural capital** is either irreplaceable or only replaced over geological timescales, eg fossil fuels, soil and minerals.

→ Renewable natural capital can be utilized sustainably or unsustainably. If renewable natural capital is used beyond its natural income this use becomes unsustainable.

→ The impacts of extraction, transport and processing of a renewable natural capital may

cause damage making this natural capital unsustainable.

→ **Natural capital provides goods** (eg tangible products) **and services** (eg climate regulation) that have value. This value may be aesthetic, cultural, economic, environmental, ethical, intrinsic, social, spiritual or technological.

→ The concept of a natural capital is **dynamic**. Whether or not something has the status of a 'natural capital', and the marketable value of that capital, varies regionally and over time. This is influenced by cultural, social, economic, environmental, technological and political factors, eg cork, uranium, lithium.

Natural capital and natural income

Natural capital is a resource which has some value to humans. Resources are goods or services that we use.

Natural income is the rate of replacement of a particular resource or natural capital (see 1.4).

In the past, economists spoke of capital as the products of manufacturing, human-made goods, and separated these from land and labour. But we now recognize that capital includes:

- natural resources that have value to us, eg trees, soil, water, living organisms and ores bearing minerals,

- natural resources that provide services that support life, eg flood and erosion protection provided by forests, and

Key terms

Renewable natural capital can be generated and/or replaced as fast as it is being used.

Non-renewable natural capital is either irreplaceable or only replaced over geological timescales, eg fossil fuels, soil and minerals.

- processes, eg photosynthesis that provides oxygen for life forms to respire, the water cycle or other processes that maintain healthy ecosystems.

So the term natural capital is now used to describe these goods or services that are not manufactured but have value to humans. They can be improved or degraded and given a value – we can begin to give monetary values to ecosystems. We may be able to process these to add value to them, eg mine tin or uranium, turn trees into timber, but they are still natural capital. The terms resource and natural capital are interchangeable.

Just as capital yields income in terms of economics, natural capital yields **natural income** (yield or harvest or services) – factories produce objects, cherry trees produce cherries, and the water cycle provides us with fresh water. The measure of the true wealth of a country must include its natural capital, eg how many mineral resources, forests, rivers it has. In general MEDCs add value to natural income by manufacturing goods from it and LEDCs may have greater unprocessed natural capital. The World Bank now calculates the wealth of a country by including the rate of extraction of natural resources and the ecological damage caused by this, including carbon dioxide emissions.

Renewable natural capital includes:

- living species and ecosystems that use solar energy and photosynthesis

- non-living items, such as groundwater and the ozone layer.

It can be used sustainably or unsustainably (See the Millennium Ecosystem Assessment, sub-topic 1.4.) If renewable natural capital is used beyond its natural income, this use is unsustainable.

Renewable natural capital can run out if the standing stock (how much is there) is harvested unsustainably, ie more is taken than can be replaced by the natural growth rate. Then, it will eventually run out. The depletion of natural resources at unsustainable levels and efforts to conserve these resources are often the source of conflict within and between political parties and countries. The impacts of extraction, transport and processing of a renewable natural capital may cause damage making this natural capital unsustainable.

▲ **Figure 8.2.1** A non-renewable resource – coal and a renewable resource – a forest

Non-renewable natural capital are resources that exist in finite amounts on Earth and are not renewed or replaced after they have been used or depleted (or only over a long timescale – normally geological scales). Non-renewable resources include minerals, soil, water in aquifers and fossil fuels. As the resource is used, natural capital or stocks are depleted. New sources of stock or alternatives need to be found.

Depending on your source of drinking water, where you live and the annual rainfall, water may be considered **renewable natural capital** (high rainfall regions where most rain is collected and used for drinking) or **non-renewable natural capital** (drier regions where underground aquifers refill slowly at rates longer than an average human lifetime).

Recyclable resources

Iron ore is a non-renewable resource. Once the ore has been mined and processed it is not replaced in our lifetime. However from iron ore we produce iron which can be cast into numerous forms and represents a significant commodity within modern societies. About 90% of a car is made from iron or iron-derived products – steel. However, steel and iron can be recycled. Old or damaged cars can be broken down. Their parts can be used to replace parts in other cars or their parts can be remanufactured into new metal objects. Therefore iron ore is non-renewable but the iron extracted from the ore becomes a renewable resource. The same is true for aluminium.

To think about

Exploiting the poles

The Arctic and Antarctic are perhaps the last wildernesses on Earth and are beautiful. Their ecosystems are fragile and contain much biodiversity found nowhere else. Any disturbance has a long recovery time as growth is slow because temperature is limiting. On land, water is also limiting as it is frozen for much of the year and so unavailable to plants.

The Arctic

Until recently humans could not exploit the resources of the Arctic on a large scale as the seas are frozen for all but a few months of the year and conditions are harsh. But there are mineral riches locked under the Arctic Ocean and surrounding land masses, especially hydrocarbons.

The world's oil supply comes from many countries. To have a national source of oil is a desire for many countries which would then not be dependent on importing oil. Some 40% of oil comes from and is exported by OPEC (Organization of the Petroleum Exporting Countries – 12 countries whose economies rely on oil exports) and they control oil prices and supply. The USA produces about 10%, Russia about

▲ **Figure 8.2.2** Map of the Arctic

13% with the remaining dispersed across a number of other countries. The price of a barrel of crude oil varies greatly. It reached over US$100 in 2008 and 2011 while it hovered around $30 a barrel for much of the 1990s and in 2009.

With climate change causing the Arctic to warm up, there are more ice-free days. High oil prices means that reserves that were once uneconomic to extract are no longer so and the Arctic could be the next goldmine or environmental disaster, depending on your environmental worldview. At 2008 prices, the estimated value of the Arctic's minerals is US$1.5–2 trillion. There are crude oil reserves under Northwestern Siberia and Alberta, Canada. There is also oil right under the North Pole. Humans have the technology to extract this oil. Why would we not?

Who owns the Arctic?

There is no land at the North Pole, it is ice floating on water. Under the international United Nations Convention on the Law of the Sea (UNCLOS), a state can claim a 200 nautical mile (370 km) zone and beyond that up to 150 nautical miles (278 km) of rights on the seabed. So it may fish or exploit the minerals exclusively in this zone and other countries may not. This distance is not measured from the border or edge of a country but from the edge of the continental shelf, which may be some distance away from the border of the country under the sea.

In August 2007, a Russian submarine expedition planted a Russian flag on the seabed at the North Pole, two miles under the Arctic ice cap. They claimed that the seabed under the pole, called the Lomonosov Ridge, is an extension of Russia's continental shelf and thus Russian territory.

Six countries — Canada, Denmark, Iceland, Norway, Russia and the United States — have Arctic Ocean coastlines and Denmark has sent its own scientific expeditions to study the opposite end of the Lomonosov Ridge to see if they can prove it is part of Greenland which is a Danish territory.

The Antarctic

Antarctica is a continent of which 98% is covered in ice and snow. In Antarctica, no large mineral or oil reserves have been found. But humans exploit the continent through tourism, fishing, sealing and whaling. About 10,000 tourists visit the Antarctica each year and

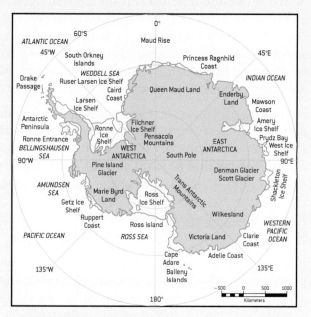

▲ **Figure 8.2.3** The Antarctic

this is increasing. No one country owns Antarctica but seven have staked territorial claims via 'The Antarctic Treaty' which was perhaps the first step in recognition of international responsibility for the environment. It was signed in 1959 by 12 countries including the US, UK and USSR who signed it in the middle of the Cold War. The treaty was strengthened in 1991 and covers all land south of latitude 60 °S. The agreement is that:

- The area will be

  - free of nuclear tests and nuclear waste
  - for peaceful purposes
  - a preserved environment
  - undisputed as a territory.

- There will be

  - prevention of marine pollution
  - clean up sites
  - no commercial mineral extraction.

- Sealing has annual limits.

- Commercial whaling is now tightly regulated.

Fishing is less of a success story with overfishing of many species which is hard to regulate in the seas around Antarctica. And this is causing the crash of many penguin and seal populations.

There is so much ice on Antarctica, that it is approximately 61 percent of *all freshwater* on Earth. If all this melted, it would add 70 m in height to the world's oceans.[1] It appears that the ice is melting and some large ice sheets

---

[1]   Kusky, TM (2009), *Encyclopedia of Earth.*

are 'calving' or breaking up and slipping away from the land. Over three weeks in 2002, a huge ice shelf, over 3,000 square km and 220 m deep, Larsen B, broke up and floated out to sea. But in other areas, the ice is getting thicker.

Questions

Why is there no Arctic Treaty?

Who should own the oceans?

Who should regulate human exploitation of the oceans?

Dynamic nature of natural capital

The importance of types of natural capital varies over time. A resource available today may not be a resource in the future. A resource available in the past may not be a resource today, or it may not have the resource value it previously had. Our use of natural capital depends on cultural, social, economic, environmental, technological and political factors. For example:

- Technocentrists believe that new discoveries will provide new solutions to old problems; for example, hydrogen fuel cells replacing hydrocarbon-based fuel, or harvesting algae as a food source.

- Arrowheads made from flint rocks are no longer in demand.

- Uranium is in demand as raw material for nuclear power by fission but may not be if we could harness the energy of nuclear fusion – the hydrogen economy.

To do

Find out about uranium as an example of natural capital.

What are its uses?

Where is it mined?

Evaluate its use as natural capital.

Examples of changing value of natural capital

1. Cork forests

Cork from the bark of the cork oak tree has been essential for centuries to seal wine bottles. But now plastic corks, screw-top bottles and plastic lids are replacing cork. Many of these are not biodegradable like cork. And they are made from fossil fuels!

Cork forests are losing their value as natural capital to humans so they are cut down and the land used for other purposes. You might think that is a good thing but it is not.

Cork oak forests in the Mediterranean region have high biodiversity, second only to that of the Amazon rainforest. In harvesting cork, the tree continues to live and only the bark is harvested by hand every 9 years.

▲ **Figure 8.2.4** A cork oak forest

2. Lithium

We use lithium carbonate batteries if we have a mobile phone, tablet or electric car. Thirty years ago, we had little idea where lithium-containing ores were in the world because we did not use much of it as a resource. Now we cannot get enough of them.

More than half the world's known reserves of lithium are underneath a desert salt plain in Bolivia. More is under the Chilean Atacama desert. China has found some in Tibet. But the annual production of lithium is not nearly enough to power electric cars if they were to replace cars with petrol engines.

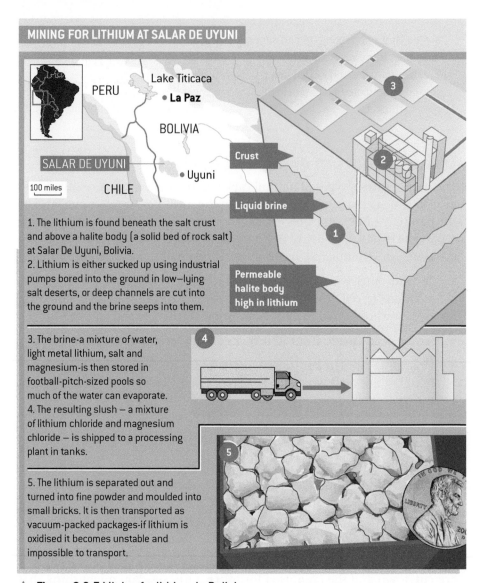

MINING FOR LITHIUM AT SALAR DE UYUNI

Lake Titicaca
PERU
● **La Paz**

BOLIVIA

Crust

SALAR DE UYUNI
● Uyuni

100 miles
CHILE

Liquid brine

Permeable halite body high in lithium

1. The lithium is found beneath the salt crust and above a halite body (a solid bed of rock salt) at Salar De Uyuni, Bolivia.
2. Lithium is either sucked up using industrial pumps bored into the ground in low-lying salt deserts, or deep channels are cut into the ground and the brine seeps into them.

3. The brine-a mixture of water, light metal lithium, salt and magnesium-is then stored in football-pitch-sized pools so much of the water can evaporate.
4. The resulting slush – a mixture of lithium chloride and magnesium chloride – is shipped to a processing plant in tanks.

5. The lithium is separated out and turned into fine powder and moulded into small bricks. It is then transported as vacuum-packed packages-if lithium is oxidised it becomes unstable and impossible to transport.

▲ **Figure 8.2.5** Mining for lithium in Bolivia

Valuing natural capital

We can divide the valuation of natural capital into two main categories:

- **Use valuation** – natural capital that we can put a price on, eg:

  - Economic price of marketable goods.

  - Ecological functions, eg water storage or gas exchange in forests.

  - Recreational functions, eg tourism, leisure activities.

- **Non-use valuation** – natural capital that it is almost impossible to put a price on, eg:

  - If it has intrinsic value (the right to exist).

  - If there are future uses that we do not yet know (science, medicines, potential gene pool).

  - If it has value by existing for future generations – existence value (Amazon rainforest).

Many people feel that the only way to make people realize the importance of these non-use valuation things is to find some way to put a price tag on them so people realize what they are worth. Others feel this may just encourage exploitation of them.

Whether a resource can be sustainably used is what we need to know. We may think that agriculture is sustainable as crops are eaten and then more are planted, but it is only sustainable if the soil fertility and structure are maintained and the environment is not degraded overall. If biodiversity is lost due to agriculture, can it be sustainable? (See 3.3 and 5.2.)

Slash and burn agriculture (shifting cultivation) or sporadic logging in virgin forest are both sustainable as long as the environment has time to recover. Adequate time to recover is dependent on low human population densities. Are we currently giving it enough time to recover?

> **To do**
>
> Review the resources of a tropical rainforest using the valuation list in the text. Identify goods and services provided by a rainforest for humans by making a table with two columns – goods and services.

To do

1. Make a list of resources that you are using. (You can do this by first writing down what objects you are using and subsequently stating which resources are required to manufacture these objects.) Don't forget transport.

2. Do all humans use the same types and amounts of resources? Explain using examples.

3. If wood became so scarce that we cannot use it for construction of houses anymore, how could we solve this problem?

4. We all use oil and oil products (fuel oil, diesel oil, chemicals, plastics). Sweden however, does not have its own oil reserves. What would happen to the carrying capacity of Sweden, if it could not import oil?

To think about

Putting a value on the environment

Consider the systems in this list:

Your school	The Sahara Desert
Your city	Lake Superior
Your home	San Francisco
Your local park or protected area	Tundra in Siberia
Tigers	Antarctica
Mosquitoes	Great Barrier Reef
Polio virus	Shanghai
The Amazon rainforest	Tokyo

In groups, put these in an order of increasing

(1) use value

(2) and then non-use value.

Write down the criteria that your group used. Compare your list with other groups and be ready to justify your decision. If you change your mind, reorder the images to your personal preference and amend your list of criteria. What are the difficulties in assessing the importance of different types of environment and what characteristics need to be taken into consideration when trying to do this? Do you think that environments can have their own intrinsic value?

Urbanization

The drift from the countryside to urban life started long ago and has continued. According to the UN DESA (Department of Economic and Social Affairs), the balance of urban to rural population worldwide is now more than 50% of us in cities. China is 50% urbanized. Some 60% of us will be city dwellers by 2030 and 70% in 2050.

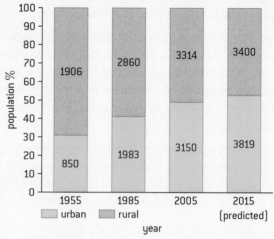

▲ **Figure 8.2.7** Ratios of urban: rural populations

Cities are not necessarily unsustainable. There are efficiencies in living in high density populations where transport costs are reduced for commuters and moving resources around, people tend to live in smaller spaces so they use less energy to heat or cool and services are nearby. But cities have to remove their waste and process it, they need a large land area to supply them with food and they create pollution. Inevitably they encroach on or degrade natural habitats.

1. Do you live in a city? (If not, select a city near to your home.)

2. What is its population and land area?

3. How much has it grown since 1955?

4. Where does the food sold in the city come from?

5. Where do the wastes (sewage, garbage) go?

▲ **Figure 8.2.6**

Which cities are these?*

* Shanghai, Mexico City, London

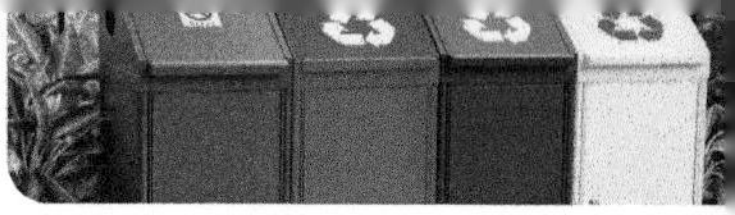

Since the early 1980s, UNEP (UN Environmental Programme) has been using a system of integrated environmental and economic accounting (or socio-economic environmental assessment – SEEA) to try to value the environment and track resource depletion. If countries would include the cost of degrading their natural resources within their GNP (gross national product), the real cost and health of the nation would be clearer to see.

From the UN Earth Summit in Rio de Janeiro in 1992 came Agenda 21 (see 1.1). An undertaking was given that local councils would produce their own plan, a local Agenda 21 involving consulting with the local community. What does your local Agenda 21 say?

To do

Fairtrade www.fairtrade.net

FAIRTRADE Guarantees a **better deal** for Third World Producers

Figure 8.2.8 Fairtrade logo

Fairtrade is an NGO charity.

1. What is the vision of Fairtrade?
2. What does it do?
3. Name three products that have the Fairtrade logo.
4. Evaluate the impact of Fairtrade on (a) the producer and (b) the consumer.

To think about

Globalization

Did you know?

- 51 of the world's top 100 economies are corporations.
- Transnational corporations:
  - control two-thirds of world trade,
  - control 80% of foreign investment, and
  - employ just 3% of the world's labour force of 2.5 billion.
- Wal-mart may be bringing 38,000 people out of poverty per month in China.

Globalization is the concept that every society on Earth is connected and unified into a single functioning entity. The connections are mostly economic but also allow the easy exchange of services as well as goods and information and knowledge.

Globalization has been facilitated by new technologies, air travel and the communication revolution. The World Trade Organization (WTO) controls the rules of this global trade. Information is one email, website, phone call away. Everyone can access the global market – if they are connected. Ebay, for example, allows someone in Europe to purchase goods from another individual in the USA.

Global trade is not new. The Ancient Greeks and Romans traded across their world. The Han dynasty in China traded across the Pacific Basin and India. European empires and the Islamic world traded via trade routes around the world. What is new is the speed and scale of the trade and the communication. Since the end of the Second World War, protectionism of markets has decreased and free trade has increased. The World Bank and the International Monetary Fund (IMF) were set up in 1944 and have influenced development and world finance, including third world debt, since then. Some think that globalization only leads to higher profits for the transnational corporations (TNCs) but there is evidence that poverty has decreased in countries with increased global contacts and economies, eg China. Ecologically, international agreements on global issues such as climate change or ozone depletion have tended to be easier to conclude with increased globalization. There is a tendency for it to westernize some countries.

Globalization is not internationalism. The latter recognizes and celebrates different cultures, languages, societies and traditions. It promotes the unit as the nation state. The former sees the world as a single unit or system not recognizing these differences. Globalization is making the individual more aware of the global community, its similarities and its differences. Such globalization is both a positive and negative force. In one instance it can make us aware of the plight of others on the other side of the globe and in another instance make us aware of what one society has and we do not have.

Many, if not most, products are now traded on a global scale. They are part of what is referred to as the 'global market' and minerals mined in South Africa or Australia are traded and shipped globally.

What do you think about globalization?

Practical Work

* Are there cultural differences in attitudes in the management of natural capital? If so, explain what causes these.

* As resources become scarce, we have to make decisions about whether to use them. To what extent should potential damage to the environment limit our pursuit of knowledge?

8.3 Solid domestic waste

Significant ideas:

→ Solid domestic waste (SDW) is increasing as a result of growing human population and consumption.

→ Both the production and management of SDW can have significant influence on sustainability.

Applications and skills:

→ **Evaluate** SDW disposal options.

→ **Compare** and **contrast** pollution management strategies for SDW.

→ **Evaluate,** with reference to figure 8.3.15, pollution management strategies for SDW by considering recycling, incineration, composting and landfill.

Knowledge and understanding:

→ There are different types of **SDW** of which the volume and composition changes over time.

→ The abundance and prevalence of **non-biodegradable** (eg plastic, batteries, e-waste) pollution in particular has become a major environmental issue.

→ **Waste disposal** options include landfill, incineration, recycling and composting.

→ There are a variety of **strategies** that can be used to manage SDW influenced by cultural, economic, technological and political barriers. These strategies include

• Altering human activity: includes reduction of consumption and composting of food waste.

• Controlling release of pollutant: governments create legislation to encourage recycling and reuse initiatives and impose tax for SDW collection, impose taxes on disposable items.

• Clean-up and restoration: reclaiming land-fills, use of SDW for 'trash to energy' programmes, implementing initiatives to remove plastics from the Great Pacific Garbage Patch (clean-up and restoration).

What is SDW?

Solid domestic waste (SDW) or municipal solid waste (MSW) is our trash, garbage, rubbish from residential and urban areas. It is a mixture of paper, packaging, organic materials (waste food), glass, dust, metals, plastic, textiles, paint, old batteries, electronic waste (e-waste) etc. (see figures 8.3.1 and 8.3.2). It is collected from homes and shops and, although it only makes up about 5% of total waste, which includes agricultural and industrial waste, it is waste that we can control.

SDW production per capita per day is about 3.5 kg in the USA and 1.4 kg in the EU. People in LEDCs tend to produce less SDW than those in MEDCs. Why is this?

▲ **Figure 8.3.1** Solid domestic waste

Type of SDW	Examples
Biodegradable	Food waste, paper, green waste
Recyclable	Paper, glass, metals, some plastics, clothes, batteries
Waste electrical and electronic equipment – WEEE	TVs, computers, phones, fridges
Hazardous	Paints, chemicals, light bulbs
Toxic	Pesticides, herbicides
Medical	Needles, syringes, drugs
Inert	Concrete, construction waste
Mixed	Tetrapaks, plastic toys

▲ **Figure 8.3.2**

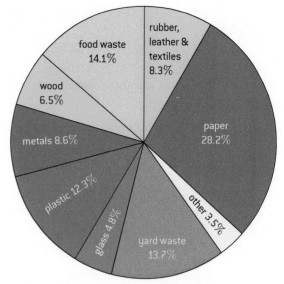

Source: Municipal solid waste in the United States: 2009 facts and figures, EPA

▲ **Figure 8.3.3** SDW proportions in the USA in 2009

To do

List (or collect) the waste that you produce in 24 hours. Do the same for waste from your household in a week. Put the waste into categories:

Recyclable

Biodegradable

Hazardous/toxic

WEEE

How much do you recycle? Who does this? Where does it go?

When is something waste?

A resource has value to humans (8.2). One human's waste is another human's resource. It depends on how we value it. That is why in many LEDCs there are whole industries set up to collect SDW. People travel round residential areas going through communal bins and taking out 'useful' stuff. In many LEDCs families live on and around the landfills just so they can trawl through the waste that arrives from the city.

Waste is material which has no value to its producer. If it is not recycled it becomes a problem and needs to be disposed of. We create waste in most of the processes we carry out – energy production, transport, industrial processes, construction, selling of goods and services, and domestic activities.

The circular economy

Most goods are produced in a linear model – 'take, make, dump'. We find the raw materials or natural capital (take) and use energy to produce goods (make). Often these goods become redundant or break down and our model has been to discard and then replace them with others (dump).

Our global economy has been built on this unsustainable premise. Earth and its resources are finite so we cannot really throw things away. There is no away. Even reducing fossil fuel use and becoming more efficient at obtaining resources only delay the inevitable dwindling of natural capital available to humans.

The circular economy[1] is a model that is sustainable.

It aims to:

- be restorative of the environment

- use renewable energy sources

[1] http://www.ellenmacarthurfoundation.org/circular-economy/circular-economy/the-circular-model-an-overview

- eliminate or reduce toxic wastes
- eradicate waste through careful design.

To do these things, the model relies on manufacturers and producers retaining ownership of their products and so being responsible for recycling them or disposing of them when the consumer has finished using them. The producers act as service providers, selling use of their products, not the products themselves. This means that they take back products when they are no longer needed, disassemble or refurbish them and return them to the market.

This model has similarities to agricultural practices in which good husbandry and soil conservation lead to sustainable growth of foodstuffs.

http://www.ellenmacarthurfoundation.org/

TOK

The circular economy is a paradigm shift. Does knowledge develop through paradigm shifts in all areas of knowledge?

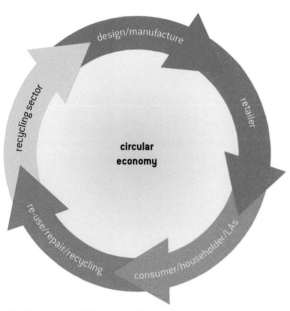

▲ **Figure 8.3.4** The circular economy

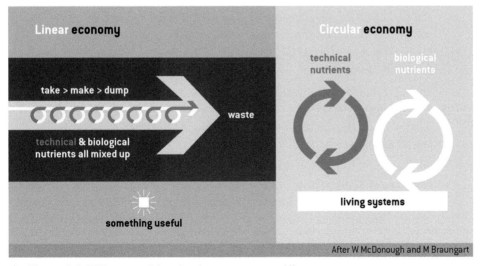

▲ **Figure 8.3.5** Diagram of the circular economy and linear economy

Principle	Agricultural sustainable practices	Circular economy practices
Design out waste	Reduce or eliminate food waste	Recycle plastics, metals
Build resilience through diversity	Manage for complex ecosystems	Build for connections and reuse of components
Use renewable energy sources	Use more solar energy, human labour, fewer chemicals and fossil fuels	Shift taxation from labour to non-renewable energy
Think in systems	Systems are non-linear, feedback-rich and interdependent, emphasize storages and flows	Increase effectiveness and interconnectedness in manufacturing
Think in cascades	Use all stages of a process. Decomposition recycles all nutrients. Burning wood shortcuts this and breaks down nutrients	Do not produce waste. Use it to produce more products

▲ **Figure 8.3.6** Applying principles of the circular economy

TOK

To what extent is emotion involved in environmental knowledge claims?

To what extent is language neutral in reference to environmental knowledge claims?

To think about

▲ **Figure 8.3.7** Shell advert

The caption on the advert reads: If only we had a magic bin that we could throw stuff in and make it disappear forever, what we can do is find creative ways to recycle. We use our waste CO_2 to grow flowers and our waste sulphur to make super-strong concrete.

Shell, a multinational oil company, was criticized by Friends of the Earth (FoE), an NGO, and others when the advert in figure 8.3.7 was produced in 2007. To show chimneys emitting flowers, according to FoE, misrepresents Shell's impact on the environment.

According to FoE, the advert implied that a significant proportion of Shell's emissions were recycled to grow flowers or reduce sulphur emissions. (Growing flowers is theoretically possible if carbon dioxide is captured and it can raise the rate of photosynthesis.) But the reality is that these are small research projects.

1. What do you think about the ethics of this advert?

2. Were FoE right to complain or should we take advertising with a 'pinch of salt'?

3. Find another advert about the environment and society that you think may be misleading.

Managing SDW

We have a choice. We can minimize waste or we can dispose of it somewhere but we can not throw it away.

1. Strategies to minimize waste

These can be summarized in three words: reduce, reuse, recycle. The best action we can take is producing less waste in the first place. See http://www.nrdc.org/thisgreenlife/0802.asp

Reduce

This is the best place to start with the 3R's and it requires us to use fewer resources. We do not have to stop our lifestyles, we just need to cut back.

- Make sure you know how to maintain your possessions so that they last longer.
- Change shopping habits:
  - buy things that will last,
  - look for items with less packaging,
  - buy products that are made from recycled materials eg paper
  - choose products that are energy efficient
  - avoid things that are imported
  - be aware of how many resources you are using in the home – water, electricity etc.

Reuse

This is where the products are used for something other than their original purpose or they are returned to the manufacturer and used repeatedly.

- Returnable bottles – take the bottle back to the shop to be returned to the manufacturer.
- Compost food waste.
- Use old clothes as cleaning rags.
- Hire DVDs – don't buy them.
- Read E-books.

Recycle

This is probably the best known R. Many towns and cities now have kerbside recycling. This is the sorting of waste into separate containers for recycling before it leaves the home.

- In Germany, for example, each household has four bins for this.
- In the UK, there is discussion about charging households more if they produce more than the standard amount of waste.

▲ **Figure 8.3.8** Recycling bins in Orchard Road, Singapore

- In India and China, very little waste is food waste as this is either not thrown away or is fed to animals.

- In MEDCs, up to 50% of waste is food waste.

Recycling involves collecting and separating waste materials and processing them for reuse. (If materials are separated from the waste stream and washed and reused without processing in some way, this is reuse.) The economics of recycling determine whether it is commercial or not and this can vary with the market cost of the raw materials or cost of recycling. Some materials have a high cost of production from the raw material and so recycling of these is particularly worthwhile commercially. Aluminium cans are probably the best example of this.

To do

Waste electrical and electronic equipment (WEEE)

The **Waste Electrical and Electronic Equipment Directive** (WEEE Directive) is a term from the European Community. It had a target of minimum recycling rate of 4 kg per capita of electrical goods recycled by 2009. It failed to meet this target but awareness was increased and more recycling of WEEE goods did happen.

Find out the fate of unwanted or broken electrical and electronic goods in your country.

To think about

Germany: what belongs where?

Adapted from www.howtogermany.com/pages/recycling.html

Many, but not all, German households are allocated three waste bins of different colours and what goes in which is carefully controlled.

Brown bin (biological waste)

Kitchen waste: old bread, egg shells, coffee powder and filters, food leftovers, tea leaves and tea filters.

Fruit and vegetables: peels, apple cores, leaves, nutshells, fruit stones and pips, lettuce leaves.

Garden waste: soil, hedge trimmings, leaves, grass clippings, weeds, dead flowers, and twigs.

Other: feathers, hair, kitchen towels, tissues, sawdust, and straw.

Blue bin (paper)

Envelopes, books, catalogues, illustrations, cartons, writing pads, brochures, writing paper, school books, washing detergent cartons without plastic, newspapers, paper boxes.

Yellow bin or yellow plastic bags (plastic, etc.)

Aluminum foil, plastic wrap, inside packaging materials.

Tins, cans, liquids refill sachets/bags, yogurt cups, body lotion bottles.

Plastic bags, margarine tubs, milk sachets, plastic packaging trays for fruit and vegetables, screw-top bottle tops, detergent bottles, carry bags, vacuum packaging, dishwashing liquid bottles.

Grey bin (household waste)

Ash, wire, carbon paper, electrical appliances, bicycle tubes, photos, broken glass, bulbs, chewing gum, personal hygiene articles, nails, porcelain, rubber, plastic ties, broken mirrors, vacuum cleaner bags, street sweeping dirt, carpeting pieces, diapers, cigarette butts, miscellaneous waste. Those households that do not have a brown bin put their biological waste in a grey bin.

To think about

Recycling plastics

Plastics are made from oil and the world's annual consumption of plastic materials has increased from around 5 million tonnes in the 1950s to nearly 100 million tonnes today. As much as 8% of the world's oil production may be used to make plastics and we throw away most of this as it is used mainly in packaging. Plastic is a difficult material to recycle as there are many different types of plastic and it is bulky and light. Some types of plastic are worth more than others to recyclers but these have to be sorted from the rest. However, plastic recycling is carried out to some extent.

A report on the production of carrier bags made from recycled rather than virgin polythene concluded that the use of recycled plastic resulted in the following environmental benefits:

- reduction of energy consumption by two-thirds,
- production of only a third of the sulphur dioxide and half of the nitrous oxide,
- reduction of water usage by nearly 90%,
- reduction of carbon dioxide generation by two-and-a-half times.

A different study concluded that 1.8 tonnes of oil are saved for every tonne of recycled polythene produced.

Recycled plastic can be made into fleeces and anoraks, cassette cases, window frames, bin bags, seed trays and a range of other products. It takes 25 two-litre plastic drinks bottles to make one fleece garment. Sadly, most plastic is used once and then put in holes in the ground.

Many LEDCs have an informal recycling sector — see earlier in this sub-topic.

 PET **Polyethylene terephthalate** – Fizzy drink bottles and oven-ready meal trays.

 HDPE **High-density polyethylene** – Bottles for milk and washing-up liquids.

 PVC **Polyvinyl chloride** – Food trays, cling film, bottles for squash, mineral water and shampoo.

 LDPE **Low density polyethylene** – Carrier bags and bin liners.

 PP **Polypropylene** – Margarine tubs, microwaveable meal trays.

 PS **Polystyrene** – Yoghurt pots, foam meat or fish trays, hamburger boxes and egg cartons, vending cups, plastic cutlery, protective packaging for electronic goods and toys.

 OTHER **Any other plastics** that do not fall into any of the above categories. An example is melamine, which is often used in plastic plates and cups.

▲ **Figure 8.3.9** Types of plastic and their uses

To think about

The success of the plastic bag

How many have you used once? Plastic bags are everywhere and are clean, cheap to produce, waterproof, convenient. It costs one cent to produce one and about four to make a paper bag. They are so cheap that stores give them away. We are so keen to use them that an estimated 500 billion to one trillion are made each year. Most are used once and then thrown away as they are so thin. A few degrade in sunlight if they are made of biodegradable starch polymer materials but most are made from oil. When we have finished with them, they may end up in landfills, trees, oceans, turtle stomachs, on deserted islands. Everywhere. Although they take up less room in a landfill than a paper bag, they take 200–1,000 years to break down. Burning them releases toxins. We only started using them in quantity in the late 1980s. Before that we carried reusable shopping bags. We can do this again.

The plastax – plastic bag tax – may be the answer. In South Africa, Ireland, Australia, Taiwan and Bangladesh, governments have acted to ban or tax plastic bags. In Ireland a tax on the bags resulted in a decrease in their use of 95%. In South Africa, thin bags were banned and the thicker ones can be reused, have to be paid for and do not float around the country. China, said to once use three billion bags a day, started a ban on free plastic bags from supermarkets in 2008. Five years on, while not universally obeyed, this had cut consumption by 67 billion bags and some 6 million tonnes of oil.

To do

The issue of plastic cups

http://www.eia.doe.gov/kids/energyfacts/saving/
recycling/solidwaste/plastics.html

Here are different viewpoints on whether we should use paper, ceramic or styrofoam cups.

A paper or a plastic cup (styrofoam) or a ceramic mug that you wash and reuse? Plastic cups are made from non-renewable oil; paper from renewable wood; ceramic mugs from non-renewable clay. Which should you choose?

A study by Canadian scientist Martin Hocking shows that making a paper cup uses as much petroleum or natural gas as a polystyrene cup. Plus, the paper cup uses wood pulp. The Canadian study said, 'The paper cup consumes 12 times as much steam, 36 times as much electricity,

and twice as much cooling water as the plastic cup.' And because the paper cup uses more raw materials and energy, it also costs 2.5 times more than the plastic cup. But the paper cup will degrade, right? Probably not. Modern landfills are designed to inhibit degradation so that toxic wastes do not seep into the surrounding soil and groundwater. The paper cup will still be a paper cup 20 years from now.

Surely it is kinder on the environment to use a ceramic mug? Well, it depends how you clean it. If you consider the energy cost of making it, the use of hot water and detergent in a dishwasher in cleaning it, you would need to use the mug 1,000 times to get down to the environmental impact that the plastic mug has.

Justify your decisions on what cups you use.

Test do

How many years?

How long do you think these objects take to break down in a landfill?

1. Disposable nappy/diaper
2. Cotton T-shirt
3. Leather belt
4. Styrofoam cup
5. Glass bottle
6. Plastic bottle
7. Paper bag
8. Banana peel
9. Aluminium can
10. Block of wood

(Turn the book upside down to read the answers at the bottom of the page.)

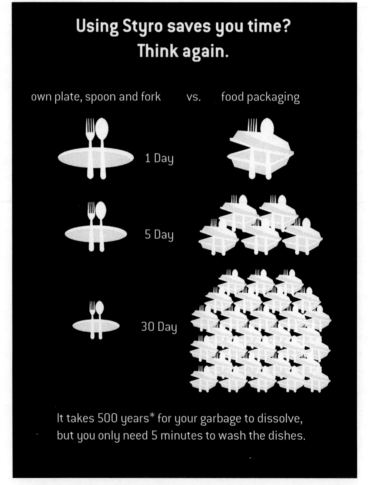

Using Styro saves you time?
Think again.

own plate, spoon and fork   vs.   food packaging

1 Day

5 Day

30 Day

It takes 500 years* for your garbage to dissolve, but you only need 5 minutes to wash the dishes.

▲ **Figure 8.3.10** Another viewpoint: styrofoam or washing up. Which produces more waste?

Answers: 1. 500–600 years, 2. 6 months, 3. 50 years, 4. 1 million years, 5. 1 million years, 6. 1 million years, 7. 2 months, 8. 1 month, 9. 500 years, 10. 20 years.

388

2. Strategies for waste disposal

If waste materials are not recycled or reused, the options are to put them in landfill sites or incinerate them, dump them in the seas or to compost organic waste.

Landfill is the main method of disposal. Waste is taken to a suitable site and buried there. Hazardous waste can be buried along with everything else and the initial cost is relatively cheap. Landfill sites are not just holes in the ground. They are carefully selected to be not too close to areas of high population density, water courses and aquifers. They are lined with a special plastic liner to prevent leachate (liquid waste) seeping out. The leachate is collected in pipes. Methane produced as a result of fermenting organic material in the waste is either collected and used to generate electricity or vented to the atmosphere. Soil is pushed over the waste each day to reduce smells and pests. New landfill sites are getting harder to find as we fill up the ones we have at a faster and faster rate.

▲ **Figure 8.3.11** A waste truck unloading in a landfill site, Wales

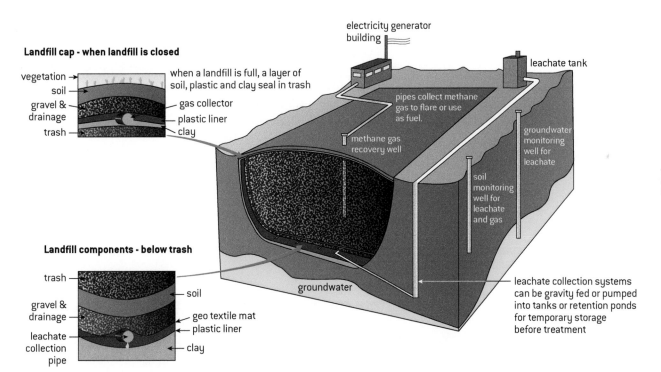

▲ **Figure 8.3.12** The technology of a landfill site

Land can be reused after the site is closed and has settled but there have been issues of subsidence or leaking gases on some reclaimed sites.

Incinerators burn the waste at high temperatures of up to 2,000 °C. In some, the waste is pre-sorted to remove incombustible or recyclable materials. Then the heat produced is often used to generate steam to drive a turbine or heat buildings directly. This is called waste-to-energy incineration. In others, all the waste is burned but this practice can cause air pollution, particularly release of dioxins from burning plastics, heavy metals (lead and cadmium) from burning batteries and nitrogen oxides. But the ash from incinerators can be used in road building and the space taken up by incinerated waste is far smaller than that in landfills. Plants are expensive to build though, and need a constant stream of waste to burn, so do not necessarily encourage people to reduce their waste output.

▶ **Figure 8.3.13** Waste incinerator in Vienna, designed by the architect Hundertwasser as something in which the city can take pride

Anaerobic digestion is when biodegradable matter is broken down by microorganisms in the absence of oxygen. The methane produced can be used as fuel and the waste later used as fertilizer or soil conditioner.

Domestic organic waste can be composted or put into anaerobic biodigesters. Composting can be done at home on a small scale or local government authorities can collect home organic waste and compost it on a larger scale and sell the composted materials which are fertilizers back to the public. On an even larger scale, anaerobic digesters (figure 8.3.14) break down the waste and produce methane (biogas) which can be used as a fuel and a digestate (the solids that are left) which is a fertilizer.

▲ **Figure 8.3.14** A modern biodigester plant for organic waste

Selecting SDW management strategies

It is not straightforward to select the best management strategy for waste. While not polluting may seem the obvious choice, in being alive we all cause the emission of carbon dioxide and greenhouse gases and produce waste. It would be impossible not to. Economies depend on production of goods and these need raw materials. Politicians have to make difficult choices which sometimes come down to a balance between employment for people or protecting the environment. Culturally, we may not be willing to change, or we may not have the finance to invest to do so.

To do

Make a table listing the advantages and disadvantages of landfill, incineration and recycling as waste disposal methods.

To do

Study figure 8.3.15 and find local examples of each strategy in the right-hand column.

Process of pollution	Level of pollution management
HUMAN ACTIVITY PRODUCING POLLUTANT ↓	*Altering Human Activity* • Reduce packaging • Recycle goods • Reuse clothes, goods, containers • Compost organic matter
RELEASE OF POLLUTANT INTO ENVIRONMENT ↓	*Controlling Release of Pollutant* • Separate waste into different types • Legislate about waste separation • Educate for waste separation • Tax disposable items
IMPACT OF POLLUTANT ON ECOSYSTEMS	*Clean-up and Restoration of Damaged Systems* • Reclaim landfills • Incinerate SDW for energy • Collect plastics, eg from the Great Pacific Garbage Patch

▲ **Figure 8.3.15** Three-level model of waste management

Practical Work

* The circular economy can be seen as a paradigm shift. To what extent does environmental knowledge develop through paradigm shifts in all areas of knowledge?

* As resources become scarce, we have to make decisions about whether to use them. To what extent should potential damage to the environment limit our pursuit of knowledge?

8.4 Human systems and resource use

Significant ideas:

→ **Human carrying capacity** is difficult to quantify.

→ The **ecological footprint (EF)** is a **model** that makes it possible to determine whether human populations are living within carrying capacity.

Applications and skills:

→ **Evaluate** the application of carrying capacity to local and global human populations.

→ **Compare and contrast** the differences in the ecological footprint of two countries.

→ **Evaluate** how EVSs impact the ecological footprints of individuals or populations.

Knowledge and understanding:

→ **Carrying capacity** is the maximum number of a species or 'load' that can be sustainably supported by a given area.

→ It is possible to estimate the carrying capacity of an environment for a given species however this is problematic in the case of human populations for a number of reasons.

→ An **EF** is the area of land and water required to support a defined human population at a given standard of living. The measure takes into account the area required to provide all the resources needed by the population, and the assimilation of all wastes.

→ **EF is a model** used to estimate the demands that human populations place on the environment.

→ EFs may vary significantly from country to country and person to person and includes aspects such as lifestyle choices (EVS), productivity of food production systems, land use and industry. If the EF of a human population is greater than the land area available to it this indicates that the population is unsustainable and exceeds the carrying capacity of that area.

→ Degradation of the environment together with the utilization of finite resources is expected to **limit human population growth**.

→ If human populations do not live sustainably they will exceed carrying capacity and risk collapse.

Human carrying capacity

Difficulties in measuring human carrying capacity

By examining carefully the requirements of a given species and the resources available, it should be possible to estimate the carrying capacity of that environment for the species. This is problematic in the case of human populations for a number of reasons.

1. Humans use a far **greater range of resources** than any other animal so it is not just a case of working out what we eat and drink and what space we need for a house.

2. We also **substitute** resources with others if they run out. We may burn coal instead of wood, use solar energy instead of oil, or eat mangoes instead of apples.

3. Depending on our lifestyles, culture and economic situation, our **resource use varies** from individual to individual, country to country. Money buys stuff so the more money there is available the more demand there tends to be for resources.

4. We **import resources** from outside our immediate environment so we cannot just look at the local environment to see how many people it can support.

5. **Developments in technology** lead to changes in the resources we use. This can mean we use less because machines become more efficient or it could mean we use more because we can exploit new resources (eg shale oil).

While importing resources increases the carrying capacity for the local population, it has no influence on the global carrying capacity. It may even reduce carrying capacity by allowing cheaper imports of food and forcing farmers to reduce their costs to compete with imports and so reduce incentives for conservation of the local environment. Plus, at the moment, it involves the use of fossil fuels in transport. If the environment becomes degraded, eg by soil erosion, the land may become less productive and so not produce food for as many people.

All these variables make it practically impossible to make reliable estimates of carrying capacities for human populations.

Ways to change human carrying capacity

Ecocentrists may try to reduce their use of non-renewable resources and minimize their use of renewable ones. Some even try to 'drop off the grid'. Meaning they become self-sufficient to varying degrees – use solar cells for their electricity, use rainwater and grey water recycling for their water supply, grow their own food.

Technocentrists may argue that the human carrying capacity can be expanded continuously through technological innovation and development. 'We shall always grow enough food, have enough water. It is just a matter of being more efficient and inventive.'

Using the remaining oil twice as efficiently means it lasts twice as long as it would have otherwise. But that is only if the population stays the same and given the UN's estimate of human population size in 2050 of 9.6 billion, efficiencies will have to increase dramatically.

Conventional economists argue that trade and technology increase the carrying capacity. Ecological economists say that this is not so and that technological innovation can only increase the efficiency with which natural capital is used. Increased efficiency, at a particular economic level, may allow load on the ecosystem to increase but carrying capacity is fixed and once reached cannot be sustainably exceeded. The other difficulty with technology is that it may appear to increase productivity (eg energy-subsidized intensive agriculture giving higher yields) but this cannot be sustainable and long-term carrying capacity may be reduced (eg by soil erosion). (See the circular economy in sub-topic 8.3.)

Reuse, recycling, remanufacturing and absolute reductions

Humans can reduce their environmental demands (and thereby increase human carrying capacity) by reuse, recycling, remanufacturing and absolute reductions in energy and material use.

Reuse: the object is used more than once. Examples include reuse of soft drink bottles (after cleaning), furniture and pre-owned cars.

Recycling: the object's material is used again to manufacture a new product.

- The use of plastic bags to make plastic fence posts for gardens or fleeces to wear.

- Recycling of aluminium. Obtaining aluminium from aluminium ore requires vast amounts of energy. Melting used aluminium to make new objects only takes a fraction of this energy, much energy can be saved by recycling.

Remanufacturing: the object's material is used to make a new object of the same type. An example is the manufacturing of new plastic (PET) bottles from used ones. See also circular economy in sub-topic 8.3.

Absolute reductions: absolute reduction means that we can simply use fewer resources, eg use less energy or less paper. Unfortunately the advantages of reductions in resource use, ie increased carrying capacity, are often eroded by population increase.

But remember that changes in birth rates and death rates do not change the carrying capacity, because carrying capacity is what the land can provide and reducing the birth rate does not change that.

Limits to human carrying capacity

In 1798, when the human population was about 1 billion, Thomas Malthus (sub-topic 8.1), an economist, wrote, 'The power of the population is infinitely greater than the power of the Earth to produce subsistence for man.' In 1976, when the population was 3.5 billion, environmentalist Paul Ehrlich warned of 'famines of unbelievable proportions' and that feeding a population of 6 billion (exceeded in 1999) would be 'totally impossible in practice'. So far these predictions of disaster have been wrong and human carrying capacity may continue to increase. Though some would say that the famines in Africa are a sign of things to come.

Ecological footprints – EF

Two researchers in Canada, Rees and Wackernagel, first published a book on ecological footprints and their calculation in 1996. Since then, the concept has become widely accepted with many website calculators designed to help you measure your footprint.

EF is a model used to estimate the demands that human populations place on the environment. The measure takes into account the area of land and water required to provide all the resources needed by the population, and the assimilation of all wastes.

Key term

An ecological footprint (EF) is the area of land and water required to support a defined human population at a given standard of living. (See also sub-topic 1.4.)

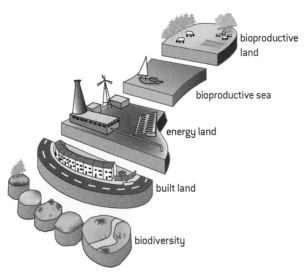

bioproductive land

bioproductive sea

energy land

built land

biodiversity

▲ **Figure 8.4.1** Types of land and sea usually used to calculate an EF

Practical Work

∗ Evaluate how EVSs impact the ecological footprints of individuals or populations.

∗ Evaluate the application of carrying capacity to local and global human populations.

∗ Compare and contrast the differences in the ecological footprint of two countries.

Where the EF is greater than the area available to the population, this is an indication of unsustainability as the population exceeds the carrying capacity of the population.

EFs may vary significantly from country to country and person to person and include aspects such as lifestyle choices (EVS), productivity of food production systems, land use and industry.

In 2012 it was calculated that the EF of all people on Earth was equivalent to 1.5 Earths or 2.7 global hectares (gha) per person. So humanity would take 18 months to regenerate one year's worth of resources that we use. We are in ecological overshoot and have been since the 1970s in that our annual demand on the natural world exceeds what it can supply.

To do

In its 2012 Living Planet Report[1], the The World Wildlife Fund has graphed and mapped the per capita ecological footprint by country, using numbers calculated by the Global Footprint Network.[2]

10 countries with the biggest ecological footprint per person	10 countries with the smallest ecological footprint per person
1. Qatar	1. Occupied Palestinian Territory
2. Kuwait	2. Timor Leste
3. United Arab Emirates	3. Afghanistan
4. Denmark	4. Haiti
5. United States	5. Eritrea
6. Belgium	6. Bangladesh
7. Australia	7. Rwanda
8. Canada	8. Pakistan
9. Netherlands	9. Democratic Republic of Congo
10. Ireland	10. Nepal

[1] http://awsassets.panda.org/downloads/lpr_2012_rio_summary_booklet_final_120509.pdf

[2] http://www.footprintnetwork.org/en/index.php/GFN/

1. Copy and complete the table below. Look at the countries in the table above and think about the lifestyles, diet, transport, industry, agricultural practices of the people. Remember EF includes water.

Reasons for high EF	Reasons for low EF

2. Discuss the changes between average EF per person in 1961 and 2008 in figure 8.4.2.

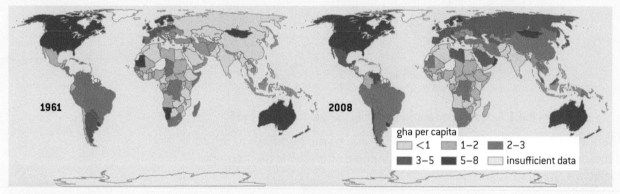

▲ **Figure 8.4.2** Global map of national EF per person in 1961 and 2008

Humans can exceed their local carrying capacity by several means including trade to import resources (see earlier in 8.4).

EF is the inverse of carrying capacity. How does the cartoon in figure 8.4.3 show this?

▲ **Figure 8.4.3** Our ecological footprint

Personal ecological footprints

Figure 8.4.4 shows a fair Earthshare for one person. A fair **Earthshare** is the amount of land each person would get if all the ecologically productive land on Earth were divided evenly among the present world population.

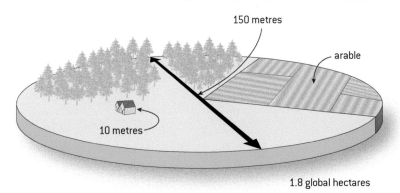

150 metres

arable

10 metres

1.8 global hectares

▲ **Figure 8.4.4** A fair Earthshare if all productive land were shared equally. Area of circle =1.8 global hectares

Do you think your Earthshare is larger? Check it by searching on the Internet for an ecological footprint calculator. http://www.footprintnetwork.org/en/index.php/GFN/page/calculators/ is a good place to start.

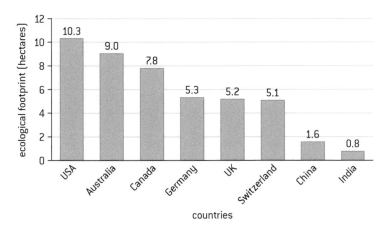

▲ **Figure 8.4.5**

On average, a Canadian's ecological footprint is 7.8 hectares or approximately the size of 15 football fields. Only the United States and Australia have larger footprints at 10.3 and 9.0 hectares respectively. To compare, the average person in India has a footprint of 0.8 hectares, China 1.6. In the United Kingdom it is 5.2, in Germany 5.3 and in Switzerland 5.1 hectares.

In 2008, if we all shared equally, there would have been 1.8 hectares available per person or 1.3 if you do not include productive marine areas. Clearly, we are living beyond the Earth's ability to provide for our consumption.

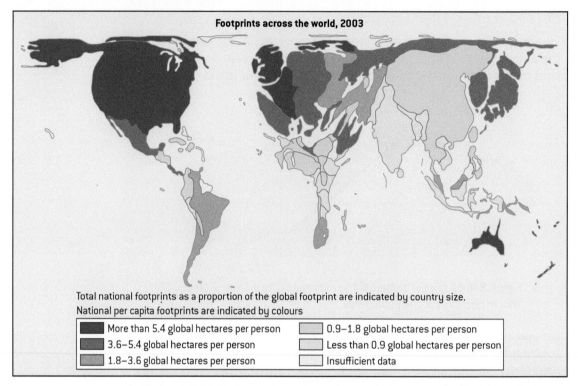

Footprints across the world, 2003

Total national footprints as a proportion of the global footprint are indicated by country size.
National per capita footprints are indicated by colours

	More than 5.4 global hectares per person		0.9–1.8 global hectares per person
	3.6–5.4 global hectares per person		Less than 0.9 global hectares per person
	1.8–3.6 global hectares per person		Insufficient data

▲ **Figure 8.4.6** World ecological footprint sizes from the Living Planet report 2006

The ecological footprint of a country depends on several factors: its population size and consumption per capita – how many people and how much land each one uses. It includes the cropland and other land that is needed to grow food, grow biofuels, graze animals for meat, produce wood, dig up minerals and the area of land needed to absorb wastes, not just solid waste but waste water, sewage and carbon dioxide. Figure 8.4.6, taken from the WWF Living Planet Report 2006, attempts to show countries as either ecological debtors or creditors. The creditors have smaller footprints than their biocapacity (living capacity or natural resources) and the debtors have larger footprints, represented here by changing the sizes of the countries in proportion. Debtors could be harvesting resources unsustainably in their countries, importing goods or exporting wastes. There is no such thing as 'throwing away' on the Earth. There is no 'away' in a closed system.

Ecological footprints of MEDCs and LEDCs

Data for food consumption are often given in grain equivalents, so that a population with a meat-rich diet would tend to consume a higher grain equivalent than a population that feeds directly on grain. Look at the data in figure 8.4.7.

Population from	Per capita grain consumption kg yr^{-1}	Local grain productivity kg ha^{-1} yr^{-1}	Per capita CO_2 emissions from fossil fuels kg C yr^{-1}	Net CO_2 fixation by local vegetation kg C ha^{-1} yr^{-1}
Africa	300	6,000	200	6,000
North America	600	300	1,500	3,000

▲ Figure 8.4.7

1. What does the high per capita grain consumption in North America suggest about the diet?

2. What does the local grain productivity suggest about the two farming methods in use?

3. Which population is more dependent on fossil fuels? Explain.

4. Why is there a difference in the net CO_2 fixation of the two regions?

These, and other factors, will often explain the differences in the ecological footprints of populations in LEDCs and MEDCs.

5. Calculate the per capita ecological footprint (food land and CO_2 absorption land only) for each region, using the two stated formulae.

$$\frac{\text{per capita food consumption (kg yr}^{-1})}{\text{mean food production per hectare of local arable land (kg ha}^{-1} \text{ yr}^{-1})}$$

$$\frac{\text{per capita CO}_2 \text{ emission (kg C yr}^{-1})}{\text{net carbon fixation per hectare of local natural vegetation (kg C ha}^{-1} \text{ yr}^{-1})}$$

6. State two differences you would expect between the ecological footprint of a city in a LEDC and that of a city in an MEDC.

7. It has been calculated that the ecological footprint of Singapore is 264 times greater that the area of Singapore. Explain what this means.

8. Assume that in a large city with a stable population, the proportion of the population that has a vegetarian diet increases. Explain how and why this change might affect the city's ecological footprint.

Assignment and summary

Use the information and examples in this sub-topic to write a summary of human carrying capacity under the title:

Human ingenuity, reduction of energy and material consumption, technical innovation and population development policies all increase human carrying capacity.

Consider the following points. Give two good examples of each.

a. Define human carrying capacity. List ways in which local human populations can exceed the natural carrying capacity of the area in which they live.

b. Define and give examples of reuse, recycling and remanufacturing. How can these lead to an increase in human carrying capacity?

c. What is the relation between technological development, resource use, carrying capacity and population growth? What consequences of these can limit population growth?

d. How can national population policies decrease population size? What cultural changes can lead to decreased population growth?

Evaluate the models used to predict population change.

To what extent are our solutions to management of SDW mostly aimed at prevention, limitation or restoration?

Explain how the EVS of a population affects the carrying capacity of its country.

BIG QUESTIONS

Human systems and resource use

Explain how a nation with a growing population can achieve sustainable development.

Our global EF is greater than the biocapacity of the Earth. Examine your predictions for the future of humanity on Earth.

Comment on the factors that may make your own EVS differ from that of others with respect to issues of population growth.

Reflective questions

→ A variety of models and indicators are employed to quantify human population dynamics.

- How can we know which indicators to use?
- How can we judge which are the most accurate models?

→ The circular economy is a paradigm shift. Does knowledge develop through paradigm shifts in all areas of knowledge?

→ To what extent is emotion involved in environmental knowledge claims?

→ To what extent is language neutral in reference to environmental knowledge claims?

→ To what extent can development be sustainable?

→ Can economic development of a country ever take place independent of other countries and international organization?

→ What factors create differences in attitudes to the management of natural capital (resources)?

→ How can one country's SDW become a global problem?

→ What are the limiting factors for the planets human carrying capacity?

→ At what point is an MEDC's ecological footprint unsustainable?

Quick review

All questions are worth 1 mark

1. Which list below contains only components of renewable natural capital?

   A. Fish, timber, cattle

   B. Methane, the ozone layer, water vapour

   C. Groundwater, hydroelectric power, solar energy

   D. Rice, whales, diamonds

2. Which could not be described as natural capital?

   A. A stand of forest on a hillside

   B. The fish stocks of a fish farm

   C. The fish harvested from the inshore waters of one country in one year

   D. A waterfall

3. Recycling of a non-renewable resource can be of environmental value because it is likely to lead to

   I. a reduction in use of the resource.

   II. a reduction in resource exploitation.

   III. an increase in natural income.

   IV. an increase in natural capital.

   A. I, II, III and IV

   B. I, II and III only

   C. I and III only

   D. II only

4. Which is a renewable resource?

   A. Soil in an agricultural region

   B. Ground water in an aquifer

   C. Fish in the sea

   D. Gold in the Earth's crust

5. Which row in the table includes examples of natural capital and natural income of a tropical forest?

	Natural capital	Natural income
A.	All harvestable timber	Market value of the timber
B.	All the trees of the forest	The capacity of the trees to reduce soil erosion
C.	A population of organisms in the forest	The total number of offspring the organisms produce in one year
D.	All minerals in the soil	The fertility of the soil for agriculture

6. If the harvesting of a fish population were to exceed the sustainable yield in one year, this would mean

   A. the natural capital had been reduced but the natural income had increased.

   B. the natural capital remained the same but the natural income had been reduced.

   C. the natural capital had been reduced and the natural income would be reduced in the following year.

   D. the natural capital remained the same but the natural income would be reduced in the following year.

7. Which statement is correct?

   A. Only items with economic value can be considered natural capital.

   B. The ozone layer is a form of non-renewable natural capital.

   C. Aesthetic values of an ecosystem can be considered natural capital.

   D. The flood and erosion control carried out by trees are considered their sustainable yield.

The graph below shows human population projections by region.

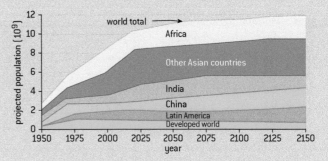

8. By approximately how many times is the world population in 2125 expected to exceed the population in 1950?

   A. 5

   B. 4

   C. 3

   D. 2

9 INTERNAL ASSESSMENT

'I hear and I forget, I see and I remember, I do and I understand.'

Confucius

The purpose of this chapter is to guide you through the requirements of the internal assessment (IA) in ESS. It will:

- Explain why there is an IA.
- Tell you what is required and how long you should take.
- Explain how much help you can expect.
- Show you how you will be assessed.

Key points

The IA:

- Is an individual investigation of a student-designed research question related to environmental systems and societies.

- Is compulsory for all students taking ESS.

- Contributes up to 25% of the total assessment (75% is external assessment).

- Is marked out of 30 by the teacher according to markband level descriptors published in the ESS guide.

- A sample is sent to the IB (after being authenticated by the teacher) for moderation.

So why do an IA?

Apart from the fact that you have to, the IA serves many purposes. You can:

- apply the skills and knowledge you have gained
- pursue your own personal interests (within the limitations of the course)
- be creative
- solve problems
- show your passion for the subject.

What does it involve?

The IA:

- is an individual investigation of a research question that has its foundations in the environment **and** society
- must be designed and implemented by you the student
- is a written a report of between 1,500 and 2,250 words long (do not exceed this or you will lose marks)
- must be properly referenced and cited
- must address the assessment criteria (see 404–408)
- cannot be the same research question as your extended essay
- should be **guided** by your teacher
- should take 10 hours – this will include
  - time for consultation with the teacher to discuss the research question before the investigation is implemented
  - time spent developing methodology
  - time collecting data
  - but **not** time in writing it up.

Your teacher will not set you adrift – they will help prepare you for the IA by:

- going through the assessment criteria; you must understand what these are so you can aim to fulfil them to the best of your ability

- going through the animal experimentation policy (if necessary)

- discussing your initial ideas with you – your teacher does not come up with the idea, you do, but they will guide you

- reviewing your progress, giving feedback and checking authenticity

- reading **one** full draft of the work – your teacher is not allowed to edit your draft but they can read it and give you advice on how to improve – this may be written or oral

- explaining clearly the significance of academic honesty (authenticity and intellectual property).

The purpose of the IA

It is important that you investigate an issue that interests you (you are spending 10 hours investigating it) because this will make it easier to use your results and apply them to the real world and suggest solutions.

In the investigation you should:

- Focus on a particular aspect of an ESS issue (you have a word limit so do not make it too broad) but make sure it is linked to a broader area of environmental and societal interest.

- Develop appropriate methodologies that will generate sufficient data. Here are a few ideas:

  - interviews

  - issues-based enquiries to inform decision making

  - ecosystem modelling (eg mesocosoms or bottle experiments)

  - models of sustainability

  - use of systems diagrams or other valid holistic modelling approaches

  - collection of both qualitative and quantitative data.

- Analyse the data and the knowledge you gain to improve your understanding of the issue:

  - estimations of NPP/GPP or NSP/GSP

  - application of descriptive statistics (measures of spread and average)

  - application of inferential statistics (testing of null hypotheses)

  - other complex calculations

  - cartographic analysis

  - use of spreadsheet or database.

- Then apply your results to a broader (local, regional or global level) environmental and/or societal context so you can suggest solutions to a problem/issue – so if you do not start with an issue/problem you cannot propose a solution. This broader discussion does not have to be in direct relation to your findings. However, the discussion is a chance for you to show creative thinking and novel solutions.

Internal Assessment criteria

For Internal Assessment, the following assessment criteria will be used:

Identifying the context	Planning	Results, analysis and conclusion	Discussion and evaluation	Applications	Communication	Total
6	6	6	6	3	3	30
(20%)	(20%)	(20%)	(20%)	(10%)	(10%)	(100%)

Assessing your IA

This section looks at the criteria you will be assessed against. These are taken straight from the ESS guide.

1. Identifying the context (20%, /6)

This criterion assesses the extent to which the student establishes and explores an environmental issue (either local or global) for an investigation and develops this to state a relevant and focused research question.

Criteria/Markbands

Achievement level	Descriptor
0	The student's report does not reach a standard described by any of the descriptors given below.
1–2	• **States** a research question, but there is a lack of focus. • **Outlines** an environmental issue (either local or global) that provides the context to the research question. • **Lists** connections between the environmental issue (either local or global) and the research question but there are significant omissions.
3–4	• **States** a relevant research question. • **Outlines** an environmental issue (either local or global) that provides the context to the research question. • **Describes** connections between the environmental issue (either local or global) and the research question, but there are omissions.
5–6	• **States** a relevant, coherent and focused research question. • **Discusses** a relevant environmental issue (either local or global) that provides the context for the research question. • **Explains** the connections between the environmental issue (either local or global) and the research question.

So to get the top marks here you need to:

- Choose a relevant environmental issue that you can:

  - Discuss and explain its global and/or local relevance.

  - Use to develop a focused/narrow research question.

  - Explain how the issue is linked to the research question.

2. Planning (20%, /6)

This criterion assesses the extent to which the student has developed appropriate methods to gather data that is relevant to the research question. This data could be primary or secondary, qualitative or quantitative and may utilize techniques associated with both experiential or social science methods of inquiry. There is an assessment of safety, environmental and ethical considerations where applicable.

Criteria/Markbands

Achievement level	Descriptor
0	The student's report does not reach a standard described by any of the descriptors given below.
1–2	• **Designs** a method that is inappropriate because it will not allow for the collection of relevant data. • **Outlines** the choice of sampling strategy but with some errors and omissions. • **Lists** some risks and ethical considerations where applicable.
3–4	• **Designs** a repeatable* method appropriate to the research question but the method does not allow for the collection of sufficient relevant data. • **Describes** the choice of sampling strategy. • **Outlines** the risk assessment and ethical considerations where applicable.
5–6	• **Designs** a repeatable* method appropriate to the research question that allows for the collection of sufficient relevant data. • **Justifies** the choice of sampling strategy used. • **Describes** the risk assessment and ethical considerations where applicable.

*Repeatable in this context means that sufficient detail is provided for the reader to be able to repeat the data collection in another environment or society. It does not necessarily mean repeatable in the sense of replicating it under laboratory conditions to obtain a number of runs or repeats in which all the control variables are exactly the same.

So to get the top marks here you need to:

- Design an appropriate way to collect enough data to study your research question.

- Explain your method so that someone else could repeat the data collection in another environment.

- Explain why you have used the methods you have used.

- Show that you have looked at any ethical issues or risks.

3. Results, analysis and conclusion (20%, /6)

This criterion assesses the extent to which the student has collected, recorded, processed and interpreted the data in ways that are relevant to the research question. The patterns in the data are correctly interpreted to reach a valid conclusion.

Criteria/Markbands

Achievement level	Descriptor
0	The student's report does not reach a standard described by any of the descriptors given below.
1–2	• **Constructs** some diagrams, charts or graphs of quantitative and/or qualitative data, but there are significant errors or omissions. • **Analyses** some of the data but there are significant errors and/or omissions. • **States** a conclusion that is not supported by the data.
3–4	• **Constructs** diagrams, charts or graphs of quantitative and/or qualitative data which are appropriate but there are some omissions. • **Analyses** the data correctly but the analysis is incomplete. • **Interprets** some trends, patterns or relationships in the data so that a conclusion with some validity is deduced.
5–6	• **Constructs** diagrams, charts or graphs of all relevant quantitative and/or qualitative data appropriately. • **Analyses** the data correctly and completely so that all relevant patterns are displayed. • **Interprets** trends, patterns or relationships in the data, so that a valid conclusion to the research question is deduced.

So to get the top marks here you need to:

- Record your data in a way that is clear to everyone else – remember the moderators reading your report do not know as much about the report as you do.
- Present the data in a clear way so that it helps you and others interpret it – remember axes titles and clear labels for everything.
- Spot the trends and patterns and describe and explain them.
- Reach a conclusion based on the data – that may not be what you expected.

4. Discussion and evaluation (20%, /6)

This criterion assesses the extent to which the student discusses the conclusion in the context of the environmental issue, and carries out an evaluation of the investigation.

Criteria/Markbands

Achievement level	Descriptor
0	The student's report does not reach a standard described by any of the descriptors given below.
1–2	• **Describes** how some aspects of the conclusion are related to the environmental issue. • **Identifies** some strengths and weaknesses and limitations of the method. • **Suggests** superficial modifications and/or further areas of research.
3–4	• **Evaluates** the conclusion in the context of the environmental issue but there are omissions. • **Describes** some strengths, weaknesses and limitations within the method used. • **Suggests** modifications and further areas of research.
5–6	• **Evaluates** the conclusion in the context of the environmental issue. • **Discusses** strengths, weaknesses and limitations within the method used. • **Suggests** modifications addressing one or more significant weaknesses with large effect and further areas of research.

So to get the top marks here you need to:

- Look at your conclusion and its link to the environmental issue.
- Evaluate what you have done – what was good, what was bad, what worked, what didn't.
- Suggest how you could improve your investigation and maybe extend it with further research. You do not have to do this but it does have to be realistic.

5. Applications [10%, /3]

This criterion assesses the extent to which the student identifies and evaluates one way to apply the outcomes of the investigation in relation to the broader environmental issue that was identified at the start of the project.

Criteria/Markbands

Achievement level	Descriptor
0	The student's report does not reach a standard described by any of the descriptors given below.
1	• **States** one potential application and/or solution to the environmental issue that has been discussed in the context. • **Describes** some strengths, weaknesses and limitations of this solution.
2	• **Describes** one potential application and/or solution to the environmental issue that has been discussed in the context, based on the findings of the study, but the justification is weak or missing. • **Evaluates** some relevant strengths, weaknesses and limitations of this solution.
3	• **Justifies** one potential application and/or solution to the environmental issue that has been discussed in the context, based on the findings of the study. • **Evaluates** relevant strengths, weaknesses and limitations of this solution.

So to get the top marks here you need to get creative and:

- Give one solution to the problem/issue you have studied.

- Explain how it would be effective and what problems it may encounter.

6. Communication (10%, /3)

This criterion assesses whether the report has been presented in a way that supports effective communication in terms of structure, coherence and clarity. The focus, process and outcomes of the report are all well presented.

Criteria/Markbands

Achievement level	Descriptor
0	The student's report does not reach a standard described by any of the descriptors given below.
1	• The investigation has limited structure and organization. • The report makes limited use of appropriate terminology and it is not concise. • The presentation of the report limits the reader's understanding.
2	• The report has structure and organization but this is not sustained throughout the report. • The report either makes use of appropriate terminology or is concise. • The report is mainly logical and coherent, but is difficult to follow in parts.
3	• The report is well structured and well organized. • The report makes consistent use of appropriate terminology and is concise. • The report is logical and coherent.

So to get the top marks here you need to think about the reader:

- Be logical and systematic.

- Use page numbers.

- Make sure it hangs together well.

The report must be correctly referenced. You will not be penalized for a lack of bibliography or other means of citation, BUT such an omission would probably be treated under the IB Diploma Academic Honesty Policy – this can mean the loss of your Diploma.

Practical work in ESS

To help you understand the concepts and skills needed for your IA, you should first carry out a range of practical activities as part of your ESS course, examples of which are given throughout this book in the boxes titled *Practical Work*. These should be an integral part of your lessons and their purpose is to:

- Reinforce the theory.

- Show you that field and laboratory work is a practical, hands-on experience.

- Allow you to practise using secondary data and modelling.

- Show you that there are benefits and limits to practical investigations.

The practical work should:

- Be about 20 hours.

- Cover most of the topics in the course.

- Include simple and complex investigations.

- Involve investigating the relationships between environmental and social systems.

What is included in practical work?

- Short labs and projects extending over several weeks.

- Computer simulations.

- Using databases for secondary data.

- Developing and using models.

- Data-gathering exercises such as questionnaires, user trials and surveys.

- Data-analysis exercises.

- Fieldwork.

What is the proof of practical work?

Your teacher records all practical work on a form – ESS/PSOW (practical scheme of work) and sends this to the IB with IA samples for moderation.

To the teacher: preparing work for moderation

You should check that the PSOW and IA are carefully prepared for moderation. Do this by ensuring the work is carefully organized and appropriate annotations (in **black** pen or pencil) are placed on the work to direct the moderator to the key points where the mark has been awarded. You are trying to showcase the students work and help the moderator understand why you awarded the marks you did and where. The aim of the moderator is to support the your marking; this is best done by following the steps identified above. Remember that the moderator does not know the case history of the work, but can only moderate what is provided.

'I was thrown out of college for cheating on the metaphysics exam; I looked into the soul of the boy sitting next to me.'

Woody Allen

About the exams

Don't be afraid of exams. If you have prepared well, they can be a satisfying experience and a culmination of your studies in secondary education.

We hope that your studies have not focused almost exclusively on the final exams (External Assessment) but that you have gained knowledge, thought critically and reflectively about that knowledge, and enjoyed at least parts of the experience.

If you reach the end of the course and think that we have degraded the planet to such an extent that you are pessimistic about its ability to recover or the human race's ability to live in balance with the resources on Earth, don't be. We are resourceful as a species and, being sentient beings, we are aware of our own existence and its impact on others. The Earth may be moving to a new equilibrium, civilization will change, but we are capable of managing our population size and resource exploitation and humans have never yet not used an innovation or discovery that we have found.

But, back to the exams. By the time you start the exam in this subject, you will have up to 25% of your final marks as IA. You will not know how many until the results are published but you will have done at least some marks before you go in to the exam.

There are two papers in the external assessment.

For both papers you are allowed five minutes reading time before you start writing.

Paper 1 – case study with questions

Paper 2 – section A short questions

– section B select 2 essays out of 4

Component addressed by this assessment objective	How the assessment objective is addressed	Weighting	Marks	Duration (hours)	Assessment objective	Approximate weighting of objectives (%)	
						1 & 2	**3**
Paper 1	Case study	25	40	1	Objectives 1–3	50	50
Paper 2	Section A: short answer questions	50	25	2	Objectives 1–3	50	50
	Section B: two essays from a choice of four		20 + 20				
Internal Assessment (see chapter 9)	Individual investigation assessed using markbands	25	30	10	Objectives 1–4	Covers objectives 1, 2, 3 and 4	

▲ **Figure 10.1** ESS assessment components

How is the course assessed?

Objectives tell you how the aims of ESS are to be assessed. The command terms tell you what you have to do. They are in the guide and in the exams. Know what they mean.

Objectives	Command terms	
1. Demonstrate knowledge and understanding of relevant: • facts and concepts • methodologies and techniques • values and attitudes.	Define	Give the precise meaning of a word, phrase, concept or physical quantity.
	Draw	Represent by means of a labelled, accurate diagram or graph, using a pencil. A ruler (straight edge) should be used for straight lines. Diagrams should be drawn to scale. Graphs should have points correctly plotted (if appropriate) and joined in a straight line or smooth curve.
	Label	Add labels to a diagram.
	List	Give a sequence of brief answers with no explanation.
	Measure	Obtain a value for a quantity.
	State	Give a specific name, value or other brief answer without explanation or calculation.
2. Apply this knowledge and understanding in the analysis of: • explanations, concepts and theories • data and models • case studies in unfamiliar contexts • arguments and value systems.	Annotate	Add brief notes to a diagram or graph.
	Apply	Use an idea, equation, principle, theory or law in relation to a given problem or issue.
	Calculate	Obtain a numerical answer showing the relevant stages of working.
	Describe	Give a detailed account.
	Distinguish	Make clear the differences between two or more concepts or items.
	Estimate	Obtain an approximate value.
	Identify	Provide an answer from a number of possibilities.
	Interpret	Use knowledge and understanding to recognize trends and draw conclusions from given information.
	Outline	Give a brief account or summary.

3. Evaluate, justify and synthesize, as appropriate:

- explanations, theories and models
- arguments and proposed solutions
- methods of fieldwork and investigation
- cultural viewpoints and value systems.

4. Engage with investigations of environmental and societal issues at the local and global level through:

- evaluating the political, economic and social context of issues
- selecting and applying the appropriate research and practical skills necessary to carry out investigations
- suggesting collaborative and innovative solutions that demonstrate awareness and respect for the cultural differences and value systems of others.

Analyse	Break down in order to bring out the essential elements or structure.
Comment	Give a judgment based on a given statement or result of a calculation.
Compare and contrast	Give an account of similarities and differences between two (or more) items or situations, referring to both (all) of them throughout.
Construct	Display information in a diagrammatic or logical form.
Deduce	Reach a conclusion from the information given.
Demonstrate	Make clear by reasoning or evidence, illustrating with examples or practical application.
Derive	Manipulate a mathematical relationship to give a new equation or relationship.
Design	Produce a plan, simulation or model.
Determine	Obtain the only possible answer.
Discuss	Offer a considered and balanced review that includes a range of arguments, factors or hypotheses. Opinions or conclusions should be presented clearly and supported by appropriate evidence.
Evaluate	Make an appraisal by weighing up the strengths and limitations.
Explain	Give a detailed account, including reasons or causes.
Examine	Consider an argument or concept in a way that uncovers the assumptions and interrelationships of the issue.
Justify	Provide evidence to support or defend a choice, decision, strategy or course of action.
Predict	Give an expected result.
Sketch	Represent by means of a diagram or graph (labelled as appropriate). The sketch should give a general idea of the required shape or relationship, and should include relevant features.
Suggest	Propose a solution, hypothesis or other possible answer.
To what extent	Consider the merits or otherwise of an argument or concept. Opinions and conclusions should be presented clearly and supported with appropriate evidence and sound argument.

Tips 1 - Read the course guide

Make sure you have your own copy of the course guide or that you have access to it from your institution.

The guide tells you what the course covers, but more importantly it gives you a great deal of guidance and advice. So, make sure you read the guide carefully and look at these sections:

1. **Nature of the subject**, aims, assessment objectives – these tell you what the course covers, how to approach it and what we hope you will be able to do when you have completed it.

2. **Check the topics and sub-topics.** Each is arranged like this:

	1.1 Sub-topic		
	Significant ideas: Describes the overarching principles and concepts of the sub-topic.		
Main ideas	**Knowledge and understanding:** • This section will provide specifics of the content requirements for each sub-topic. **Applications and skills:** • This section gives details of how you can apply your understanding. For example, these applications could involve discussions of viewpoints or evaluating issues and impacts.	**Guidance:** • This section will provide specifics and give constraints to the requirements for the understandings and applications and skills. **International-mindedness:** • Examples of ideas that teachers can easily integrate into the delivery of their lessons. **Theory of knowledge:** • Examples of TOK knowledge questions. **Connections**: syllabus and cross-curricular links.	Tells you what depth is required IM and TOK ideas – see reflective BQs

Left-hand column can all be assessed.

Tips 2 - What else should you read

Past papers and mark schemes are published by the IB and your teacher should have some. Towards the end of the course, you should look at and do some of these. Practice getting your timing right so that you do not run out of time in the exam. Read the mark schemes carefully so you can see how the examiners award marks.

Subject reports are written by the senior examiner after every exam session. They report on each question, and how it was answered, and on the IA.

In the exam

To state the obvious, be prepared. By this, we mean:

- revise thoroughly (start well in advance)
- reach the exam room in good time,
- have all the equipment you need (and some spares) with you.

Although you can bring revision notes to the door of the exam room, **do not** take them inside. You do not want to be found guilty of malpractice because you forgot you had some notes in your pocket. Always check.

Once at your desk:

- Listen to the person starting the exam.

- Check that you can see the clock and have all that you need.

- Use the five minutes reading time to:

  - Paper 1 – focus on the questions as you are not allowed to use a highlighter during the reading time and many of you will need to do this in the resources book.

  - Paper 2 – scan section A (the short answers) but focus on which essay questions you will attempt. Remember – check the big mark questions and make sure you can do them well.

When you are allowed to start to write, do not be rushed if you see your colleagues writing furiously. You can score high marks with a few well-chosen words or could write a great deal and not answer the question.

Exam technique can be learned and you can get better at it. You will practise answering exam-style questions over the course and, in the end, do timed practice papers.

- Remember, you cannot get a mark for knowing something, only for writing it down. You cannot get a mark if you do not answer the question.

- You may write a brilliant answer but if you 'describe' and you were asked to 'explain', you do not get the marks.

- Always answer the question and answer the right number of questions. You must write on two of the four possible essays in Section B, Paper 2. Do not omit one essay, for example, or you immediately lose 20 marks out of the 65 for Paper 2. The time allocated allows you to write two essays well and address the questions fully. So do not attempt three essay – you do not have time to write three effectively.

Check the time that you have. If it looks like you will run out of time, make a list, submit your essay plan, answer the easier parts – do anything to get the marks that you can. But, before you reach that point, practise timed essay writing to make sure you do have enough time in the exam.

Examiners are humans too and want you to do your best. But they have to mark according to the mark scheme and markbands and they need to be able to read your writing. They will do their best to read what you have written and do not penalize you for spelling or grammar as long as the meaning is clear.

So practise writing clearly and quickly.

- If it is not written down, you cannot get the mark. So, in calculations, show your working. Remember to put the units. Do not assume the examiner knows you know.

- You can answer the questions in any order so you may do the ones you are most confident about first. Never leave a gap: have a guess and write something as you are not penalized for trying.

- The command terms are vital to your understanding of what you should do. Check that you know which term it is in the question and that you note any other clues, eg 'Justify, giving two reasons ...' – so give two, not just one, 'State, giving a named example, ...' – always have named examples ready in your mind.

- Sometimes you will be asked for a diagram or annotated diagram. Sometimes you will not be asked for this but you know it would help you answer the question. So do draw one if you think it will help you.

- Use the terminology of the subject and technical language correctly.

- DO NOT write outside the boxes you are given – the papers are scanned and some of your work may not show. If you need more space ask for spare paper and leave the examiner a note so they know where to find the extra information.

In Paper 1 and Section A of Paper 2, there are spaces in which you write your answer and the mark allocated is also given. The number of lines for your answer is worked out carefully and most candidates do not need to write more than this. There are usually two lines per mark. You are not penalized if you do write more on an additional sheet but it is not often that the longer answer gains more marks than the concise one. Think before you write the first thing that you think of. The number of lines and mark allocation also give you a clue. If it asks you to describe and evaluate and there are 4 marks allocated, you might guess that 2 of these are for describing and 2 for evaluating. It does not always work like that but get into the habit of thinking about this.

In Section B of Paper 2, there are four essay-style questions of which you answer two of your choice. Most of these questions are divided into three subsections and mark allocations are given. Make sure you divide your answer up in the same way as the questions are asked – eg parts a, b and c. The first sections of each essay will be marked according to a mark scheme. The final section of each essay will be marked using markbands (see ESS guide p78).

Some students like to do an essay plan for both essays at the start of the exam. Then they add to this as new thoughts come to them. If you run out of time you can even just hand this in as it will probably gain some marks.

EE

- Compulsory part of the Core of the IB Diploma
- 4,000 words
- About 40 hours work
- Area of study from one of your 6 Diploma subjects
- Or for World Studies EE, from 2 of the 6
- Independent research
- Topic of personal interest to you
- Guidance from your supervisor in school
- Sent away and marked by an IB examiner.

If you are studying for the full IB Diploma, then you must write and submit an Extended Essay (EE) as part of the requirement for the Core. This chapter gives you some advice on how to approach this task and to see it as something that will give you research skills that you need in tertiary education.

Along with your Theory of Knowledge marks, the EE contributes up to three core points to your total IB Diploma final point score. That may not seem like very many points for the amount of work involved but, if your EE is a good one, it may make the difference between getting into the university or college course of your choice or not.

Why the EE is a good thing

Researching and writing an EE may be the first time that you take real control of your academic study. Most of the time in education, you do what your teachers tell you to do to achieve the highest marks you can or to pass exams. But you do not get much choice of the subjects which you study. Here is a chance, at last, to select a topic which you find of interest and which you can research yourself. In writing the EE, you will develop research skills that will be of benefit later on and develop skills of critical thinking and systematic research. We all hope that you also enjoy it and gain satisfaction from the intellectual discovery of the process.

Tips 1 - Before you start

Read these essential documents[1]

IB Extended Essay guide

IB Academic honesty guide

IB Animal experimentation policy

Recent Extended Essay subject reports

Examples of other Extended Essays

10 tips for your EE

1. Before you start read the EE guide and the ESS subject specific details. These are full of helpful advice and information which will be very useful to you.

2. Keep it under control and do not allow your other work to suffer because of your EE. You are not likely to win a Nobel Prize for this piece of work nor change the path of environmental research.

3. Your supervisor is there to direct you and help you but not to write your EE or do the practical work for you. Arrange plenty of meetings with them and make the most of their support.

4. With a large project, you need markers so you can see your progress. Many schools have interim deadlines to help you through the process – stick to them. If they do not, make your own.

5. Pick a topic area and research area that interests you – you have to write 4,000 words.

---

[1] Available from your school

6. The title is crucial as regards:

    a. subject area

    b. topic

    c. research question.

    The title must be:

    a. relevant, ie is about an environmental systems **and** societies issue

    b. do-able within the word count

    c. ethically sound

    d. a question to which you do not know the answer at first. That is why you are doing the EE. Be patient. It takes time to find the right question.

7. Many students find it easiest to include primary data – data you have collected through fieldwork, experimentation or observation. Some may not be so lucky and must work with all secondary sources. Both are fine but you need to justify your choices.

    a. In practical work that means why did you choose the methods you chose?

    b. In secondary research that is looking at the validity of the data you are using.

8. Read and cite a range of sources from some of these: academic papers, journals, textbooks, personal conversations, reputable websites, books and reviews. If you use internet sites keep a note of the URL as you go.

9. Keep all your notes and drafts until you have your results. Your supervisor and the IB have to be sure this is your own work. These are your evidence. You do not have to write your draft essay in the order in which it finally appears.

10. Be safe and ethical. Do not do an EE in an area that inflicts damage on an organism or the environment or may endanger you or anyone else.

Common errors

-   You have not got a research question that focuses on environmental systems AND human societies.

-   Your EE is not open to analytical argument – it is only a description or narrative.

-   You are unethical or unsafe in your choice.

-   Your research question is too broad or unfocused.

-   Your research question cannot be answered in 4,000 words.

-   Your sources are not relevant to the research question.

-   You cite websites uncritically.

-   You have a one-sided argument and no counterclaims.

▲ **Figure 11.1** Your EE?

Your supervisor

How your school determines EE supervision is a matter for your school but you must have a supervisor who is a teacher within the school. Your supervisor may or may not be a subject specialist in your chosen subject but you should expect your supervisor to help you by:

- advising and guiding you on your research question and on research skills and resources

- ensuring that your research question is manageable and ethically sound

- giving you the assessment criteria

- reading and commenting on your first draft (within IB guidelines)

- confirming that this is your own work

- submitting a predicted grade (A–E) to the IB.

Your supervisor will probably spend about 3–5 hours with you and must conduct a short concluding interview (viva voce).

Your supervisor does not write your EE for you, edit your EE or do the practical work for you if there is any.

How to start

1. Decide if this subject area is the one in which you want to write your EE.

2. Ask yourself these questions:

   Do I enjoy it?

   Do I read around the subject anyway?

   Can I write 4,000 words?

3. Then take some trips – to a local park or protected area, a local farm, reservoir, stream, pond, zoo, school grounds, your back yard or garden, a site that has construction going on or has been built, a landfill site, etc. Go anywhere where organisms live and humans are having an impact.

4. Observe what is going on. Make some notes and ask yourself some questions. What is there? How many are there? Where are they? What are they doing? Why? The 'they' may be plants, animals or human animals or all three. What has changed? Why has it changed? What impact have humans had here?

5. Come up with a list of possible questions that you think you could investigate.

Then look again at this list and ask yourself:

- Can I find an answer to the question?

- Is the answer too obvious eg, 'Is the environment harmed by the new trunk road?' (You would want to change the question to ask how it is altered.)

- Does the answer need investigating by either an experiment, a questionnaire, a literature review or a combination of these?

- Can I try to find the answer within the time I have for this? If it involves you comparing ecosystems that are on different continents because you are visiting the other one in the summer vacation, or growing plants, that will take months. Think carefully about this. You always need to build in time in case things go wrong and you need to redo them.

Although you do not have to do it, an experimental or observational approach may be more rewarding to you than a pure literature review.

Keep it simple. Don't be overambitious in your choice of title or need for equipment. You do not need complex equipment to undertake an excellent essay.

What you must include

This subject looks at the systems of the natural world and the effect of humans upon these. So you need to include both in your EE. You do not need to give equal weight to environmental systems and societies but should include both to some extent. If you investigate an ecological principle such as zonation or succession, then include the interaction of humans with this. If you decide to research environmental philosophies, you also need, in your EE, to have an element of natural systems.

- Take the systems approach in your EE. If you are only looking at one organism in isolation, the EE may be better submitted as a Biology EE.

- You cannot just have a descriptive EE, it must have an analytical element. This means you will score poorly if you just describe a farm or a zoo or ecosystem. But you could compare it to another or to itself in the past or under different conditions and evaluate the relationship.

- On the other hand if you choose to look at the ecological footprint or the carbon footprint make sure you look at the impact on the environment.

- You need an argument to which there is a counter-argument(s).

- Keep the focus sharp. If it is too broad it can be superficial. See the EE guide for examples of focused and broad topics in the subject guidelines.

When to stop

Although an important component of your Diploma and a core requirement, the most points that an EE and TOK (Core points) can gain you is 3 out of 45. So keep it under control. There is a minimum standard that you must reach (figure 11.3) but do not let your study of other subjects suffer because you spend too long on your EE. Know when to stop. Being a perfectionist is not always helpful when you need to manage your time wisely in taking the Diploma.

Figure 11.2 Know when to stop your EE

Important information about the old and new EE guides

For all EEs submitted for May/November exams in 2015, 2016 and 2017, the current EE guide applies. For students taking their IB Diploma from May 2018 onwards (so starting the course in August – November 2016), there are NEW criteria in a new assessment model. See the new EE guide from the IBO when it is published.

The diploma points matrix

		Theory of knowledge					
		Excellent A	Good B	Satisfactory C	Mediocre D	Elementary E	Not Submitted
Extended essay	Excellent A	3	3	2	2	1 + Failing condition*	N
	Good B	3	2	1	1	Failing condition*	N
	Satisfactory C	2	1	1	0	Failing condition*	N
	Mediocre D	2	1	0	0	Failing condition*	N
	Elementary E	1 + Failing condition*	Failing condition*	Failing condition*	Failing condition*	Failing condition*	Failing condition*
	Not Submitted	N	N	N	N	N	N

▲ **Figure 11.3** Diploma points matrix for TOK and EE

28 points overall are required to be eligible for the diploma if you attain an E grade in either EE or TOK.

A grade A in one of these earns an extra point even if the other is a grade E.

A grade E in both the Extended Essay and Theory of Knowledge is automatic failure.

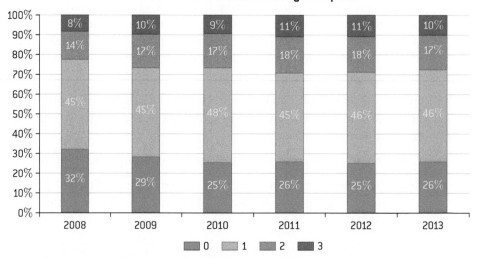

Distribution of candidates scoring bonus points

	2008	2009	2010	2011	2012	2013
3	8%	10%	9%	11%	11%	10%
2	14%	17%	17%	18%	18%	17%
1	45%	45%	48%	45%	46%	46%
0	32%	29%	25%	26%	25%	26%

Legend: 0 1 2 3

▲ **Figure 11.4** Percentages of candidates scoring 0–3 TOK and EE points from 2008–2013. Note that nearly half of all candidates gain 1 point

Academic honesty

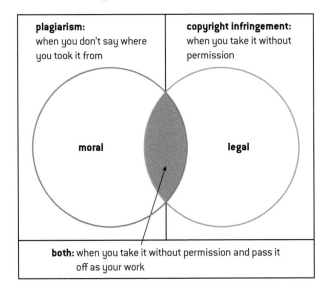

plagiarism: when you don't say where you took it from	copyright infringement: when you take it without permission
moral	legal

both: when you take it without permission and pass it off as your work

Academic honesty is a set of values and behaviours of personal integrity and respect for the integrity of the work of others.

The IB requires your school to have an academic honesty policy and you should have seen this and discussed what it means in practice. In the widest sense, being academically honest means that you do not use the ideas, creative works, words, photos or diagrams of someone else without acknowledging that they came from that source. Of course, you build upon the ideas of others as we do not live isolated from the rest of human experience and knowledge, but it is only right that you do not imply that it is your authentic work if it is not. It is only too easy to cut and paste words from websites or to scan materials. There is nothing inherently wrong with that but attempting to persuade others that this is your own material, words or ideas without stating the original source is deemed to be plagiarism by the IB. It is a fine line to tread – if in doubt GIVE THE SOURCE.

It is your responsibility to ensure that the EE is your own work and you have to sign a cover sheet when you hand it in to say that this is the case. Your supervisor also has to sign to say that they believe this to be your own work. Referencing and citations tell the reader where the information was found and show that you have acknowledged your sources. You must set out a reference so that the reader can find this information for themselves. You might reference a book or paper, magazine or newspaper, internet site or interview. Failure to acknowledge a source is plagiarism, whether intentional or by accident. Avoid it at all costs by giving citations to the facts and quotations you use unless they are your own or are common knowledge. If your school uses Turnitin use it to its fullest potential – view it as a tool to help you avoid malpractice problems.

Collusion is another form of malpractice and this is when you support the malpractice of another candidate. It may be working together with someone else on a piece of work for assessment when it should be done individually or taking your work from or giving it to another for the purpose of submitting to the IB. This carries the same penalties as plagiarism as do duplication of work for assessment and cheating in exams. The IB does not ask the question 'Who originated the work?' It is all malpractice.

While it would be easy if plagiarism fell into a black (yes you did) or white (no you did not) box, there is a grey area in the middle. This is when ideas or words may now be common knowledge (eg the Earth is not flat, gravity is a force pulling us into the Earth) yet were once the 'intellectual property' of one person or when you paraphrase someone else's words to the extent that they become your own. The best practice you can adopt is to keep track of all material that you read or note for your EE and other work and **if in doubt, cite**.

Keep your notes, drafts and copies of any papers or print-outs that you use until after the exam session closes. Then you can be sure that you have a paper trail to show that you did the work. Your supervisor may ask to see these notes as they build up a picture of how you worked. They are your evidence.

The 10–15 minute long **viva voce examination** is a chance for you to talk about your EE with your supervisor and show what you have learned.

What exactly is plagiarism?

We emphasize the positive aspects of academic honesty – know how to reference correctly – but, sadly, plagiarism is the most common type of malpractice that the IB finds in work submitted for assessment. In many cases, this is due to negligence or ignorance and is not intentional but you are still guilty of malpractice and the penalties range from no mark being awarded for that component to no mark awarded for the subject. It is your responsibility to get this right.

The following article is by Chris Willmott and published in the **Journal of Biological Education (2003) 37 (3) pp 139–140**.

Taking someone else's words or ideas and presenting them as your own work is known as plagiarism. But how much do you need to change something before it becomes a legitimate re-working?

The paragraph below is taken from Pharmacology (4th edition, 1999) by Rang, Dale and Ritter. Study the essay extracts in the table and decide whether or not you consider the author of the work to be guilty of plagiarism – some may be more obvious than others!

> **During the last 60 years the development of effective and safe drugs to deal with bacterial infections has revolutionised medical treatment, and the morbidity and mortality from microbial disease have been dramatically reduced.**

Essay extract	Plagiarism?(✓ or ✗)
1. During the last 60 years the development of effective and safe drugs to deal with bacterial infections has revolutionised medical treatment, and the morbidity and mortality from microbial disease have been dramatically reduced.	
2. During the last 60 years the development of effective and safe drugs to deal with bacterial infections has revolutionised medical treatment, and the morbidity and mortality from microbial disease have been dramatically reduced. [Rang et al., 1999]	
3. "During the last 60 years the development of effective and safe drugs to deal with bacterial infections has revolutionised medical treatment, and the morbidity and mortality from microbial disease have been dramatically reduced." [Rang et al., 1999]	
4. In the 4th edition of their textbook *Pharmacology* (1999), Rang, Dale and Ritter state that: '*During the last 60 years the development of effective and safe drugs to deal with bacterial infections has revolutionised medical treatment, and the morbidity and mortality from microbial disease have been dramatically reduced.*' Such a bold assertion understates the ongoing threat posed by microbial infection. It is estimated, for example, that worldwide there were over 8 million cases of tuberculosis in 1998 (WHO, 2000).	
5. The development of safe and effective drugs to deal with bacterial infections has dramatically reduced the death rate arising from microbial diseases.	
6. During the post-war years, the development of effective and safe drugs to deal with bacterial infections has transformed medical treatment, and death and illness resulting from microbial disease has been dramatically reduced.	
7. The availability of antimicrobial compounds has transformed healthcare in the period since the second world war. People are far less likely to die or even be seriously ill than they had been prior to the introduction of these drugs.	

Plagiarism – the answers

1. The first version listed is an 'ice-breaker'. It is clearly a verbatim account and is thus seriously guilty of plagiarism. The fix is easy – put it in parentheses and quote it.

2. The second version is marginally better, but is still not acceptable. The original work has been acknowledged as a source of ideas and information, but no indication has been made that the text itself has actually been used. Again, put it in parentheses and quote it.

3. In this case the addition of quote marks makes an important distinction from the previous versions. The author is clearly acknowledging that both the ideas and the word order have come from the textbook. It is not therefore guilty of plagiarism. We include this version to highlight a different weakness, namely that stringing together a series of quoted 'chunks' of text is a poor way to construct an essay and work written in this way is therefore likely to score low marks. But it is not technically plagiarism, just poor practice.

4. This version of the essay is fine. The quotation is indicated and is used in an appropriate way; it is being critiqued by the author and contrasted with a view supported by a second reference. Not plagiarised.

5. Here we get to the crux of the matter. The fifth and sixth versions of the essay are illustrations of practice that undergraduate students

early in their studies consider acceptable but we do not. They are derivatives of the original work with only cosmetic alterations. The wording and sentence construction of version 5 bears a very close relationship with the source and is guilty of plagiarism.

6. Similarly, this is a 'thesaurus-ed' or word-swapping version of the same text. A few words have been replaced with synonyms but this is not sufficient to be considered new work.

7. The author of the final essay has made a serious attempt to produce a novel account of the subject. It is still not perfect – lined up as it is here with one original source document, there are still echoes of the thought processes within the work and we would ideally want the student to draw on a number of sources in order that the essay has genuine originality. Nevertheless, significant effort has gone into bringing freshness to the text and we would consider that this is not guilty of plagiarism.

Some ESS EE titles

While your research title will be specific to your interests, some of these and a few comments about them, may help you in clarifying your ideas. All these titles are real, past titles.

High scorers

- Evaluation of water quality in Ramallah district springs.

- Is the decline in wild salmon in Vancouver due to the increase in number of farmed salmon?

- A comparison of a dairy farm in the Netherlands to one in Tanzania.

- Which ecosystem within the UK suffered the worst ecological fallout as a result of the Chernobyl disaster?

- An investigation into the soil erosion rates and effect on people and habitat in two areas of the lower slopes of Mt Kilimanjaro, Tanzania.

- Comparing inputs, outputs and efficiencies in an organic and non-organic sheep farm in New Zealand.

- What are the environmental and social impacts of building the Three Gorges Dam in China? (Although based on secondary data, some siltation experiments were carried out.)

Too general

All these are far too broad and global or not meeting the Systems and Societies requirement and cannot be addressed within the EE to meet the criteria. Avoid such general titles.

- Tourism in Antarctica.

- Global dimming. (But this could be a good idea if it were experimental and addressing light and/or heat transmission through different levels of atmospheric pollutants and their effect on the growth/germination of plants.)

- Oceans and their coral reefs. (But investigating a specific coral reef bleaching or death over time and comparing primary and secondary data would work.)

- The greenhouse effect. (Don't go there unless you do specific experiments. If thousands of scientists are working on this, what would you say in 4,000 words?)

- The effects of oil spillage on marine life. (Again, if you made this specific to one oil spill, or a comparison of two and their clean-up and restoration, you could have a good title.)

- The nuclear winter.

Too one-sided or just not right

- The wisdom of Lynx reintroductions in Colorado. (The title suggests the conclusion – that they are not wise – and there is a danger in only putting one side of the argument. Better to keep the title open as 'Lynx reintroductions in Colorado; success rates, methodology and evaluation'.)

- Temporary habitats for aspiring Martians. (Perhaps a little too flippant as a title. It was about terraforming but, again, too general to score well.)

- The effects of the Chernobyl catastrophe 1986 on Germany and the Bavarian woods 19 years after the event. (How could you compare then and now?)

Too vague

- Do the positive consequences of destroying the Amazon rainforest to benefit humans outweigh the consequences or are the negative consequences so critical that there is nothing worth risking them?

- The study of one key aspect to determine success of a community-based conservation project.

- Effect of fertilizer run-off on freshwater ecosystems.

- The qualities of water and their impact on plants.

Not addressing the subject

- Antimicrobial resistance: a growing ethical dilemma between economics and the environment.

- Heavy metal contamination in dietary supplements.

- Does the frequency of noticeable earthquakes off the west coast of S California and the Gulf of Alaska correspond to Southern Oscillation Index Values?

- An investigation of teenage noise pollution.

Referencing and citations

As the IB EE guide states, academic referencing may follow one of several styles. It does not matter which one you select but you must be consistent and it does matter if there is a full-stop or a comma in the right places. Be accurate. Find a guide to referencing on many university websites by putting 'academic referencing' into your search engine.

These pages contain resources and website addresses for further reading. It is not exhaustive and is a personal choice of the authors. But it is a start. We cannot be sure that the addresses will not change but most are of established organizations and should be there for some time. As things change so rapidly on the Earth, the data in this book may well date quickly and we recommend that you look at more recent data if necessary. Many university websites contain lecture notes and presentations on topics in this book. Happy reading.

Websites

General

http://about.greenfacts.org	Belgian non-profit organization, aims to bring complex scientific consensus reports on health and the environment to the reach of non-specialists
http://earthtrends.wri.org/	Online database maintained by the World Resources Institute
https://www.cia.gov/library/publications/the-world-factbook/	The CIA World Fact Book with detailed facts about the world and countries
http://news.bbc.co.uk/	Links to updates on environmental issues as do daily newspaper and other media channel websites
http://communities.earthportal.org/	Comprehensive resource for objective, science-based information about the environment. Includes the Encyclopaedia of the Earth
www.nasa.gov	A huge site from the US space agency, Earth sections are full of data, photos are fantastic
www.newscientist.com	UK weekly science magazine with reputable articles and reviews, eg on climate change, seas, biodiversity
www.noaa.gov	A US federal agency, full of info on climate, oceans and coasts
www.peopleandplanet.org/	Global review and internet gateway with very readable articles on many topics
www.unep.org	UN Environment Programme
http://www.unesco.org/new/en/natural-sciences/environment/ecological-sciences/man-and-biosphere-programme/about-mab/	UNESCO's Man and the Biosphere programme
www.worldwatch.org/	A 35-year-old research organization that reports on global environmental events
www.wri.org	A 25-year-old environmental think-tank with great articles and publications about the state of the Earth
www.iiasa.ac.at/	International research organization based in Austria
www.oxfam.org.uk	Oxfam International – charity and NGO that works around the world

www.foei.org/	Friends of the Earth International – a large grassroots organization with 2 million members
www.greenpeace.org/international/	Greenpeace International – a global campaigning NGO
www.unc.edu/~rowlett/units/index.html	A dictionary of units of measurement
www.oneworld.org	International civil society network with the aim of building a more just society
http://www.gapminder.org/ Gapminder	various statistics on a range of topics
http://www.breathingearth.net/	global population statistics and carbon emissions
http://puzzling.caret.cam.ac.uk/	games and simulations for variuos topics
http://www.abc.net.au/science/	great articles, games, puzzles and quizzes

Worldviews and environmental philosophies

| http://plato.stanford.edu/entries/ethics-environmental/ | Stanford University |
| www.worldchanging.com | Ideas for a better future |

Ecology

www.envirolink.org/	Links to hundreds of environmental sites
www.globe.gov	The GLOBE programme coordinates hands-on science projects between schools around the world
www.epa.gov/	US Environment Protection Agency
www.epa.gov/students/	Has some good links
www.enviroliteracy.org/	All things environmental. Put 'ecology lecture notes' into your search engine to find university course notes.

Biodiversity

www.panda.org	Worldwide Fund for Nature main site
www.iucn.org/	IUCN main site
www.iucnredlist.org	The entry point to the IUCN Red Lists of endangered species
www.eol.org	Encyclopaedia of life

Climate change

| www.ipcc.ch | IPCC website with their reports and views |
| www.epa.gov/climatechange/ | US Environment Protection Agency site |

Human populations

www.unfpa.org/	UN Population Fund, international development agency that promotes the right of every woman, man and child to enjoy a life of health and equal opportunity
http://esa.un.org/wpp/unpp/	UN Economic and Social Affairs Department searchable population database
www.census.gov/ipc/www/idb/	US government census site with dynamic population pyramids

http://www.census.gov/population/international/	World population clock
http://www.undp.org/content/undp/en/home/mdgoverview/	UN Development Programme Millennium Development Goals

Resources

www.earth-policy.org/	Dedicated to building a sustainable future
www.iisd.org/	International Institute for Sustainable Development

Ecological footprints

www.footprintnetwork.org	Ecological footprint calculator
www.carbonfootprint.com/	Carbon footprint calculator
www.ecologicalfootprint.com/	Simple calculator

Energy resources

www.bp.com	British Petroleum site, one of the world's largest energy providers. BP reports are full of information
www.shell.com	Another big energy company with some renewables information on the site

Water resources

www.peopleandplanet.com	Global review and internet gateway with very readable articles on many topics
www.wateraid.org	Find practical solutions to water issues
www.ifpri.org/publication/world-water-and-food-2025	Water resources booklets
www.worldwatercouncil.org/	International, multi-stakeholder platform for water issues

Soil resources

http://soil.gsfc.nasa.gov/	Soil made exciting by NASA again
http://soilerosion.net/	Just what it says

Food resources

http://faostat.fao.org/	UN Food and Agriculture Organization database
www.iwcoffice.org	If you care about whales
www.ifpri.org/	International Food Policy Research Institute – good on food as well

Pollution

www.atm.ch.cam.ac.uk/tour/	Cambridge University ozone hole tour: very informative
www.eia.doe.gov/kids/energyfacts/saving/recycling/solidwaste/plastics.html	Everything you need to know about recycling plastics
www.wasteonline.org.uk/	UK solid waste site full of facts on waste disposal

INDEX